Carl R. Moore
775540

Introduction to Engineering Design

McGraw-Hill Book Company New York San Francisco St. Louis Toronto London Sydney

INTRODUCTION TO ENGINEERING DESIGN

Department of Engineering, University of California, Los Angeles **Thomas T. Woodson**

To my wife, Marjorie, and to Loren, Craig and Claire who together have been obliged to share me with this book.

Introduction to Engineering Design

Printed in the United States of America

Library of Congress catalog card number: 66-18215

67891011 HOCZ 754

ISBN 07-071760-5

Foreword

The education of the "new breed" of innovative engineers is increasingly claiming the attention of engineering leaders, in and out of engineering colleges. More widely acknowledged is the institution's obligation to teach engineering design and to do so more and more in an authentic context, sharing, as professionals, the medical practice of teaching through direct experience.

Yet, the student learns engineering design less by the word and more by the favorable stimulating environment. The traditional approach to design in engineering education not only failed in this element of environment, but committed the more serious error of formalizing procedures to a dull, unimaginative, routine level. Today we recognize that a very different approach is necessary, and that the most capable, most impatient and creative students can be attracted rather than repelled.

The student must have access to the comprehensive domain of resource materials. His basic training in science and mathematics must be thorough; and he must be informed in many areas of economics, business administration, sociology, and public affairs, which provide so much of the basis for engineering decisions.

The scientist carries through his observation and correlation of natural phenomena; the technician relies on his handbook and standardized procedures; but the engineer must create, to satisfy society's need, those things which have never existed before.

Mr. Woodson's book captures very well the new spirit of education in engineering design. Both the student and the instructor will be inspired in their joint effort to learn by involvement in an authentic design project.

NEWMAN A. HALL
Executive Director
Commission on Engineering Education

Preface

This is a textbook for the engineering student who is beginning his design experience. He may be at any stage of his educational career, but it is assumed that he has had little instruction in integrating interdisciplinary material or synthesizing engineering systems. In addition, the student is probably questioning what the engineer really does and how he goes about it.

In brief, the book's mission is to assist the instructor in guiding the student through an authentic, if miniature, design project. We insist upon starting the student at the very beginning, at his validation of the need. Following this, he identifies and defines the problem and synthesizes alternative solution concepts. Then he analyzes, optimizes, tests, and iterates. The practicing engineer must necessarily be associated with his project from "cradle to grave."

Certainly engineering is more than design, but many believe that design is the essential, unique hallmark of the profession. Research, analysis, synthesis, experimentation, decision-making, communication, and persuasion may all be area concentrations within engineering, but no one or two of these *comprise* the practice of engineering.

In the midst of compartmentalized education, the student should live through at least one, and preferably several, "whole" experiences. One of these experiences should come early in his formal education, to add reason and motivation for ensuing courses. But, even if not early, some one experience should start, and eventually conclude, in the real world.

> "Education is an admirable thing, but it is well to remember from time to time, that nothing really worth knowing can be taught." *Oscar Wilde*

We need to sow the seeds of engineering judgment and of willingness to take responsibility. The engineer must take the risks of handling power—

huge blocks of physical power as well as the subtle personal power of irreversible social change. All the engineer's important problems are new; he must learn to venture.

The educational institution cannot compete with business and industry in providing the coherent, concentrated work experience. The school can, however, more than offset this disadvantage by carefully selecting the size and variety of project, by focusing upon the pedagogically "critical incidents," and by extracting the full value of preparation and generalization.

We strongly urge that the course be complemented by semester-long student projects. Long projects are preferable because of the time required to collect satisfactory kinds and amounts of information. Help in the crucial choice of these projects is offered at the end of Chapter 1 and in Appendix B. We work on the hypothesis that all students have a useful store of general knowledge which may serve as a starting point. It is not unrealistic to compare this to the engineer's starting on a project in a field new to him. We expect the student, as the engineer, to use the tools he already has, and to add more as he proceeds. However, the important point is not the information or the tools, but what the student expects to do with them and how he goes about doing it. A small but thorough interdisciplinary project may be just as valid as a large one, and possibly more instructive because of relatively less clutter. Additional project suggestions and/or appropriate case studies are available from a variety of sources.*

The book is organized chronologically, so the project can be started immediately. The chapters are arranged so that clusters can be selected for quarter, semester, or full-year courses. Further, material is presented so that the instructor may emphasize either the creative, the quantitative, or a blend of both aspects. The supporting disciplines of graphics, computation, and communication are brought into discussion as they become relevant. Communication in the form of progress-report writing should be injected at intermediate intervals.

For adaptation to quarter or semester courses, we suggest eliminating problems related to Chapters 7, 8, 9, 11, 14, and 16; although the text of some of these chapters may be assigned as survey reading. For nonquantitative emphasis, Chapters 12, 13, 14, and 16 should be excluded.

The book has been written in a realistic tone, because engineering is performed by humans—time-taking, error-ridden, subjective beings that we usually are. To delete the spicy and bitter flavors would rob engineering of essentials, and overlook the reasons for checks and safeguards. Engineers, as all creators, must struggle for their conquests; they must iterate their attempts. In sheer honesty we must not present ice-cold, simon-pure

* One of the best sources for case studies and project ideas is the National Science Foundation project at Stanford University. Write: Department of Design, Mechanical Engineering, Stanford University, Stanford, California.

derivations, pretending that no fumbling preceded the inspiration. Innovation has never happened that way, and probably never will.

Engineers develop strong feelings and close identification with their projects, and students should glimpse engineering as they will in truth later experience it.

A truly professional engineering project consumes many months or years of full-time effort. Since neither a full project case narrative nor actually very many days of student time may reasonably be expected, a substitute has been supplied: A series of design incidents from a particular assumed project (a transit-mix concrete batching plant) which is within the subject familiarity of most readers. This vehicle should provide cohesiveness, bridging many apparent gaps in continuity from chapter to chapter, and it should demonstrate the wide variety of problems to be discovered in a relatively simple project.

Acknowledgments

This book is the outcome of work in the development of Engineering 4B at UCLA, initially sponsored by the Educational Development Program and partially funded by a grant from the Ford Foundation. So many persons and institutions have contributed ideas and criticisms that few can safely be attributed. The author is finally responsible for the words and opinions herein expressed.

Professor Morris Asimow was originally a co-author, and I am grateful for the strong start he gave the book. Professors Allen Rosenstein, Benjamin Ostrofsky, Winfield Heinz, and Henry Fuchs have been especially influential as supporters, contributors, and critics. I would also like to acknowledge many valued discussions with Robert McKee, Melvin Kline, Melvin Lifson, Russell Strom, Peter Serrell, Wilton Stewart, Milton Lewis, Harry Buchberg, Joseph McCutchan, John Lyman, John Bowen, and other instructors of Engineering 4B at UCLA. A great deal of stimulation was derived from conversations with Llewellyn Boelter, Francis Shanley, Louis Grandi, Myron Tribus, Alex Levens, Margaret Jones, William Schimandle, John Bennett, Peter Bulkeley, and Newman Hall.

Countless hours of care have been invested by Estelle Dorsey and her Reports Group, and by our indispensable secretaries, Carole Fogen and Harriet Palk.

Professor William D. McIlvaine, of California State College at Long Beach, contributed Chapter 8, "The Engineer and Money."

And finally, over the past five years, the students have, by the high quality of many of their projects, justified our very high expectations.

THOMAS T. WOODSON

Table of Contents

v Foreword

vi Preface

One

1 **The Engineering Problem Situation**

Designing for Need, 1. Design, 3. Role of the Engineer, 4. Varieties of Engineering, 5. Branches and Fields of Engineering, 5. Kinds of Engineering Work, 8. Disciplines Necessary for Success in Design, 12. Summary, 14. Book Project Example, 14. Exercises, 16. Student Project Suggestions, 17. References, 19.

Two

21 **Design Project Organization: Planning a Design**

Organizing to Produce Results, 21. Participation of the Engineer in Goal Definitions, 23. Time Sequence of Design Project: Structure of Design, 24. The Design Process (A Brief Glossary), 29. End Objective, 31. Book Project Examples, 32. Exercises, 34. Student Project Exercises, 35. References, 35.

Three

37 **Information and the Needs Analysis**

The Engineer and Information, 37. Information Flow Diagram, 39. The Real Problems about Information, 41. Information and the Production-Consumption Cycle, 42. Varieties of Information, 43. Criteria of Rationality, 45. Numerical Information, 47. Starting Point in the Needs Analysis: The Primitive Statement of Need, 48. Variety of Needs, 50. Translating the Needs into the Statement of Goals, 51. Summary, 53. Book Project Example, 53. Student Project Exercises, 55. References, 55.

Four

56 **The Feasibility Study: Identification of the Problems**

Is It Feasible?, 56. Elements of a Feasibility Study, 57. Tools to Aid in Clarifying Goals, Stating the Problem, Formulating Criteria, 59. From Goals to the Engineering Problem, 62. Book Project Example, 65. Student Project Exercises, 66. References, 66.

Five

67 **Information Sources; Laws and Effects**

Where Can I Find Out?, 67. Book Project Example, 68. Many Sources, 68. Economic Basis, 69. Types of Sources, 69. Suggested Procedures to Obtain Information, 71. Specific Situations, 71. Laws and Effects, 72. Cause and Effect Chart (Input/Output), 73. Book Project Example, 74. Student Project Exercise, 74. Idea Scanning Table, 75. References, 83.

Six

84 **Synthesis of Alternatives: Creativity**

Re-checking the Problem Starting Point, 84. Synthesis, 85. Creativity, 87. Hindrances and Aids to Creativity, 89. Patents and the Inventor, 90. Original Idea Disclosure or Design Record, 92. Summary on Creativeness, 93. Book Project Example, 94. References, 95.

Seven

97 **Estimation and Order-of-Magnitude Analysis**

Confidence from Experience, 97. Estimation, 98. Order-of-Magnitude Analysis, 99. Estimation: Examples, 99. Order-of-Magnitude Analysis: Examples, 103. Summary, 107. Student Project Exercises, 107. References, 108.

Eight

109 **Engineering and Money**

Why Money?, 109. Fundamental Relationships between Money and Engineering, 111. Gross Analysis of the Engineering Project, 112. Conclusions about the Economics of Engineering Projects, 116. The Business Unit, 117. The Business Unit and the National Economy, 120. Summary, 122. References, 122.

Nine

124 **Preliminary Design**

Alternate Usages, 124. Stages in the Preliminary Design, 125. Illustration: Pole Support, 125. Illustration: Conveyor Belt, 130. Variables, Parameters, Constants, 134. Sensitivity, 134. Conclusions, 134. General Discussion: Variables, Parameters, and Constants, 135. References, 136.

Ten

137 **Engineering Problem Modeling**

Perfect Solutions?, 137. Values of a Model, 138. Limitations and Cautions, 139. Aspects of Solutions, 140. Variety of Models, 143. The Iconic Model, 143. The Analog Model, 146. Analogies between Technical Fields, 153. The Analog Computer, 153. The Symbolic Model, 156. Summary, 158. Book Project Examples, 158. Exercises, 159. References, 160.

Eleven

162 **General Aspects of Graphics: The Iconic Model**

Ubiquitous Graphics, 162. Communication and its Purposes, 163. Capability of Eye-Mind Team, 165. What Can Be Done by Graphics?, 166. Specification, 166. Visualization and Synthesis, 168. Determination of the Dimensions, 172. Computation by Graphical Methods, 173. Showing Relation and Emphasis, 181. Persuasion and Selling, 183. Recording, 183. Iconic Models: Scaling, 184. Skills Necessary, 186. Summary, 187. Exercises, 187. References, 189.

Twelve

191 **Conceptual Representation: Symbolic Models and Modeling**

Introspection, 191. System Description, 192. Physical vs. Logical Laws, 194. Laws and Implicit Answers, 195. Symbols, Expressions, and Equations, 196. How to Get Started, 197. Characteristics of a Good Model, 198. Book Project Example: Mathematical Model for a Water Tank, 198. A Basic Design Equation, 202. Decisions, the Value System, and the Criterion Function, 204. Working with the Criterion Function, 206. Revision of the Tank Example, 209. A Quantitative Solution, 211. Summary: Check List for Assembling Criterion Function, 212. Exercises, 213. References, 213.

Thirteen

214 **Expansion of the Criterion Function**

Résumé, 214. General Form of the Function, 215. Cost Plus One Physical Variable, 217. Multiple Criteria, 219. Permutation of Alternatives: Educational Toy, 227. Summary, 237. Exercises, 238.

Fourteen

239 **Checking in Engineering Design**

The Role of Checking, 239. Necessary but Not Sufficient, 240. Checking by Logic: Assumptions, 241. Checking by Mathematical Rules, 243. Checking by Units and Dimensions, 245. Checking by Symmetry, 249. Limiting-Value Checking, 252. Checking Against Experience, 254. Summary, 255. Exercises, 256. References, 257.

Fifteen

258 **Optimization**

Meanings, 258. Systems vs. Subsystems, 259. Optimum vs. Optimal, 260. Criteria, 261. When to Optimize, 262. Methods of Optimization: Subjective, 263. Methods of Optimization: The Use of General Principles, 263. Methods of Optimization: Graphics, 264. Methods of Optimization: Analytical, 267. The General Analytically Optimized Problem, 267. Book Project Example: Bucket Conveyor Optimization, 270. Invention for Optimizing, 276. Optimizing with Regional Constraints: Linear Programming, 279. Review of Linear Programming, 282. Summary, 283. Exercises, 284. References, 286.

Sixteen

287 **Digital Computers and Design**

Introduction, 287. The Antecedents, 288. Essential Blocks and Operation, 289. How Numbers Are Set to Work between Input, Output, and Memory, 292. Special Uses and Varieties of Computers, 297. Use of Computers in Design, 299. Computerized Solutions to Problems in Design, 300. Massive Computation, 300. Iterative Solution, 303. Differential Equations, 314. Summary, 317. Book Project Example: Algorithm, 317. References, 320.

Seventeen

321 **Communication and the Engineer**

Communication and Information, 321. Information Theory, 323. Information Measures, 323. Communication and Reports, 325. The Communication Diagram and Engineering Reporting, 327. The Nature of the Report, 329. The Preparation of the Report, 334. Presenting the Report Conclusions, 343. Summary, 344. References, 346.

Eighteen

347 **Oral Reporting**

Engineering Presentations, 347. Advantages of Oral Presentation, 348. Comparison with Written Communication, 348. Special Opportunities, 349. Preparation, 350. Visual Aids, 351. The Room, Facilities, and Rehearsal, 353. The Speaker, 354. The Presentation, 355.

Nineteen

357 **Open Challenges to Engineering**

Appearances, 357. Realities, 358. Energy, 359. Raw Materials, 363. Urban Environment, 365. Transportation, 368. Information Handling, 369. Biotechnology, 369. Health, Medicine, 370. Education and the Developing Nations, 371. Summary, 373. References, 373.

Appendix A

374 **Common Engineering Reference Data**

Work, Power, and Heat Conversions, 374. Physical Properties and Costs of Materials, 376. Common Engineering Reference Sources, 378. Items in Construction Job Costs, 379. Elements of Cost in Factory Mass Production, 380.

Appendix B

382 **Student Design Project**

Criteria for Choosing a Project, 382. Suggested Project Areas, 383. Supplemental Information on Projects at End of Chapter One, 383. Illustrative Information Concerning Project L, 394.

Appendix C

397 **Information Sources**

Library of Congress Classification System, 397. A Few Useful Sources, 398. National and International Engineering and Technical Societies, 399.

Appendix D

402 **Units and Dimensions of Quantities**

Dimensions of Mechanical Quantities (Fundamentals: L, M, T), 403. Dimensions of Thermal Quantities (Fundamentals: L, M, T, plus t, Temperature), 403. Dimensions of Electrical Quantities (Fundamentals: L, M, T, plus K, Electric Permeability, and μ, Magnetic Permeability), 403.

Appendix E

404 **Summary of Graphics Essentials**

Appendix F

416 **Use and Care of the Engineering Notebook**

General Suggestions, 416. Essential Rules for Keeping Lab Notebooks, 417.

Appendix G

419 **Flow Charting for Digital Computer Solutions**

To Solve $Ax^2 + Bx + C = 0$, 419. "Interpretive Language," 421. FORTRAN Symbols, 422.

Appendix H

423 **Large System Modeling: Supersonic Aircraft (SST) for 1970-80**

427 Index

The Engineering Problem Situation

One

"Man is the measure of all things."
Protagoras

Designing for Need

Picture one of our unnamed ancestors of 20,000 years ago laboriously poling his small raft up a stream along the shore. He notices another man up on the bank leaning against the force of the wind while trying to rack up a drying animal skin. Tired though our raftsman is, he realizes the significance of the other man's struggle. We don't know how the sailing boat was conceived, but this imagined situation carries all the ingredients of invention: An urgent need, and eyes that can see in an "unhabitual way."

We move to 300 B.C. and watch the construction of one of the battering rams of Alexander the Great. This ram is reported to have been made of a log 180 feet long and provided with a heavy head. It was hung from a wooden framework, roofed over to protect the accompanying soldiers. It was said that no walls could long withstand repeated punishment from this large weapon.

Coming up to 1770, we look in on English businessmen buying and installing Richard Arkwright's spinning frame, which would produce four threads with the work of a single pair of hands. Arkwright claimed that his machine would be of great utility to manufacturers and to the public "by employing a great number of poor people in working said machinery."

Finally we can look back just a few years from the present and recall the incredible flights of the Mariner spacecraft, one grazing past the planet Venus and the other past Mars. The National Aeronautics and Space Administration had spent several hundred million taxpayers' dollars simply to record data from these flying packages of scientific equipment. The stated purposes of these expensive undertakings were to measure the interplanetary radiation, the magnetic fields and the surface temperature of the two planets, and to photograph Mars; laudable but highly speculative objectives. Obviously the world's opinion of the United States was at stake also, and these Mariners met a number of psychological needs.

The common element of the preceding illustrations is the use of technological design as a means of satisfying human need. There are about half a dozen fundamental needs of man, and many ways of satisfying them. We shall discuss here only the technically obtained means of satisfaction. Engineers, and the proto-engineers and craftsmen who preceded them, have planned and provided these satisfactions ever since Homo sapiens emerged 50,000 years ago. Anthropologists[1] believe that, beginning half a million years ago, tools, the use of tools, and the brain of the primate evolved interdependently together, producing a race that could adapt both its environment and itself to its needs. But, what were these needs? Our illustrations have brought up the examples of transportation, military power, low-cost textiles, and political status; widely divergent but typical historic needs. In fact, the dominant military role of the first acknowledged engineers gave rise to the distinction of "civil" engineers to describe the later men concerned with nonmilitary public works. Needs, then, whether factual or fancied, moral or amoral, present or future, important or unimportant, have been and are the starting points in design.* Note that now, as then (1) the designer has only a limited supply of resources to work with, and (2) upon his own internal motivation he seeks to make the best use of them by choosing from among several alternatives, and (3) at that moment he must be a pioneer in the use of "natural law." He may be "engineering," but he is not *designing* when he knowingly repeats embodiments that were earlier established by someone else.

In the distant past, a person who had become skilled in the art of design had first been trained as an apprentice to a master and then as a probationer

[1] All numerical superscripts refer to the respective numbered references at the end of the chapter.

* As with any citizen, each engineer chooses his subject of concentration and his employer or client based upon his own cultural value system, and his own ethical standards, whatever these may happen to be.

in the unsympathetic grip of experience. Years later he became a master himself. In the last century the physical sciences and mathematics progressively have become such invaluable tools that the embryonic engineer now develops competence in them before essaying the field. Also, new understanding of design as a creative development process allows us now to describe and teach it as a body of knowledge, even though engineering design still retains some mystique as an "art." Thus, currently, learning time can be shortened, and professional attitudes and relatively new tools can be introduced to the student and applied by him earlier in his career.

Design

We can combine some of the foregoing ideas to develop what we mean by *engineering design*. It is "an iterative decision-making activity to produce the plans by which resources are converted, preferably optimally, into systems or devices to meet human needs." There are other ways of defining design and of performing it, but this description seems to reflect the experience of engineers who have spent years successfully *doing* engineering design. Design work can be thought of as consisting of separate, identifiable stages or phases, called different names by different groups (as we shall see later), but still having a shape and purpose. Design is one of the many difficult-to-define tasks of society (like government or commerce or parenthood). The need and performance come first, the precise name and description second.

Let us expand upon our definition to clarify our meanings of the words and to distinguish our usage from the broader dictionary definition, which is based on the Latin *designare*, which means to "mark out."

Iteration. Engineering design is marked by repeated attempts which are generally necessary to accomplish a single phase. Information developed during the initial study is employed in a second trial, and more if necessary, for finally arriving at a satisfactory conclusion to a given phase. The result of a given phase is practically never acceptable the first time attempted.

Decision-making. In engineering design we encounter many alternatives, some of which are more or less equally applicable to the final outcome. Faced with these alternatives, the engineer must decide which specific problem he must choose to solve and which solutions among many he will accept. He bases his decision upon various working criteria and upon various ends or values which are exchanged from case to case. These exchanges are called "trade-offs."

Conversion of resources. All engineering design is performed within the constraints of limited supplies of time, money, talent and natural resources.

These resources are always finite, and their nature and availability varies from situation to situation.

Plans for system or device. The plans that are the destined output of the engineer may be for a new organization or a new operating method; or they may be for the more frequent result: New hardware. In scope, engineering "products" range from safety pins to world-wide telephone networks, from mining methods to manufacturing systems. However, the engineer seldom if ever produces the hardware or puts the system to work; he produces only the plans. He continues as a teacher, leader, or supervisor for those who do the actual work.

Human need. Needs can be satisfied only by paying the various prices that are exacted. *Function* requires materials and energy. *Safety* usually requires complications in design or operation. *Lightest weight* frequently requires costly materials. *Reliability* requires more costly manufacture. The optimum, or best value for a given situation, requires a thoughtful balance between first cost, function, appearance, convenience, maintainability, and life. These must be traded off, one for the other, in differing degrees depending upon the assumed priorities. Of course, a real optimum can be attempted only on a *whole* system; sub-optimization is the best that can be expected elsewhere.

Role of the Engineer

In consultation with the end-user, the design engineer or (as he is sometimes called) the systems engineer is the person most likely to apprehend both the technology and the situation; and it is upon him that the responsibility for design decisions frequently rests. To be sure, others set policy and participate, including government agencies and financing people as well as the customer. However, at the risk of claiming too broad an empire, we assert that the professional engineer, as a designer, is the key man in integrating all factors into single solutions. Incidentally, it is easier to make an engineer into a generalist, than a generalist into an engineer; and the generalist is the indispensable one in carrying out a project.

The purpose of this book is to unfold the concepts of design activity as summarized above and to introduce and illustrate the use of some of the tools of design. A particular project will be carried along through the book so that characteristic examples can be found naturally, to illustrate the various methods and concepts as they become appropriate. Not every project uses all or even the same tools of design, just as a doctor does not always use a blood-pressure gauge in diagnosis or always prescribe aureomycin in treatment, even though they may be exactly what are needed in certain cases. Pertinence and suitability will be our guides.

Varieties of Engineering

We speak of engineering design as if it were a single activity. In a sense it is; and yet, it is composed of many activities and done by different people when performed at the professional level. To avoid confusion we shall present the several titles that are conventionally applied to engineers. Then we will list fields of engineering and functions of engineering, and then attempt to relate our concepts of engineering design to them.

First, the titles that are usually given to engineers, considering education, age, experience and other qualifications, are:

Years out of College	*As an Individual Contributor*	*As a Supervisor*
0–2	Engineer-in-training (becomes oriented)	
1–3	Junior engineer (assists an engineer)	
2–5	Assistant engineer (given first responsibilities)	
5–15	Engineer	Project engineer, Group leader
15–25	Senior engineer	Section engineer, Division engineer
20–35	Consulting engineer	Chief engineer, Manager of engineering

Recognize that in the supervisory titles above, the man's responsibilities may vary widely. For instance, he may be the project engineer responsible for developing a new supersonic aircraft, or he may be the chief engineer of a small broom factory! Do not depend upon title alone in judging responsibility. Find out the annual budget he controls: $10,000 or $10,000,000?

Branches and Fields of Engineering

The "fields" of engineering are narrower than the "branches" and are shown under the branches as subdivisions below. The major branches and fields are given here in approximate chronological order of their historic development.

1. *Civil:*
   Highways, Structures, Water, City Planning, Airports, Pipelines.
2. *Mechanical:*
   Machines, Heat, Fluids, Automotive, Refrigeration, Gas Turbines.
3. *Architectural:*
   Residential, Industrial, Governmental, Scientific.
4. *Mineral and Mining:*
   Fuel, Ceramics, Metals, Petroleum.
5. *Electrical:*
   Communications, Illumination, Power, Control, Computers, Space-Telemetry.
6. *Industrial:*
   Manufacturing, Planning, Safety, Cost, Management.
7. *Chemical:*
   Basic Chemicals, Food, Paper, Pharmaceuticals, Plastics, Electrochemical.
8. *Aeronautical:*
   Aerodynamics, Structures, Control, Navigation, Propulsion.
9. *Nuclear:*
   Weapons, Power, Instrumentation.
10. *Astronautic:*
    Spacecraft, Propulsion, Control, Life Systems.

Other branches: Agricultural, Marine, Military, Acoustical, etc.

The kind of work that an engineer finally does and the satisfaction he derives are decided more by what he does (his *function*) than by where he does it (his *field*). This is true because the function more clearly relates to the fundamentals of his psychological make-up and basic motivations, while his field of interest is more related to the "accidents" of his reading and personal contacts. In most cases, his choices mutate and evolve throughout his career.*

In all the above-listed fields, natural differences occur between engineering for products and engineering for processing (e.g., electric motors vs. cement). In mechanical and electrical engineering, for instance, the manufacturing facilities are engineered by people other than the product engineers. However, by contrast, in chemical engineering the man who engineers

* One of the most important designs an engineer can undertake is that of *his career*. This, obviously, deserves his best efforts, but surprisingly, it seldom receives it.

the product is usually the same man (or very close to that man) who engineers the process.

We shall list below the functions of engineering under the broad heading of product-related activities, in a conventional sequence.

Function Title	*Description of Duties*
Research	Obtain knowledge and understanding.
Development*	Establish technical feasibility or improve performance.
Design	Decide among alternatives and specify for manufacture.
Production* (synonymous with quality or reliability engineering)	Improve quality and reduce cost.
Construction	Supervise erection of capital equipment or buildings.
Industrial*	Plan and/or operate manufacturing plant.
Evaluation-Testing	Measure performance and life.
Marketing-Sales	Persuade customers and transmit information.
Operation	Operate the product or system.
Service-Maintenance-Repair	Service the customer, by the manufacturer.
Education	Provide general or specialized training.

*The asterisked functions (as well as the Design function) may occasionally perform design. There is little or no design activity in the other categories.

You will notice that this listing generally follows the time sequence associated with a product or process development. However, it only roughly approximates the chronological pattern of evolution of a specific project or a broad-scale undertaking, as we shall later see.

There are also engineers who work on the set-up, operation, and analysis of systems and relationships. The functions in these latter cases include such titles as industrial engineers, manufacturing engineers, operations research engineers, and management consultants. Of course, there are many engineers who have left the field and yet retain their engineering orientation in such occupations as city managers, patent attorneys, bankers etc. As we have said earlier, engineers rarely have assignments which are purely restricted to one field, although generally the assignments are closely limited to one function.

We mentioned earlier the individual contributor as contrasted with the supervisor. Engineers who have broad interests and specific leadership capabilities tend to move in the direction of project responsibilities, supervision, or the professional practice of engineering.

In this book we are considering "engineering design" in its broadest sense and identifying it with the essential activity of professional engineering,* and the highest level of true engineering accomplishment.

Kinds of Engineering Work

The U.S. Department of Commerce[2] lists the income from all classifications of national economic activities in Table 1.1. This study can be considered a typical comparison for the at least partially-developed areas of the world, although the actual figures will vary from country to country.

Table 1.1 Total U.S. Annual Income and Engineers Employed.†

	U.S. National Income (billions of dollars)			*No. of Engineers Employed*
	1957	1960	1963	1961
Agriculture, forests, fisheries	16	17	19	100
Mining	6	5	5	16,700
Contract construction	20	22	25	34,900
Manufacturing	112	121	137	459,400
Wholesale, retail trade	60	68	77	23,700
Finance, insurance, real estate	35	43	49	1,700
Transportation	17	18	19	8,000
Communications, public utilities	13	17	19	37,600
Services	40	49	59	68,900
Government, government enterprises	43	52	64	800
	362	412	473	651,800

† For a table of the full interaction between U.S. economic sectors, see *Scientific American*, pp. 1–11, April, 1965.

Most of the "industries" listed in Table 1.1 provide some kind of an employment opportunity for the young engineer. But what kind of a job will he get when he first goes to work? It is safe to say that if he is hired with any idea of permanence he will first be given orientation and indoctrination

* To a professional, his occupation is more than a vocation; it is a way of life by which he serves himself, his client, and society. He professes skills in a specialized body of knowledge and accepts responsibility for the performance of his creations. Acknowledging his debt to the practitioners of the past, he contributes through his own original work. In his relations with the public and with other engineers he considers himself bound by the obligations of a professional code of ethics (from W. L. Orr). See Smith, in the references listed on page 20 at the end of this chapter, on the subject of professional ethics (p. 39).

for a period of weeks or months (engineer-in-training). This orientation may or may not be associated with a productive job assignment, though it usually is. If employment is in the usual manner, the first assignment is generally that of making tests or measurements on small devices, or obtaining information or making analyses in very small projects. This probably would involve some small element of a process or a single mechanical part. The engineer would report to his supervisor, another engineer, along either of two types of organization pattern.

The first chart, Figure 1.1, delineates a product-oriented engineering organization and the second chart, Figure 1.2, shows a field- or specialty-oriented organization. These charts show forms of the relatively large engineering organizations generally associated with technical manufacture or consulting. In such organizations the low-echelon engineers find themselves isolated from the major direction and policy-making functions of the group or business. Here the engineer (along with other newcomers) sees only his department and often develops the feeling that his function is supreme and the business is in operation to do the bidding of just the engineers.

Such an attitude is rare in the small organization, where there are only one or two engineers, and the dominant purposes of the operation are only too clearly brought to a focus for the employee. Very quickly the young engineer in these latter circumstances learns that the boss, customer, or client is only interested in prompt, accurate answers to specific questions. None of these people is very much interested in the feelings, preferences, ambitions, or personal problems of the young employee. The same impersonality is suspected in the large organization; but more time passes before discovering the real purposes and procedures of the enterprise.

With all these obvious constraints, what freedom has the young engineer to exercise any creativity, any independence of choice, any of the significant suggestions for basic improvement that we hint he should make? The boss gives him a specific problem to do, and that's that.

In spite of possible raised eyebrows and shrugs of hesitation, we are introducing the all-embracing subject of engineering design early in the life of the engineering student because the broad-gauge attitude is precisely what will help the alert young engineer to be of most value to himself and to his organization. Let us illustrate: The chief in a refrigerator engineering design office needs a modified freezer compartment designed. He calls in the young engineer and gives him the size of the freezer, size of the refrigerator, the maximum and minimum temperature, the type of insulation, the method of support, etc. This appears to be a narrow technical situation in which the assignee should not talk back to the boss; the engineer simply should bring in the results of his best efforts based on the data supplied. On the other hand, without belaboring the point too much, consider the

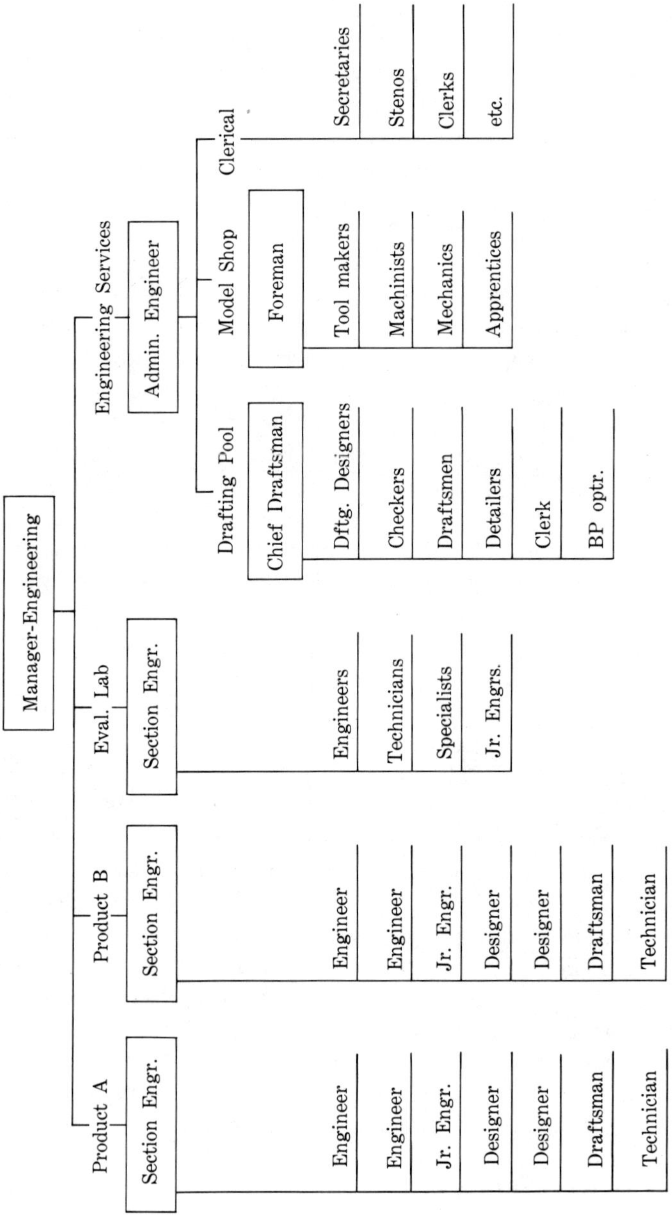

Fig. 1.1 Product-oriented engineering organization.

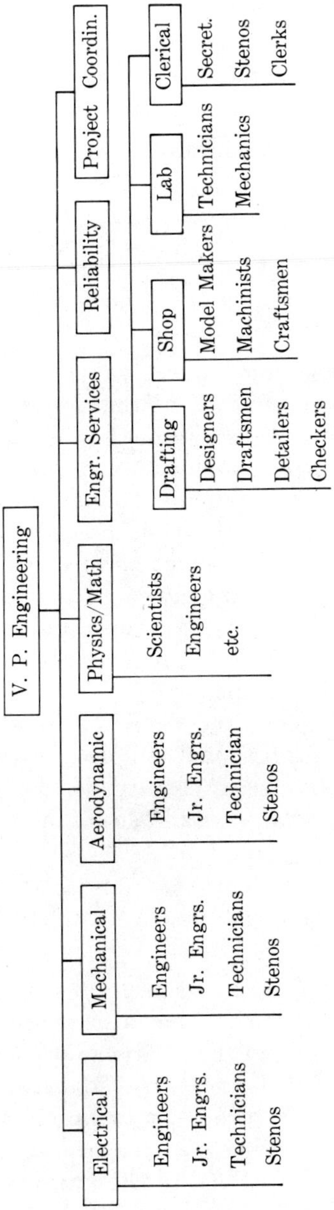

Fig. 1.2 Specialty-oriented engineering organization.

inquiring young engineer who after some hard thinking and research comes back to the boss with discreet questions (and proposals) on such subjects as:

Should the freezer compartment be enlarged this year?

Should we consider larger ice-cube storage?

What future foods will be available in 10 years, while the refrigerator is still in use?

Could we not tool it more economically with another wall material?

Should different access doors (no hands) be provided?

How important is defrosting to the housewife?

How accurately regulated does the temperature really have to be?

Has market research been done on the impact of central storage lockers?

Should the freezer compartment really be *under* the fresh food compartment?

The young engineer becomes more valuable when he thus places himself progressively in the positions of the designer, the manufacturer, the salesman, the housewife, the serviceman, the stockholder. He should then take as many of the logically indicated resultant steps as he can by himself and ask permission to take more. These all lead to better performance for himself, for his boss, and his boss' boss. Their objectives generally revolve around total competitive product performance, time schedule, and dollars. The total engineering group, down to the last man, is employed to reach the organization's objectives. Simply taking pride in some elegant engineering accomplishment is an unjustified luxury, except when it is indulged in after work, or over some coffee.

In the discussion throughout the remainder of the book we will thus take the viewpoint of the professional engineer, rather than that of the student or junior engineer. We trust each reader will adapt the respective conclusions to his own particular case and situation.

Disciplines Necessary for Success in Design

Most students are resigned to the necessity of thorough training in mathematics, physics, and chemistry. They also assume that the engineering curriculum will provide training in thermodynamics, electricity, electronics, mechanics, materials, experimental science, and whatever other technology is deemed necessary. Our purpose here is to broaden his expectations by urging the study of optional subjects in the life sciences, the humanities, the social sciences, and the arts. Even in the areas of his own technology, the engineer is now being forced to take living processes more and more into account, so that biology, physiology, and allied subjects are more frequently called upon.

In working with man both as a part of the machine and with him as an employer or customer, the engineer has much to learn from the life and social sciences. As the engineer provides more and more fundamental services, his value system must be guided by various aspects of his culture, and these are the areas with which philosophy deals. Consider, for instance, a freeway whose projected path will cut through and split the community of houses of a local minority group. Should the freeway be directed around the area? Also, how much extra money is it worth to make the freeway safer by an expected one less traffic fatality per month (or two or ten)? Although the engineering designer is not an architect, he must be similarly sensitive to the arts, because his cooperation can make a difference in the degree to which aesthetic values are incorporated in a given project. Examples can also be cited in other fields ranging from agriculture to aircraft, illustrating the sharp, persistent relevance of the humanities, social sciences, and life sciences in their various disciplines.

Finally, there is the discipline of self. Here we use "discipline" in another and more important sense. The awareness and control of self comes from the conscious development of realistic attitudes about what oneself is worth, what one's capabilities and limitations are, what one's over-all goals are, and one's expectations as an engineer. The importance of attitudes can hardly be overstated, because primarily attitudes rather than skills distinguish the professional from the commercial man, with all due respect to the latter. It is worth much self-examination and objective measurement to come to satisfying conclusions about oneself, because these conclusions are the premises upon which are based not only one's education but the first major directions of adult life.

Another subtlety of attitude revolves around job expectations.[3] It is something of a shock for the young engineer in industry to find his group is *only* a service in the way that Accounting or Personnel are services. Whatever the facts may be, the company considers that its money is made through manufacturing and sales (buying and selling), not by designing the conversion of the resources, because the company designs only "once," whereas it repeatedly and continually converts. Since the company can generally manufacture whatever quantity it chooses, sales is the bottleneck in getting the goods to the customer. Hence, many believe that sales gets the attention, the compensation, and the glory. This varies, naturally, from industry to industry. In the consulting business, the engineer himself constitutes the money-maker and therefore is at the focus. However, even this engineer is kept working by the sales effectiveness of the head of the consulting firm.

In any case, the engineer performs a service and the value of it is judged by a boss, whether immediate or remote, or by an associate, a customer, or a client. Incidentally, it must be kept in mind that the final judge of an engineering work is a layman, just an ordinary person, whereas, another equal or better scientist is the only one qualified to judge the value of a scientific contribution. Thus, the engineer is paid for results, and in order to perform adequately he must be goal-oriented. The method is definitely secondary; any productive method is acceptable. He must continually ask himself: What is the real, stripped-down, essential need of my client? How can I meet this need or help him meet it most economically and in the shortest time?

SUMMARY

Thus, engineering design may concern itself with a whole enterprise, with a major sector or a system within the enterprise, with a smaller subsystem, a component or a part, and each of these in turn, may be considered from many viewpoints. The engineer starts with a need, frequently vague and nebulous, and proceeds through an identifiable process to plan the optimum use of resources to meet that need. He decides among available alternatives, judging on the basis of accepted criteria, iterating until a satisfactory solution is reached.

BOOK PROJECT EXAMPLE*

We can extract a variety of illustrative design problems from a typical project: A transit-mixed concrete company. It is simply one project from among many which could provide manageable design problems. The project pertains to subjects in common experience and further information is usually within easy reach of most persons. Let us assume that the reader is a young engineer now employed by a successful transit-mix company which operates about 25 mixing trucks. The company has several loading plants, such as illustrated by Figure 1.3. It operates its own sand and gravel supply system and it serves customers who are located over a relatively large area. In the past, these customers have been able to receive engineering services from the

* Book Project Examples, interpolated at frequent intervals throughout the text, are displayed in this fashion and are enclosed by rules at the beginning and end. Such examples are simple vignettes or one-act plays lifted from the running drama of a continuing project effort.

company in certain technical areas including limited structural concrete design, materials recommendation, specialized pouring, curing schedules, and so on.

The first group of illustrative situations simply spreads the table, shows the variety of diet, and prepares for consumption and digestion in later chapters.

Fig. 1.3 Transit-mix concrete batching plant. General view of dispensing hopper and control room. (Courtesy Consolidated Rock Products, Inc.)

First a customer comes in who is planning a simple two-car garage. He wants the sketches and the costs for the wall footings and the floor slab. This is essentially a *single-answer problem* and you, the engineer, will make a sketch with plan and side views and compute the volume and cost using the minimum slab thickness and standard sizes specified by the building code. In half an hour the customer is on his way with the information he wanted.

The next illustration concerns a problem brought to you by the plant foreman. A new belt conveyor is being installed and the foreman wants to know how far the sand will be thrown when the belt is running at its full speed, and what the size and shape of the pile will be as it builds up under discharge spout. This is essentially a *single-answer problem* also. You can answer the foreman's first question by considering the sand as a liquid with initial horizontal and vertical velocity components, and a known distance to fall in a parabolic trajectory. We shall defer work on this problem until later on, in Chapter 11.

The next illustration is a problem which comes from a customer who owns an ungraded corner area which he wants to convert into a parking lot by grading it flat and erecting a retaining wall. His question is, what size and shape of the retaining wall will allow him the greatest annual income, considering the cost of the wall as a part of the cost of operating the lot? This is a problem with *many answers*, a bona fide design situation where you will be concerned with how you arrange the cars and the entrance aisles, how you would buttress the wall against cave-in, how much of the parking traffic would come from one street versus the other, and the average occupancy of the parking lot. We will resume the parking lot design in Chapter 7.

Each of the preceding examples relates to a different problem model and these will be expanded in Chapter 12. For the moment the examples show the varieties with which an engineer is faced, and suggest that his approach must be consistent with the needs, the timing, the priority, etc.

EXERCISES

1.1 A small borax mine has run into an underground vein of water and needs to add a pump to bring the water to the surface. What information should be assembled before attempting to specify the pump?

1.2 Your employer, a small manufacturer, has been disturbed by news of recent burglaries. He has asked you to design and supervise the installation of an alarm system for the building. How would you go about it?

1.3 Your company uses a large number of electric and electronic meters. In the past these have been sent out periodically for calibration and repair. The company believes it could save money and time by setting up a special standardizing section internally. They have asked you to investigate. How would you proceed?

1.4 You work in the research department of a large "broad line" appliance manufacturing company. The government has approached your boss with a request to design and build the food-supply system for a two-man, one-week orbital flight around the earth. What factors would you take into account in preparing for such a development?

Student Project

For a selected one of the following projects, outline the problem situation, covering end-user and his need, the project goal, the purchaser and his interest, other interested persons or groups, limitations in the resources likely to be available to an engineer if it were to be produced, limitations in resources available to you as the designer, various fields of knowledge likely to be involved, how much of the project you would expect to build as a prototype in a working or visual model, and your expected schedule. (See the Appendices for a limited amount of pertinent information on each subject.)

A. *Small Article Retainer for Gemini (two-man) Spacecraft** (Zero gravity, high acceleration).
Astronaut John H. Glenn, Jr., reports that using string to restrain binoculars, light meter, and camera caused serious tangles. He accidentally bumped an untied small container of film, "and it floated out of sight behind the instrument panel." Lift-off and re-entry should be considered.

B. *Isotope Shipping Container* (Nuclear radiation).
A manufacturer wants to build and sell a line of isotope shipping containers to the Atomic Energy Commission (AEC), generally covering the range from 1 curie to 300 curies. The containers must meet the limits of radiation intensity specified by the Atomic Energy Commission publications and conform to the regulations of the Interstate Commerce Commission. The radioactive materials are enclosed in stoppered glass bottles of approximately 4-oz size.

C. *Automatic Titrator* (Chemical handling-measuring).
A chemical laboratory makes frequent use of titration equipment as a part of a research program. The general accuracy desired is within 0.2 per cent beyond the endpoint, over an acid-base concentration range of $0.1N$ to $0.5N$. There are a number of ways of measuring endpoint including: Color, conductance, chemical EMF. How would you make this measurement automatic using simple inexpensive equipment?

D. *Pick-up Device for a Paraplegic* (Medical-mechanical).
The reach of a paraplegic (legs paralyzed) is to be extended so that objects on the floor (home or office) or higher overhead can be grasped and brought to within reach of the hands. The typical array of common objects is to be anticipated.

E. *Blueprint Paper Sensitizer* (Chemical processing).
To be able to carry out a home hobby, it is desired to make $8\frac{1}{2} \times 11$ blueprint paper in a simple device at home, using the ferric oxalate process. The device should take freshly mixed solution and clean white paper, and immerse, dry, and then store the paper in a dark container.

* *Aviation Week and Space Technology*, April 23, 1962, p. 59, quoted by John R. M. Alger and Carl V. Hays, "Creative Synthesis in Design," Prentice-Hall, Inc., Englewood Cliffs, N.J., 1964, p. 54.

F. *Layout of Low-Cost Home for Underdeveloped Country* (Fundamentals of construction).
A serious need exists for the layout and initial estimates for a low-cost home to be built and sold in underdeveloped regions, such as Brazil. The climate is sub-tropical like that of Los Angeles, and basic local materials are available: Cement, clay tiles (wall and roof), pressed-wood sheets, etc., but little structural wood can be obtained. Secondary supplies and equipment must be imported from distributors several hundred miles away: Kitchen, bathroom, and plumbing fixtures and fittings, metal pipe and rod, electric fixtures and wiring, etc. See Appendix B for additional information, including costs. Assume three bedrooms and a small bathroom (shower only) and provide for only the simplest basic needs within 1000 sq ft (90 sq meters). Can a house be made for $500 to $1000?

G. *Burglar Alarm* (Electronic-optical).
A need exists for a low-cost, reliable burglar alarm for the rear entrances of commercial establishments, particularly where there is a walled-off area at the rear also. The alarm may employ an optical or radio method, and use conductive circuits or others, but it should be fail-safe.

H. *Low-Cost "Bull Horn" Speaker* (Electro-acoustic).
A portable voice amplifier is used in many outdoor situations (bands, beaches, large picnics, etc.). Assume you need to speak to 100 or 200 persons over normal background noise. The parts should be obtainable at a radio-electronics dealer. The device should be easy to carry and use.

I. *Educational Toy* (Electro-mechanical-physical).
In the large field of educational toys the interests and pleasures of children can be used to reinforce their education. This field can make use of many devices and cover many subjects: Arithmetic, spelling, geography, simple physics and chemistry, etc. It is understood that such toys would be designed for mass-produced manufacture.

J. *Jar-Top Opener for the Home* (Mechanical).
Whether they contain mayonnaise, olives, or fruit, screwtop jars are difficult to open when they are new. A very useful device would be one that allowed a woman or child to grasp it and apply enough force to unscrew the lid.

K. *Freeway Connections* (Graphics-highways).
Here is an exercise in combining many constraints. The aerial photograph, Figure 1.4, shows the center line of a freeway under construction. One bridge has been built and many houses have been moved or demolished. Based on your estimate of the best over-all results, decide which streets should be blocked off, bridged or tunneled, and which connected to the freeway. Assume the necessary entrances and exits. See Appendix B for standard practice on lanes, turning radii, etc. Draw a layout of the streets and approaches in your final recommended form. Present the reasons for your decisions.

L. *Water Recovery System* (Fluid flow).
Water effluent from a sewage treatment plant is substantially ready for return to lakes and ponds in the hills where it will re-enter the water-table, help com-

bat forest fires, conserve trace minerals, etc. Consider the source to be a treatment plant in the area of suburbs about ten miles from some appropriate canyon reservoirs located in the nearby National Forest. Propose a water transport system to connect source and termini and examine the economics of it. See Appendix B for basic data.

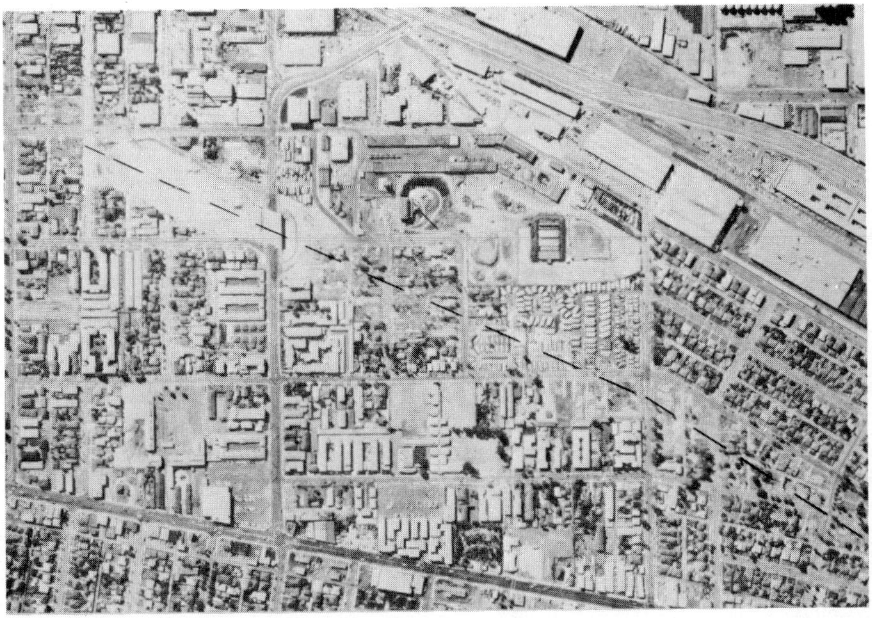

Fig. 1.4 Freeway planning. Aerial view of construction work in process along right-of-way. (UCLA photo)

REFERENCES

1. The Human Species, *Scientific American*, September, 1960.
2. *The World Almanac*, 1965, p. 742, New York World-Telegram, New York.
3. Pamphlet, W. J. King, The Unwritten Laws of Engineering, Published in May, June, July, 1944, by the American Society of Mechanical Engineers, New York.

Additional References

Whinnery J. R.: "The World of Engineering," McGraw-Hill Book Company, New York, 1965.

Krick, E. V.: "An Introduction to Engineering and Engineering Design," John Wiley & Sons, Inc., New York, 1965.

Wilson, W. E.: "Concepts of Engineering System Design," McGraw-Hill Book Company, New York, 1965.

Mantell, M. I., M. E. Reeder, and J. A. Stevens: "Orientation in Engineering," 2nd ed., Prentice-Hall, Inc., Englewood Cliffs, N.J., 1963.

Smith, Ralph J.: "Engineering as a Career," 2nd ed., McGraw-Hill Book Company, New York, 1962.

Johnson, L. H.: "Engineering, Principles and Problems," McGraw-Hill Book Company, New York, 1960.

Design Project Organization: Planning a Design

Two

"The shortest way to do many things
is to do one at a time."
Cecil

"Life is not long,
and too much of it must not pass in idle
deliberation how it will be spent."
Samuel Johnson

Organizing to Produce Results

The engineer should be goal-oriented, as was observed in the last chapter. The engineer must hence maintain an organized approach to his work. The logical development of a task far surpasses a haphazard, intuitive one, because of the crucial importance of countless items of detail in the final success. Every connected relay, switch, vacuum tube, and wire must be operating for a telephone call to go through; and every microscopic leak must be avoided in order to keep a refrigeration system in service.

"Problem-solving" is the name given to procedures suggested for organizing the smaller units of engineering work. We list a variety of these procedures in Table 2.1. The larger units of engineering work are projects, usually including several or many problems, the total projects consisting of the technical portions of some entrepreneurial or governmental undertaking. Let us point out again that, in our context, a "problem" is not the clear, incisive, thorough statement of numbers and factors usually associated with physical or mathematical exercises. Our problems might

Table 2.1 The Problem-solving Process: A Variety of Descriptions.

Design Process[1] *(Asimow)*	*Thought Process*[2] *(Wallas)*	*Professional Method*[3] *(Ver Planck and Teare)*	*Engineering Method*[4] *(Smith)*	*Problem-solving*[5] *(Buhl)*	*Scientific Method*[6] *(General Electric)*	*Law (Ballantine)*[7]	*Management (General Electric)*[8]	*Military (Staff College)*[9]
Analysis	Preparation	Define problem	Preliminary analysis	Recognition Definition Preparation Analysis	Collect existing facts List of missing facts	Reject irrelevant material	Plan	Recognize, gather data
Synthesis	Incubation Illumination	Plan treatment	Statement of question	Synthesis	Develop hypothesis	Translate and restate	Organize Integrate	List possible solutions
Evaluation				Evaluation			Measure	Test possible solution
Decision			Solution			State legal issues		
Optimization	Elaboration	Execute plan			Design and conduct experiment			Select best possible solutions
Revision		Check as a whole	Check		Revise hypothesis			
Implementation		Learn and generalize	Interpretation	Presentation	Develop theories which prove valid			Apply

Note: Superscript numerals refer to references at end of chapter. The horizontal alignment does not really imply functional correspondence; only an approximate stage-wise agreement with column one is attempted.

better be called "unresolved technical needs," "perplexing situations," or "messes."

We include Table 2.1 to allow the engineer to make a composite suited to his tasks. The table also points out some of the subtleties of mental processes with the inclusion of such words as illumination, integration, and synthesis. The different columns do not purport to characterize exactly the same things, but they are close enough to be useful.

One other comment is appropriate here. Solutions to problem situations come to the mind through *understanding*, not simply by the superficial or fortuitous fitting of a few facts together or by the substituting of some numbers into a formula. Such insight or understanding comes from a deep penetration and an encompassing of the many aspects and facets of the original situation. Desirable solutions come more from the application of persistence than flash, of care than energy, of wisdom than wit.

What is the difference between a problem and a project? The answer is simply that of degree. The border between the two is shaded, but the design projects we shall describe below are generally dignified by importance, risk, cost, and complication. Systems, subsystems, components, and parts are all involved. The individual case will determine the name. A project usually encompasses man-months or man-years. In all cases, however, the design effort is expected to produce viable ideas and then detail drawings, specifications, instructions, and predictions.

Participation of the Engineer in Goal Definitions

Strange as it may appear, the engineer must defend himself, his reputation, and his future by entering as early as possible into the setting of goals, and the defining of problems. This is particularly important when his work is only a part of the whole or when the client has done some investigating on his own before asking for bids or for further professional help. Only as the engineer can understand what is the ultimate goal or technical objective will he be able to search along fruitful avenues. Consider, for example, the re-entry heat shield on the Project Mercury space capsule. The temperature just ahead of the leading face was 14,000°F as measured by wind-tunnel tests during simulated re-entry decelerations. The problem was originally stated: "What materials can we find or develop to withstand 14,000°F for many minutes?" The ensuing search produced no known or foreseeable suitable materials. However, when the question was restated as: "How can the capsule structure be protected and the man inside safety insulated?" a method was developed promptly. The solution consisted of using the exterior heat to boil out gases through a porous ceramic shell, forming a protective moving gaseous interface at only 5000°F. This "ablation" method was used for all the later successful orbital flights.

The engineer must start his thinking at the "primitive statement of need," the simple English statement of what is wanted, frequently even before numbers enter consideration. The later in the process an engineer enters as a participant, the more restricted his alternatives must be, and the less able will he be to compete successfully. This is as true in industry as in consulting, and applies even to the very well-qualified engineer.

A natural question would be, how does an engineer influence the original goals, when for one reason or another he enters late? One answer is, of course, that he may be too late to do anything; another more constructive answer is that, by studying the situation thoroughly, he can confidently make his own proposals leading back to the beginning. These proposals, of course, must offer benefits evidently better than any issuing from the progress to data; but this is frequently not too difficult. In companies soliciting United States government contracts today, considerable effort is expended to keep up with the thinking of various agencies. This is done so that, at strategically early moments, an appropriate proposal can be made for such long-term expected projects as a new supersonic airplane, a new weapons system, or a new salt-water purification plant, which have been under extended consideration.

Time Sequence of Design Project: Structure of Design

The problem-solving elements described earlier would seem to constitute a unit sequence of events for some kind of a stage or module. However, a whole design project should be granted a grand strategy, a time sequence having a distinct pattern for fitting together successive, widely-differing phases. Although occasionally the same designer is in charge during all of these phases, as time passes he performs different functions and judges the results on different bases, as the ruling circumstances change from creativity to performance, to cost, to deadlines.

We will call the collection of time phases of differing steps the *structure** or morphology of design. The following names given to the major subdivisions or phases of the project structure are more or less self-explanatory:

Phase	*Function*
1. Feasibility Study	To validate the need and produce a set of possible solutions.
2. Preliminary Design	To quantify the parameters so as to yield the optimum solution.

* The study of the forms and structures of design is called the *morphology of design*, from the Greek *morphe*, form.

Phase	*Function*
3. Detail Design	To reduce the best solution to a description for manufacture.
4. Revision	To produce an improved or acceptable design, based on experience with a manufactured and tested system.

Before we examine these phases more closely, let us identify them as applied to the design of a modified bicycle, made to be easily interchangeable for boy or girl. Under the *Feasibility Study*, the need would be stated and validated. Such a statement of need might be:

"We need a new bicycle frame with a movable center bar and increased structural rigidity, the drive gear ratios are all right, but the gear shift must be changed to require lower hand or finger forces."

Taken alone, these might be considered problems, but taken together and requiring manufacture, they qualify as a "project." The *Feasibility Study* would be carried on to produce sets of plausible ideas for these new components. The *Preliminary Design* would select the group of ideas that best fitted together, establish the necessary sizes and other relationships, and attempt to predict the performance. The *Detail Design* would produce manufacturing drawings with all dimensions, tolerances and specifications, down to the last nut and cotter pin. The same sequence, of course, would be carried out for a large system such as an oil refinery, for example. The component words would be different, there would be many more substeps, and vastly greater problems, but the steps of the procedure would apply directly.

We said that the attitude of the designer changes from phase to phase. At first during the Feasibility Study, he explores widely and makes only rough order-of-magnitude checks on cost and performance. Then in the Preliminary Design he proceeds carefully to weed out the inaccuracies and the less valuable alternatives. In the Detail Design he can let no detail go unconsidered. Every number, every dimension and every material specification must be checked and double-checked. Time and cost are important here, both for the project and the product. Thus, the activities and viewpoints of the designer follow the mood of the project, up to and beyond the point of manufacture. Naturally, the procedure holds whether the outcome is simply a design report or paper study, or whether it finally eventuates in hardware.

There are several ways of describing what a designer does. Sandor,[10] for example, discusses the seven stages of design. Buhl[5] presents his version of seven steps in designing. Our particular way of charting the various steps within each phase is shown in Figures 2.1 and 2.2. At this time we will men-

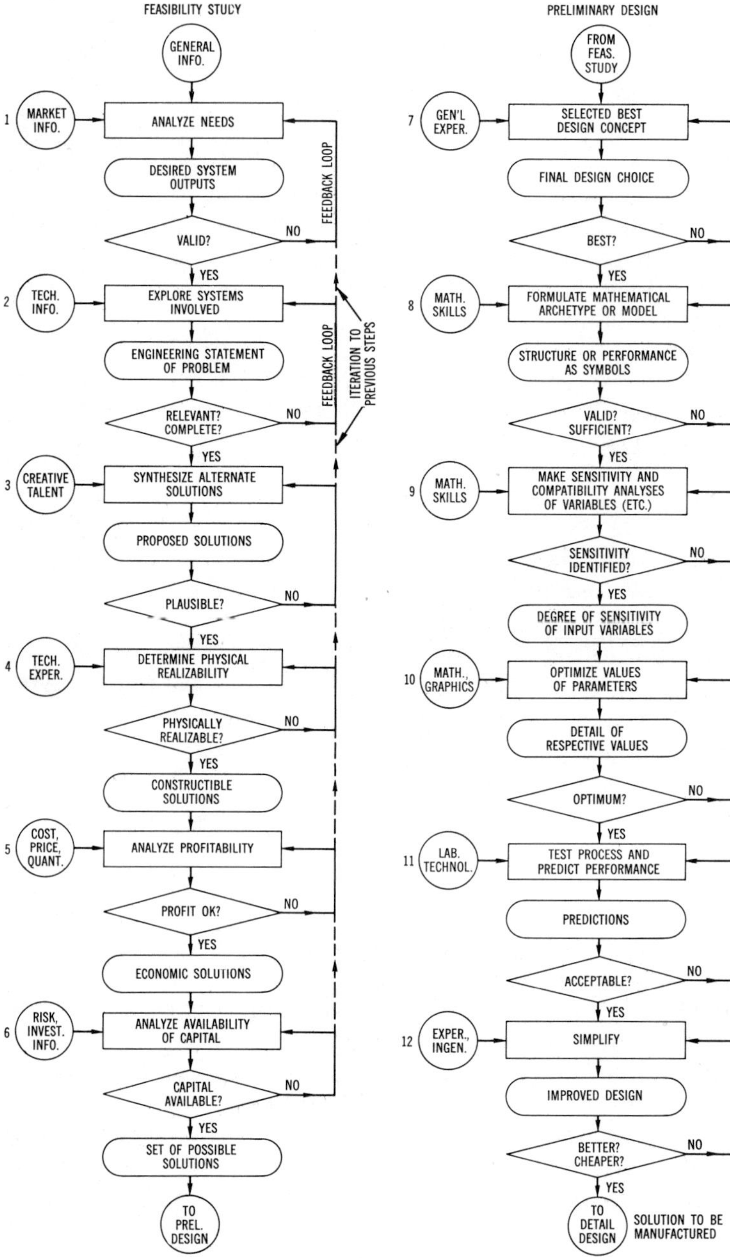

Fig. 2.1 Structure of design. Feasibility study and preliminary design.

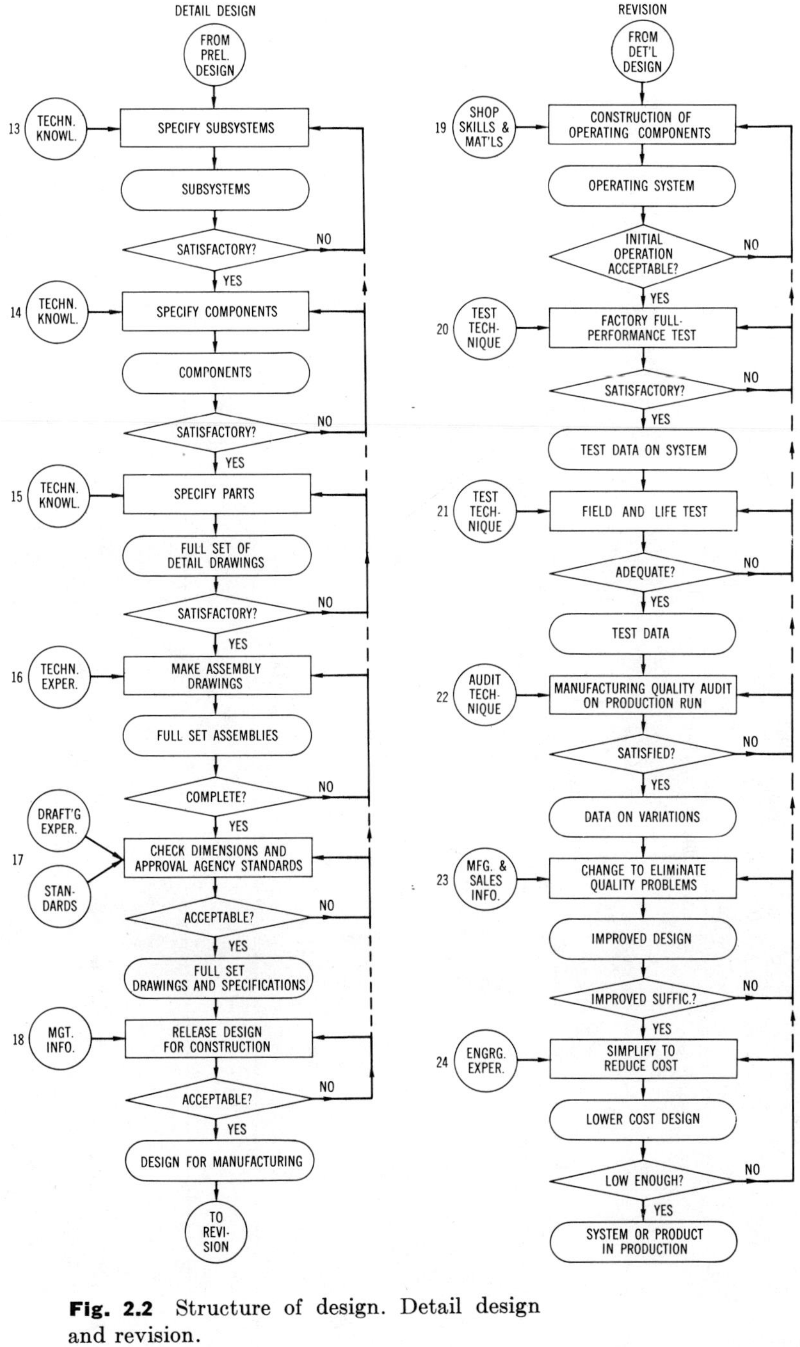

Fig. 2.2 Structure of design. Detail design and revision.

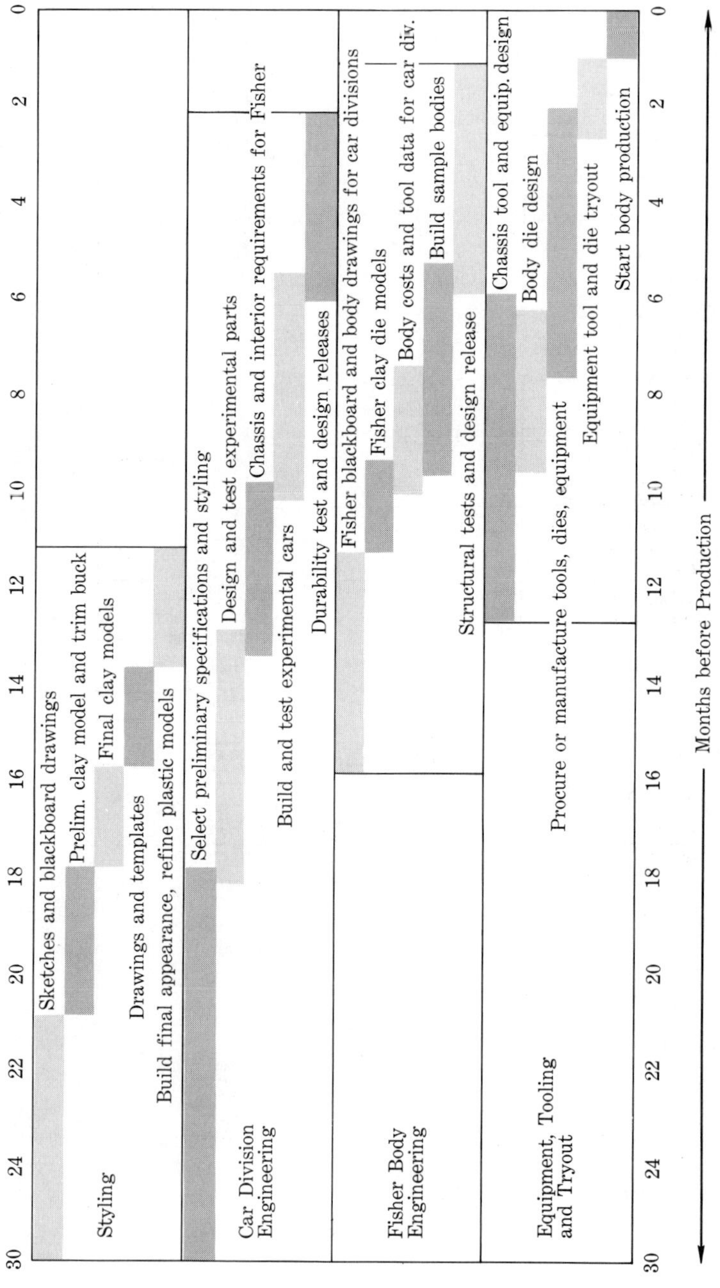

Fig. 2.3 New automobile body model timetable. (Courtesy General Motors Corp.)

tion their interrelations only briefly. A more detailed discussion is given in Chapter 4 for the Feasibility Study and in Chapter 10 for the Preliminary Design.

The reader may well question whether all the 24 steps shown in Figures 2.1 and 2.2 are really employed in practice. The answer is yes; these and more are used on any sizeable project, whether or not they are known or named this particular way. We refer the reader to any conventional development bank loan application form, to see the penetrating examination to which any new venture is subjected. Another illustration in depth is shown in Figure 2.3, the new model automobile body production schedule.

We trek through the following Glossary of Design so the reader can understand and profitably use those portions which fit his own needs, from case to case.

The Design Process (A Brief Glossary)

Note that each of the major phases in Figures 2.1 and 2.2 is composed of a number of similar graphic units. Asimow[1] calls this unit stage the "Design Process," as shown in Figure 2.4.

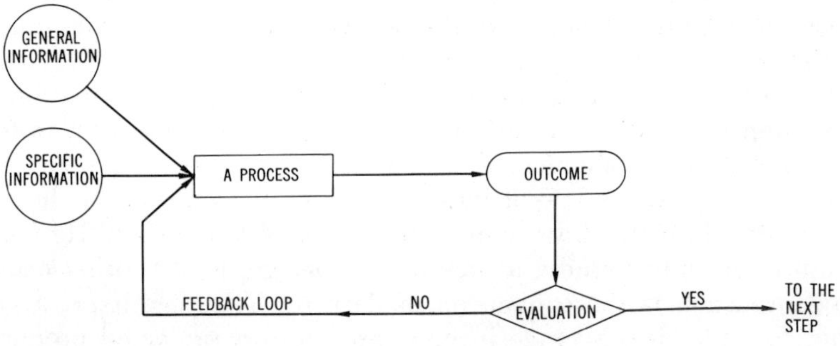

Fig. 2.4 The repeated graphic module used in Figures 2.1 and 2.2.

These diagrams all represent the flow of information and decisions. Except for the "Feedback Loops," all information would come into each "Process" and flow straight through to the end of the design. These return-path loops enable experiences gained through a given activity to enter into reconsideration, if the first results prove unsatisfactory. The concept is based upon the essential premise that the designer seldom arrives at a finally acceptable solution the first time he attempts it. Further thought inevitably uncovers weaknesses in his proposals and surprise factors he had overlooked. With the new information, he can now "iterate" the process, this time making a much closer approximation to the needed answer. The results usually converge satisfactorily after only a few iterations.

The various words in the latter flow chart, Figure 2.4, have been given the meanings and uses defined in the next paragraphs. Again, we stress that these are conceptual aids. They are not the actual processes, nor are they the only way of describing the process. Here are the meanings we attach to the key words:

General Information includes all types of common knowledge, whether common to the layman or to all engineers. It includes all varieties and all degrees of certainity, all fields of endeavor, and all sources at the fingertips of the engineer or his associates.

Specific Information is that knowledge directly pertinent to the immediate problem. It comes generally from specialized sources such as experts, libraries, or manufacturers of special equipment. It can also involve such "non-engineering" subjects as market, finance, personnel, product, or any field pertinent to the design step at hand.

All such information must be acknowledged to be within the limits of certainty and uncertainty which every person knows: From that absolute certainty which the mind knows about its own feelings, to the remote uncertainty about the probable happenings of tomorrow. Self-confidence must be developed to deal effectively with such a wide range of possibilities.

The *Process* is a transformation of the motley input information into integrated useful output information. For example, the engineer transforms information on the mechanical and magnetic properties of sheet steel, information on Faraday's law of induction, plus information on the needs of a $\frac{1}{4}$-hp motor, into the drawing of a motor stator lamination. He may do this within a feasibility study, a preliminary design, etc. Another example of a transformation is the conversion of data on living habits of a group of persons, sample size, average income, etc., into a projected production rate for a new product. The engineer uses the Process, therefore, to convert raw input information into necessary output information, useful for the next step(s) in the design.

The *Outcome* is the useful information, presumably ready for the next step.

Evaluation is the review and decision process by which the effectiveness of the step is adjudged acceptable or not (yes, no). If *yes*, the next step is put in motion; if *no*, the need for further effort or more information is implied. This evaluation itself points up the deficiencies which are returned as added input information along the "Feedback Loop." This iterative return is repeated as often as necessary to yield a *yes* finally at evaluation.

If the next step is undertaken with unsatisfactory information (i.e., no), it simply is a waste of time and must later be repeated when critically reviewed.

Such successive approximations, or repeated trials, may at the outset seem an inappropriate, even a sloppy way for an engineer to proceed. However, it is a very realistic method. The only tasks that can be done "in-line," straight-through, are either those which have in essence been done before or those whose process rules permit no exceptions. It will be found upon investigation that *all* new creations of the mind are the result of trial and error, whether in the field of mathematics, logic, physics, law, life science, etc. Engineering design is the creation of something new, something that has never before been put together in this way for this purpose, even if it is assembled from "old" parts. A most pertinent illustration is the digital computer. Before 1945, no one had thought of using relays, switches, vacuum tubes, and counters in a stored-program machine. It took the creative genius of three men, Burks, Goldstine, and Von Neumann* at Princeton to put together in a new way these and other available "old" components. As a result, they increased man's computing-speed capability by a factor of a million.

Thus, iteration, or going back over previous stages when necessary, is one of the prime characteristics of the design structure. In Figures 2.1 and 2.2, iteration not only includes parts of the same step or stage; it may include much earlier steps. Consider the situation in which a design is predicated on the use of stainless steel for a given member. Later it is found that a new plastic, polypropylene, is now commercially available and has the same corrosion resistance but considerably less cost. Naturally the structural design must be completely changed to permit the use of the new material. Wall thickness, stiffness, and color possibilities all are radically different.

End Objective

The use of iteration is dependent on the designer. Does he need to return and redesign, or should he go on ahead? Only the designer's sense of suitability (or his client's), his judgment of how well he has met the objectives and how much time he has left, only these can govern the iterations. Another matter of choice is the observance or the skipping of the various steps. *The outline is not a cookbook recipe to be slavishly followed.* It is a check list for thinking.

* A. W. Burks, H. W. Goldstine, J. Von Neumann, Preliminary Discussion of the Logical Design of an Electronic Computing Instrument, *Datamation*, September, 1962, pp. 25–31.

The individual case determines the usage and the emphasis. One can think of cases where optimization would be a waste of time. But in the design and manufacture of a kitchen refrigerator, careful optimization is vital; it makes the difference between success and failure commercially. On the other hand, when a very newly conceived system is built, only the principles can be optimized; optimization of detail cannot be attempted seriously; that must wait until later. Likewise, when only one embodiment of the system is to be manufactured, like the Glen Canyon Dam, one does not first make a model-shop sample for lab test! The activities must make good sense. Each step the engineer takes should be based on his evaluation of need, timing, cost, and alternatives.

Another point should be noted. Step 8, Mathematical Model, is highly desirable because of the power of analysis it gives the engineer. But there are many cases where no such model can be constructed either because of high cost or insufficient knowledge. Then the steps 9 and 10 cannot be done mathematically; they must be done by logic and common sense. Even if a variable cannot be exactly quantified, it can be evaluated as to great or small effects.

The desired qualities and features of the end result of the design must be continuously kept in the mind's eye. This will guide the choice of procedures, of methods, of materials, and of geometries that are selected along the way.

BOOK PROJECT EXAMPLES

The first example of the previous chapter concerned a request for a floor slab design and cost. Let us illustrate the application of the structure of design to this example.

First, let us recognize that in this case the engineer's experience allows him to go immediately to the Detail Design phase and start with the components. The following constitutes the total job for him. (The item numbers for the various steps correspond to those in the flow chart of Figure 2.2.)

Step	*Engineer's Action*
Specify Components (14)	1. Determines that reinforcing rod is necessary. 2. Ascertains information on floor drains, footing lag bolts, condition of ground base, delivery of material, etc.
Specify Parts (15)	1. Specifies reinforcing rod on the detail-assembly drawing.

Step	*Engineer's Action*
	2. Specifies drains, bolts, etc. on the above drawing.
	3. Specifies slab and footing dimensions and mix on the above drawing.
	4. Computes cost estimate.
Checks (17)	1. Checks Building Code Handbook.
Release (18)	1. Releases design to customer.

The second example concerned the discharge from a sand conveyor. Here the engineer started with a math model and needed only a few more steps. (See Figure 2.1.)

Step	*Engineer's Action*
Formulate Math Model (8)	1. Select a liquid equivalent, with the density of sand: 95 lb/ft^3.
	2. Assume flow concentrated along its median line.
	3. Assume no adhesion to the belt.
	4. Write the equation of motion using Newton's Laws.
Check Sensitivity (9)	1. Check the effect of air on the trajectory.
	2. Check the effect of the moisture content of the sand.
Predict Performance (11)	1. Make numerical computations based on the above assumptions (e.g., parabolic path).

The last example from Chapter 1 concerned the design of a retaining wall for a parking lot. This, an extensive project, uses all steps in the Feasibility Study, and most of those in the Preliminary Design and Detail Design. Let us illustrate only with the Feasibility steps.

Step	*Engineer's Action*
Analysis of Needs (1)	1. The engineer may roughly estimate the cost and value of the wall relative to other uses of the land (i.e., store, office, filling station, etc.) to be able to discuss alternatives.
	2. Traffic loads, directions, timing, parking alternatives are explored.
	3. Assuming that a parking lot is decided upon, criterion is established to be the maximum return on investment. Only \$5000 is available from owner. Balance must be borrowed.

Step	*Engineer's Action*
Systems Involved (2)	1. Parking lot operation (automatic vs. attendant). 2. Convenience of access vs. maximum number of cars. 3. Wall materials (concrete, steel, wood). 4. Wall methods (posts, piles, braces, flat wall, internal buttress, etc.).
Alternate Solutions (3)	1. Various geometric combinations of flat area, entrances, wall, operation, etc.
Physical Realizability (4)	1. Check wall designs for their capability of supporting earth. 2. Check availability of automatic entrance gate equipment.
Economic Worth (5)	1. Compare estimated income with the total expenses, both initial costs and operating expenses.
Plan Financing (6)	1. Ascertain where to obtain money to pay for the initial improvements. Does the owner know a source or will he have to ask a commercial bank for a loan? If so, what rate of interest will he have to pay?

EXERCISES

The following situations are offered to help further the reader's understanding of the use or non-use of the formal structure of design. In each case, identify the step by number and write a short phrase describing what you believe should be the appropriate actions of the design engineer.

2.1 Design a counterweight system for your overhead garage doors, given that its spring system is not effective.

2.2 Submit a design for a hand-operated brake-pedal attachment for a conventional car, for use by a crippled person having both legs paralyzed.

2.3 A family desires the wall removed between two downstairs rooms in a two-story house.

2.4 The sales department submits a request for a remote-controlled mechanism to turn a home television roof-antenna to the position of maximum signal strength.

2.5 A client asks for the calibration of a dip stick to be used in measuring fuel in a horizontal-axis cylindrical tank having a diameter equal to one half the length and rated nominally at 1000-gal. capacity.

2.6 Assume a conical sand pile built up by a conveyor discharging onto its vertex. The water content of the sand ranges from zero to ten per cent by weight. What is the formula or graph for the volume of the sand pile in terms of the height and the percent water content? (*Hint:* Where can you find the angle of repose as a function of water content?)

Student Project Exercises

Organize your thinking concerning the project you have selected. Outline the steps in operational form for your own use, including as many of the phases of the design structure as you presently estimate will apply usefully. Be realistic; don't include a step unless you know it must be included or it seems necessary and you have some idea about how to start it. From Chapter 1:

A. Small-article restrainer for spacecraft
B. Isotope shipping container
C. Automatic titrator
D. Pick-up device for paraplegic
E. Blueprint paper sensitizer
F. Low-cost subtropical home
G. Burglar alarm
H. "Bull Horn" speaker
I. Educational toy
J. Jar-top opener
K. Freeway interchange
L. Water recovery system
M. Other

REFERENCES

1. Asimow, M.: "Introduction to Design," Prentice-Hall Inc., Englewood Cliffs, N.J., 1962.
2. Wallas, G.: "The Art of Thought," Harcourt, Brace and World, Inc., New York, 1926.
3. Ver Planck, D. W., and B. R. Teare, Jr.: "Engineering Analysis," John Wiley and Sons, Inc., New York, 1954.
4. Smith, R. J.: The Engineering Profession, College Training in Engineering, The Engineering Sciences, in "Engineering as a Career," 2nd ed., McGraw-Hill Book Company, New York, 1962.
5. Buhl, H. R.: "Creative Engineering Design," Iowa State University Press, Ames, Iowa, 1960.

References 6–9 are Special References from Alger, J. R. M., and C. V. Hays, "Creative Synthesis in Design," p. 9, Prentice Hall, Inc., Englewood Cliffs, N.J., 1964:

6. "Professional Management in General Electric," General Electric Company, Schenectady, N.Y., Book III, p. 139, Figure 10, 1954.
7. Ballantine, H. W.: "Ballantine's Problems in Law," 3rd ed., pp. 5–7, West Publishing Co., St. Paul, Minnesota, 1949.
8. "Professional Management in General Electric," General Electric Company, Schenectady, N.Y., Book III, p. 14, Figure 1, 1954.
9. "Air Command and Staff College," ASC-1, Maxwell Air Force Base.
10. Sandor, G.: "The Seven Stages of Design," American Society of Mechanical Engineers, New York, paper 63-WA-113, 1963.

Information and the Needs Analysis

Three

"For your information,
let me ask you a question."
Sam Goldwyn

The Engineer and Information

"The engineer earns his salary by organizing, improving and transmitting information." This, as first heard, may seem to be a facetious, unappreciative description of the engineer's mission. Yet, it doesn't seem quite so warped when one recalls that the client or supervisor really seeks *only* information from the engineer; pertinent, reliable information in usable form: Drawings, specifications, performance predictions, bills of material, technical advice. The engineering department may produce an operating prototype; but the engineer himself produces only information. The engineer, therefore, must collect whatever types he may find from whatever sources within access, and convert them into the desired form. To be sure, other persons in an enterprise also process information, but the engineer uniquely creates systems and products based on his information.

Let us examine a case of "converted information." Some years ago a company, which manufactured electric power transformers, analyzed the competitive market prices of a transformer line to deduce what portion of the product cost could be allocated to the fabrication of the iron core. In other words, the engineer converted market information into numerical information which signified potential cost reduction. Also, information about the improved magnetic qualities of along-the-grain steel sheets versus cross-grain sheets, costs of these materials, and so forth, was brought together and assimilated. Under the pressure of the project, these were synthesized into a revolutionary invention, coiled-core transformers, illustrated in Figure 3.1. "Spiracore" was the name given to this ingenious,

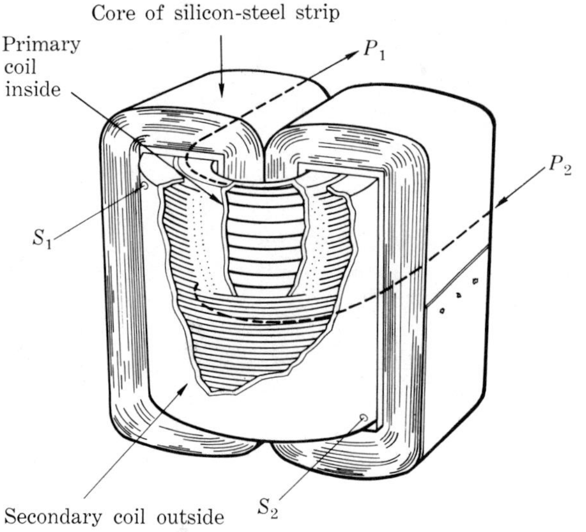

Fig. 3.1 The "Spiracore" transformer.* Pre-rolled magnetic strip core threaded through finished winding assemblies. (Courtesy General Electric Co.)

highly efficient, lower-cost method of coiling the cores from strip steel instead of stacking them from punched flat sheets.* Thus, common, readily procured information was converted by purpose and ingenuity into a profitable new method of manufacture. If the costs and the magnetic quality information had not been prepared there would have been no stimulus to invent.

A parallel case is the original invention of the Bendix auto-engine starter, shown in Figure 3.2. The beginning information here included the exertion and hazard of hand-cranking the engine, the terminal voltages

* Steel strip is tightly rolled into hollow cylindrical cores. Then these coils are opened up and threaded through the pre-assembled winding coil-forms. The core coils are re-rolled closed and cinched tight in the original compact shape and spotwelded secure. (General Electric Co. "Spiracore," Reg. T.M.)

and other electrical data on the different types of battery cells, the parameters of motor design, gear life, etc. These were all converted by engineering, into the now familiar gear-disengaging starter, which only slightly modified has served the auto industry for over 40 years.

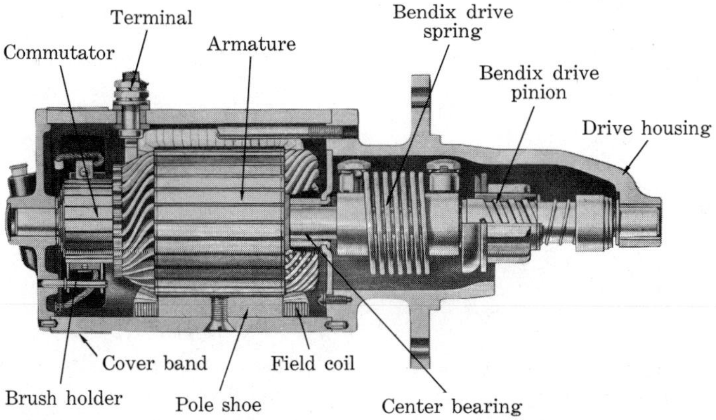

Fig. 3.2 The Bendix automotive starter. This is the early form used for many years after the initial conception. (Courtesy Delco Remy Division, General Motors Corp.)

These preceding examples illustrate how the engineer generates and controls the flow and conversion of information. In most cases, in addition to the control, he actually performs the conversion or transformation himself.

Information Flow Diagram

The following basic diagram of flow symbolizes a variety of flow processes which occur in engineering: Flow of goods, money, energy, *and* information.

Fig. 3.3 Basic diagram of flow.

This diagram, Figure 3.3, is in fact the fundamental symbol from which is evolved the Design Process (Fig. 2.3), and in turn the stage framework of design (Figs. 2.1, 2.2).

To generalize our thoughts so far, it is the engineer's responsibility to transform nebulous, incomplete inputs of information into useful, specific, and complete information outputs.

The information flow diagram for our case of the engine-starter may be shown as in Figure 3.4.

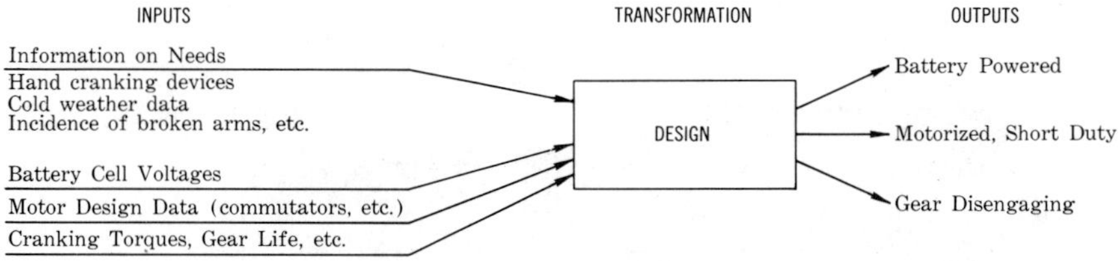

Fig. 3.4 Example of nonphysical transformations.

One can accept this kind of transformation conceptually without understanding exactly how it takes place. This mental transformation is usually followed by a physical transformation; and we can better visualize the former by going on to the latter. Consider an electronics factory where the physical transformation is producing radios by changing the shape and location of raw materials and components.

For the manufacture of our radios, the Input/Output diagram takes the simple shape shown in Figure 3.5. Note, however, that the manufacturer knows in advance exactly what his end product will be, and through what processes he must go to arrive there.

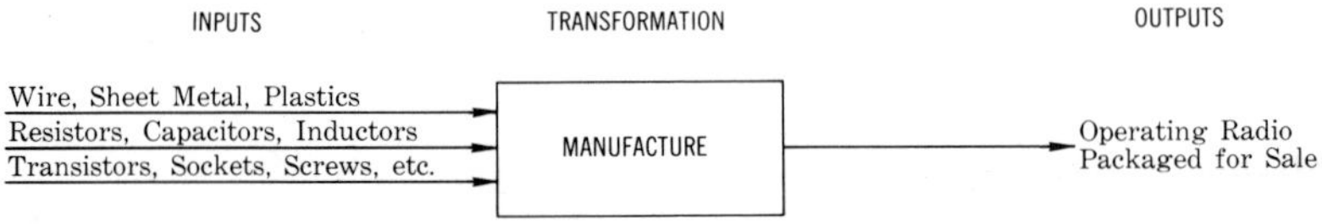

Fig. 3.5 Example of physical transformations.

If we apply this illustration to the engineer transforming information, we can picture him choosing iteratively from many unidentified, seemingly random materials and components, from which to build a partly-conceived device or system, which will be carefully scrutinized and evaluated, but which as yet has been only approximately specified. The engineer selects alternative assemblies and projects their cooperating performances, to see if he can accomplish what he thinks he should.

The engineer designing the radio went through work, shown in Figure 3.6, similar to that done by the engineer for the automotive starter, shown in Figure 3.4.

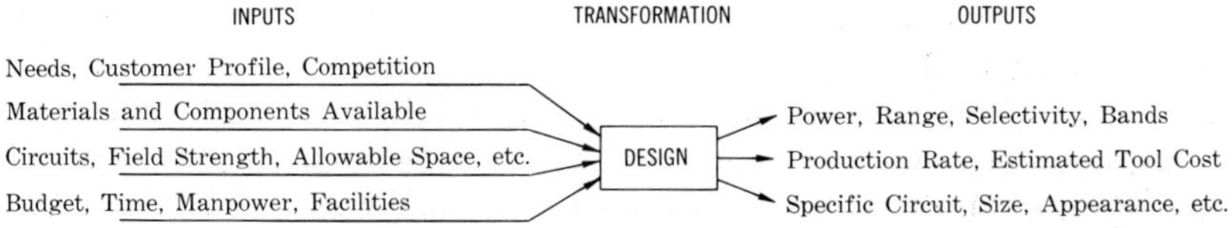

Fig. 3.6 Example of information transformation in a radio project.

The Real Problems about Information

It is fairly obvious from the foregoing discussion that the engineer needs a great deal of information, whether he has it in his mind already or whether he must get it from other sources. As yet, however, we have only hinted at what real problems he will encounter searching for information.

The engineer's real problems about information generally are these:

Where can I find it? (Availability, location of sources, nature of sources.)
How can I get it? (Accessibility, cost, directness, delay.)
Can I believe it? (Credibility, authenticity, relevance, accuracy.)
How do I interpret it? (Meaning, applicability.)
Is it enough? (Sufficiency of amount and variety.)
What is the decision as a result of it? (Yes, no, maybe, later?)

The engineer does not need new information at all stages of his design; but where he does, the above questions dog his tracks and affect his progress.

At the beginning of a design, the Needs Analysis, as we shall discuss, focuses the engineer's attention on "Can I believe it?" "Is it enough?" Later on, during the concentration on creativity, the engineer seeks a wide variety of ideas, and then he is most interested in "How can I get it?" As we now pursue these subjects, let us define what we mean by information, and describe some of its varieties that the engineer will encounter. We will go into further detail about sources in Chapter 5.

We are using "information" here in its general sense as facts, data, unorganized knowledge, or intelligence. It may be verbal, graphic, or symbolic. A mathematical theory of information has been developed for the field of communications systems in which "information" is used in an entirely different sense.* Here, however, as we use it, information covers all aspects of the problem situation, the materials, the principles, the

* In the mathematical theory, information comprises not what you *did* say but what you *could* say. If there was only one possible message, the information is zero. Information is quantified by measuring the probability of the occurrence of the symbols (letters) available.

economics that the engineer works with; it includes the governmental laws and regulations, the whims, preferences, prejudices and pressures of all those persons involved in the project; and finally it includes the necessary conclusions, recommendations, and descriptions he finishes with. If the necessary and desired information were immediately available, the engineer's job could be done in a tenth of the time. This explains why the efficient engineering office is normally well equipped with catalogs, reports, files, data books, operating samples, and all types of handbooks and other sources.

Engineering experience bears out the following generalizations about "information" and the project:

1. A small amount is available at the outset, but it is never complete; much more is necessary, but what to seek is not then known.
2. Time is required to find out what else to search for, and more time is required to find it.
3. Information items offered are frequently conflicting; also much of it is not necessary or useful.
4. Information costs time and money to obtain.

Information and the Production-Consumption Cycle

In this chapter we shall concentrate on information flow and its validation during the first phase of a design, the Needs Analysis. For background on where the engineer's information originates, we shall consider the total production-consumption cycle[3] of man's socio-economic system. Figure 3.7 portrays the various stages that material goods pass through in their cycling from raw materials around to raw materials again, after use and discard. In man's economy we find frequent instances of no material recovery; all such consumption results in waste. In the natural world, the equivalent cycle is closed; waste becomes a raw material.

In the schematic diagram of Figure 3.7, *goods* flow counterclockwise with the outside arrows, while *information* necessary for Production generally flows clockwise with the inside arrow. That is, the consumer tells the distributor what he wants or needs, and he in turn tells the manufacturer. On the other hand, the purchaser of large capital goods negotiates directly with the manufacturer concerning needs and preferences. The consumer-goods purchaser usually communicates simply by his (or her) buying habits. Occasionally the individual consumer reports directly to the manufacturer, in which case he is generally registering a complaint, rather than stating a need.

The information about Needs from various parts of the cycle is of crucial importance to the Design Engineer, because even slight changes in assumptions can cause major changes of design direction. Consider, for example, the design of a water-evaporation-type home cooler. Will it ever be subjected to sudden freezing temperatures? This makes a *major difference* in the design. All persons in the cycle of Figure 3.7 are interested in the "goods," but each is interested from his own special viewpoint. The customer wants them functional, good-looking, long-lasting, and inexpensive. The distributor wants them with sales appeal, with no service problems, safely shipped, and with lots of profit. The manufacturer wants them easily made

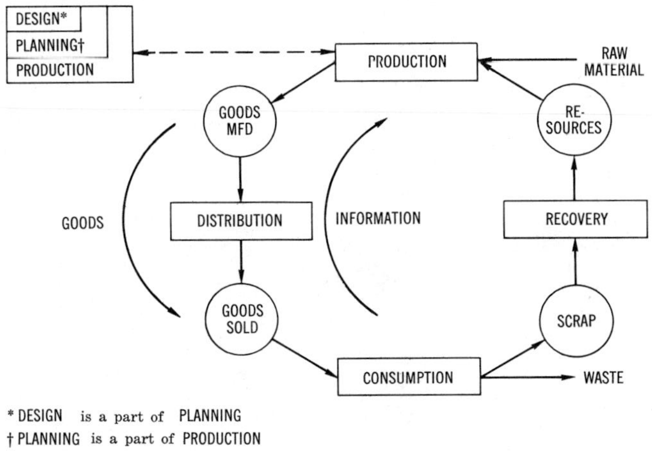

Fig. 3.7 The production-consumption cycle, showing the place of engineering design.

by low-cost labor and easily procured materials, with no seasonality, and with high demand. These extremes of particular preferences may seem incompatible, but they represent strong feelings and desires on the part of those who must be considered in a Needs Analysis. These varieties of viewpoint apply equally to all engineering projects; to building construction, to large industrial machines, to military equipment, etc., because *persons* are still the makers, sellers, and users.

Varieties of Information

The foregoing discussion dealt with general sources and kinds of information. Here we shall examine certain informal but important *categories* of information, especially those having to do with reliability and usefulness.

Uncertainty in the engineering situation is due partly to uncertainty about the information itself. Some information is reliable and we know it; some is represented as reliable, but actually is unreliable or incorrect and we do not know it—possibly never can—during the time we need it. As an example of misinformation, the Wright brothers* reported that their early machine would fly only in winds of 25 mph, although they had used data from Lilenthal's air-pressure tables indicating it should have flown at 16 mph. A year later, using a new machine, the tables and results still didn't jibe, and they decided to run some homemade wind-tunnel tests. Finally data from those tests did agree with their flight results, and the originally-accepted published data were shown to be in error. Engineers frequently encounter the same trouble: Error in published data that shows up only upon a belated proof-test.

This example leads us to an informal way of classifying information into "hard" and "soft" categories as well as "necessary" and "dispensable." The classifications are suggested as urgent warning against treating all information in the same way.

"Hard" information is verifiable, unambiguous, permanent, documentable, numerical, checked by several sources, or it has some combination of these attributes. It has maximum reliability when properly transmitted. All information can suffer in transmittal, even photographs of text material, but we are referring to the basic nature of the information itself. This is the type of information the engineer depends upon most confidently.

"Hard" information includes such matters as:

Principles	Data on present systems
Laws	Contracts
Quantities	Physical relations
Standards	Drawings, photographs

"Soft" information may be equally or more important, but it is generally nebulous, qualitative, verbal, transient, not necessarily verifiable, or it has some combination of these characteristics. Having this nature, it is even more likely to suffer in transmittal, particularly if the communication is verbal. This type must be doubly verified, because its importance frequently beguiles the engineer into uncritical acceptance.

"Soft" information includes:

Opinions	Situations
Market surveys	Hearsay
Recommendations	One-run statistics
Ideas, proposals	Projected data on future systems

* Orville Wright, "How We Invented the Airplane," David McKay Co., Inc., 1953; reported in Case Study EDP #2-64, UCLA Engineering Department, Los Angeles, 1964.

"Soft" information cannot generally be "proven," so it is tempting psychologically to stretch it, even if slightly. Given the same conditions, "hard" information is likely to be more reliable than "soft," simply because of the salutary effect of verifiability.

"Soft" information ranges from the sublime to the ridiculous, from flying saucers to infallible business ideas, and frequently one cannot be easily distinguished from the other. Particularly insidious are generally accepted assumptions, recognizably false upon investigation, but still considered good information.

For instance, the assumption that Americans would not continue to buy compact cars cost the automobile industry two years of sales in the late 1950s. Similarly, unwarranted assumptions also cost the Ford Motor Company millions of dollars in the ill-advised Edsel. To be most valuable on his job, the engineer must buck the crowd occasionally and quietly harbor reasonable doubts until the evidence is all in.

"Necessary" vs. "dispensable" information is more or less self-explanatory. "Hard" and "soft" may be either necessary or dispensable because the choice depends on a value judgment of *use*, not of intrinsic *nature*. The value-importance is related to the risk of the pending decision. If the risk is found to be high, or the dollars or time strategic, then the missing information suddenly becomes necessary.

When asking whether certain information is indispensable or not, the engineer considers the following factors:

"Necessary Information"	*"Dispensable Information"*
High economic risk without it.	Interesting but not mandatory.
High safety risk without it.	Only slightly reduces the cost.
High performance risk without it.	Can wait.
No other alternative.	Not everyone agrees is necessary.
Timing is strategic.	

These various kinds of information intermix in all stages of Design. There is no predictable balance between them in any phase or activity. Laws, opinions, numbers, ideas, measurements, situations all inextricably interact in the actual evolution of a design. The engineer is thus dependent upon his own resources, and relies on his own judgment to choose which information to believe—which horse to bet on. We believe there is a rational approach, and with its use he is more likely to win.

Criteria of Rationality

Some hints can be extracted from the practice of our lawyer friends when they weigh the value of evidence. In presenting evidence to a jury, the lawyer gives highest priority to that which carries proof with the presentation. Next is that where proof is shown to be available, next is circumstantial proof, then expert testimony and eyewitness testimony. Hearsay, as second-hand information, is practically never allowed in court. We list in Table 3.1 an example of evidence submitted for proof of "hard" information, comparing hypothetical legal and engineering situations.

Table 3.1 Example of Evidence from Legal and Engineering Viewpoints.

Legal (Evidence that Mr. A contracted to perform X for Mr. B.)	*Nature of the Evidence*	*Engineering* (Evidence that a motor design A has a 10,000-hr bearing life under given conditions.)
The original contract itself.	Proof inherent	Engineer witnesses tests; examines the parts (being familiar with motors).*
A photocopy of the contract (which is not immediately available).	Proof available	Engineer reviews data of tests run by his employee.*
X is being and has been regularly performed by Mr. A for Mr. B.	Proof circumstantial	Motor design A has been sold elsewhere for similar 10,000-hr duty.
Expert testified Mr. A has accomplished X.	Expert testimony	Consultant in motor field states he knows motor design A has passed 10,000-hr tests.
Eyewitness Mr. C testifies seeing Mr. A perform.	Eyewitness testimony	Non-expert, who observed the tests at a distance, reports the results.
Mr. D testifies he heard Mr. A. performed X.	Hearsay	Man on next project heard that motor design A passed.

* The two situations are similar, but not the same.

This comparison is apt for radical new developments, where experience is not widespread and where you as engineer must finally depend upon your own judgment. In many cases, however, a supplier (manufacturer) will guarantee performance, so you can pass the responsibility over to him for that particular component. You still must accept responsibility for the total system, of course, as well as the component's suitability in the system.

Thus, in summary, the Design Engineer deals in the currency of information: Some valuable, some worthless; sometimes plenty, sometimes far too little. He must judge, case by case, which items he needs, which he accepts and which he rejects, and where he can obtain more.

Numerical Information

Whereas the uncritical acceptance and use of verbal information certainly is hazardous, it may seem that the use of numerical information is safe and obvious, and hence needs no discussion. This is true enough if someone tells you which numbers are significant and which ones ought to be disregarded. In the ordinary situation, however, no one tells you these things, and this section will attempt to help you do it for yourself.

In the Needs Analysis stage of design, all information has a basically statistical nature, whether the range of variation is wide or narrow. Let us give an illustration or two.

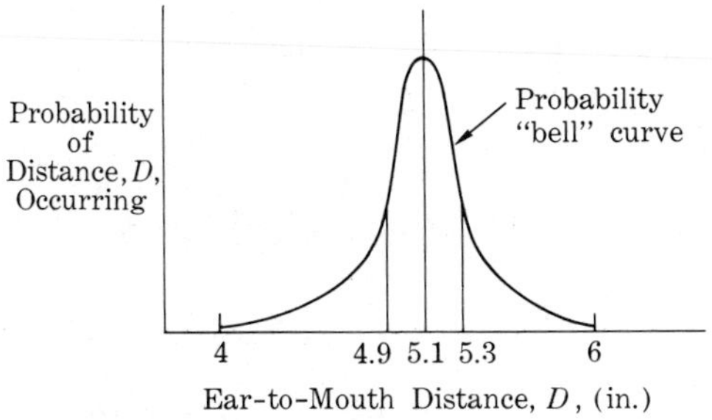

Fig. 3.8 Ear-to-mouth distances for a telephone handset.

Suppose you are designing a telephone handset and are, at this moment, concerned about just one dimension: The optimum distance between mouthpiece and earphone. If you look up the published anthropometric measurements,* you will find them distributed according to a curve such as in Figure 3.8. The curve indicates that most of the people (say 60 per cent) have a measurement near 5.1 in.; 10 per cent are less than 4.9 in. and 10 per cent are more than 5.3 in. What single dimension should you choose, since you can choose only one? This is typical of engineering problems: The solution depends on the circumstances. People using a handset that is too short would be uncomfortable; those using the long one could speak a little louder, or simply move it up along the ear. Thus the indication here is to choose the long dimension. (The General Telephone Company's handset measures 5.7 in.)

Some information may hinge around just one fixed number—no statistical variation seemingly involved—like the U.S. Standard Railway Gage,

* W. E. Woodson, "Human Engineering Guide for Equipment Designers," University of California Press, Berkeley, California, 1954.

for instance: 4 ft-$8\frac{1}{2}$ in. between rails. But notice that even though this is a dimensional standard the minute that one attempts to fit car or locomotive wheels to the rails, he must consider some statistical diversity; because all existing sets of rails are not exactly 4 ft-$8\frac{1}{2}$ in. apart. A distribution curve, such as that of Figure 3.8, applies, although probably with a narrower range of variation, say 4 ft-$8\frac{1}{4}$ in. to 4 ft-9 in. (The American Association of Railroad's standard distance between wheel flanges is 4 ft-$7\frac{1}{2}$ in.)

The same variability applies to strengths of materials, to chemical reactions, to electrical properties, and so on. No dimension, measure, or characteristic that the engineer uses is exact or fixed, no "one number" can be depended upon absolutely; the dispersion is only a matter of degree.

Thus even innocent-appearing numerical information hides some quicksand of uncertainty requiring the engineer to use caution and judgment. Each problem case determines which option will give the desired results: Should the engineer choose the mean (or average) value? Should he take the range of values, or the maximum value, or the standard deviation,* or some combination of these? These questions have no simple answers.

In the real case, the complexities are further aggravated by the interactions of the several given numbers with each other. And, whether they represent mechanical, economic or psychological interactions, this interplay forces the engineer to attempt a more basic understanding of the situation he is working with. Only then can he select the "best" numbers among those offered.

Starting Point in the Needs Analysis: The Primitive Statement of Need

Let us now go back to the original topic of the chapter, the needs analysis, and discuss what kind of information is necessary to analyze and validate the need for the project. Referring to the production-consumption cycle of Figure 3.7, notice that there are four operating groups in the total loop. Each of these groups is made up of subgroups, and a typical list is shown opposite. Any particular product will draw into action its own array of subgroups.

* The statistical "standard deviation" is σ in the equation for the normal distribution curve of Figure 3.8:

$$y = \frac{1}{\sigma\sqrt{2\pi}}\,\epsilon^{-(x-\bar{x})^2/2\sigma^2}$$

Thus σ is the distance on either side of the mean abscissa, $\bar{x}$, where $y = (1/\sigma\sqrt{2\pi})\epsilon^{-1/2}$ max. height of curve. See Reference 1.

Operating Group	*Subgroup*
Production	Planners Raw material suppliers Component suppliers Capital equipment suppliers Fabricators
Distribution	Wholesale trade Retail trade Transportation Finance Product service
Consumption	Mass consumers—indirect consumption Individual consumers—direct consumption
Recovery	Salvage Collection Conversion

Note that each constituent subgroup member receives money or value for his transaction, but that the real motivation comes from the consumer. He is the only one putting in *money* for goods or services received. All the others put in *labor* (or *materials* in the extractive industries). All the others *extract* money; they do not *contribute* money. The full motive power for the economic cycle hence comes from the consumer. And of course, he may be an individual citizen, a company, a city or state or other governmental agency, or some market combination of these. Electric light bulbs, for instance, are sold to all of these consumers; but rag-paper for currency is sold only to national governments.

It is easy to see, then, why the consumer, the end user is so important to the engineering designer. It must be to meet some consumer need that any design project receives its first feeble, or possibly energetic push toward life.

Whether it is Production or Distribution that makes the first move, the product must sooner or later meet with consumer acceptance, or the project is doomed to failure and abandonment. *Some* customers' acceptance must have been solicited for even a large, motorized street-sweeper, a television transmitter, a rocket to the moon, an oil well or even a highway system. In each of these cases, somebody said, "What we need is a good street-sweeper," a long-range TV broadcast station, or whatever. These simple English-word sentences comprise the primitive statement of need. The words fall into the category of "soft" information: Ideas, proposals, recommendations. The person who first said these words may be anyone interested; but someone in authority must repeat them to the Design Engineer before that engineer can start any serious work on them. Of course, the words are merely starting points, because not until they are examined and validated can the engineer really be sure they represent what the consumer wants badly enough to pay for. Observe that in governmental purchases, the consumer is the department that pays the bill. If, after a time the taxpayers are not pleased, they may change administrations; but the purchases will have been decided upon long ago.

Thus, as the engineer starts at the simplest statement of ultimate need, he uses all reasonable types of information to validate and clarify this need. He seeks to learn the facts, of course, rather than simply confirm a prejudice. He inquires of all the affected links in the supply chain. He checks the past, the current needs, the likely future. Because time is involved, he must project some realistic expectations of future demand, or of stimulated demand if it doesn't exist now. How much of a similar product is used now? What are the costs, the limitations, the competitive products? Why does the user say he wants it? What could he better use? Each bit of information must be tested and evaluated before it is given a place in the logically built-up case for the project. Questionable or redundant data are appropriately discarded or shelved. Modern tools are now included in market surveys: Depth interviews, testing of psychological reactions, careful sampling methods, correlations, etc. All of these are known well by public relations offices, advertising agencies, and public opinion firms. Again, notice that such careful consumer analyses apply to engineering at the highest technical level as well as to the mass-produced products of low technical sophistication.

Variety of Needs

We shall digress for a moment to recall that elementary psychology textbooks say there are two sorts of human drives. One is the *biological*, stemming from "tissue hunger," and the other is *psychological-social*, part being innate (inborn) and part being acquired from the culture since infancy (environmental). The biological needs and drives of the race give the engineering designer his basic work areas: Food, shelter, clothing, transportation, employment. However, the psychological needs have a way of continually intruding into all personal and public relations and must be faced and dealt with by the engineer. These psychological needs motivate people to seek:

Security
Response to and from others
Group approval and status
New experience and knowledge

Other apparent motivations, such as power-hunger or certain fears, are believed relatable to one or more of those just listed, or they may be included as neurotic drives.

We refer to these motivations only as reminders that at the base of all consumer needs is a variety of consumer drives or motivations; they may or may not be visible or even acknowledged, but they are still there.

The engineer and his associates, in searching for the basic needs, must not project their own needs into the picture. It is an easy trap to imagine the interests of other people as coincident with your own. Not only are other personalities involved, but so are different social behaviors, different economic levels, and business climates. In projecting future needs, many factors, including time changes, competitors, legislation, and political situations all enter into consideration. We must also answer the questions, "Is this a transitory need, like a fad which will soon die away?" "Is this an imagined need which is attractive to people now, but which is really the expression of another momentarily unsatisfied want?" The engineer seeks to get the maximum assurance from or about the customer, concerning those needs at the time when the product will be available.

Translating the Needs into the Statement of Goals [2]

One of our first cautions is to urge recognition that further analysis of the early statements of need brings surprising changes. What was first thought to be a bona fide need has suddenly transformed itself into another one. What was thought to be the technical statement of the problem evolves into something entirely different. The situation has not changed; the goal is the same; the apparent need simply becomes different as the engineer understands it better. There are many illustrations of how the needs "change" with increased understanding. Here are a few:

1. It was thought that machines to iron clothes would save hand labor. A more rational approach has produced drip-dry and wrinkle-free textiles. Ironer sales have dropped off markedly. (Not better ironers—only less labor.)
2. As central-station electric generators became larger, the conductor cooling problem apparently called for larger air blowers. The solution actually used smaller blowers and less power; hydrogen, with its higher heat conductivity and lower losses, was sealed in as the cooling fluid. (Not faster air flow—just more heat dissipation.)
3. Spring steel was customarily given a very high polish to avoid cracks that reduced the fatigue-cycling strength. It was found

that if the surface was put into compression by "shotpeening," much higher fatigue strength resulted and the degree of smoothness made no difference. In fact, the surface became pebbled. (Not smoother surfaces—simply higher surface-fatigue-strength.)

In addition to changes in the needs, further work on the project frequently changes the engineering nature of the problem. For example,

1. The problem in mass producing nylon was not in the chemical reaction but in the unusual new machines required to pump the molten mass and spin it to threads.
2. Safe entrance of incoming vehicles onto a freeway was not obtained by slowing down or stopping the main traffic but by providing parallel side lanes to speed up the entering traffic. This acceleration lane is frequently misused by uninformed motorists, even today.
3. Decreased flight resistance on the 100 series supersonic fighter plane was accomplished by bulging the nose and narrowing the fuselage waist rather than making the fuselage smooth and the noise pointed.

Of what, then, does the initial statement of the goals of the problem consist? It will naturally vary from situation to situation, but in general it should:

a. start with rather well-defined needs, a primitive statement of need,[3]
b. identify the resources that are expected to be available,
c. quantify the boundary conditions (the constraints) as far as possible,
d. present some criteria, or methods by which to judge results.

Let us examine the difference between the scientific-problem statement and the engineering-problem statement. As we have mentioned before, science is devoted to acquiring knowledge and understanding; engineering is directed toward the creation of new systems or products. Thus, for example, the initial wording of a specific scientific problem (even in an engineering project) might be: "What is the effect of moly-disulfide in lubricating the given sleeve-bearing?" After some work, it might be restated: "Does the oil film break, and hence need supplemental lubrication?" On the other hand, a related engineering statement might be, "Will moly-disulfide eliminate the sleeve-bearing troubles?" Further investigation might convert the question entirely to: "Shouldn't we use ball-bearings instead?" Each engineering question probes toward a satisfactory meeting of the original need. What primitive question should guide the designer of the "good street sweeper" briefly mentioned earlier in this chapter? Simply this: "What would be a really cheap, effective way to clean the streets?"

SUMMARY

To analyze the need for a proposed project or product:

1. Research the past, present and future market.
2. Concentrate on the needs of the end user directly.
3. Eliminate the prejudices and preconceptions of the analysts, insofar as possible.
4. Measure the needs against expected ability to satisfy them.
5. Translate the needs into a statement of project goals.

BOOK PROJECT EXAMPLE

In the preceding chapters we have assumed that our engineers were working for a large, successful concrete transit-mix company. However, they have been looking around and have found an area in the city where no transit-mix service is readily available. They had been wanting to go into business for themselves, anyway, so this seems like an opportunity.

They are considering forming a new company, calling it *The Econocrete Co.*, using their knowledge of the business, and, for starting capital, adding to their savings by borrowing from banks.

Starting an independent small business raises the most fundamental *Needs* questions of all. In the case of a branch of a business, or just a new product line, the questions would not be so personal, so basic and direct to the prospective owners. The first questions relate not to a particular business, but to the matter of anyone's going into business at all. These, then, are the questions they must ask themselves:

What are our personal needs?

Is our new company likely to satisfy them?

Can we make the necessary commitments?

Have we validated the data upon which we have argued thus far?

Will the plan work if the contingencies of market, business, finance, deliveries, etc., all go to the worst limits?

Are any new material or processes on the horizon that will decrease the demand for concrete, or reduce the price that customers are willing to pay?

What are the important unknowables (the risks) in this situation?

How late can we retract and back out?

Assuming we (or they) can satisfactorily answer the foregoing questions, the next *need* to study concerns the plant:

Do we need a plant design? Or, should the plant be bought outright? New? Secondhand?

Should the design be contracted to consultants, or ought we to attempt it ourselves?

Instead of starting a company "from scratch," should we lease, rent, form a syndicate, or make a deal with another rock products corporation?

If these are answered by sound reasons supporting a plant-design proposal, then we ask:

What do we expect the plant to do?

How much output? What varieties of concrete?

What is the minimum level of sales that will earn a profit?

How far from customers?

Should we produce products other than concrete-mix?

Are there new ways known now (but not used) for mixing concrete?

Should the plant be manual, automatic, or a combination: How much of each?

How much profit is expected?

These questions must have acceptable answers before the enterprise can proceed. We shall assume the needs have been validated in the minds of the engineers-entrepreneurs, and that they have concluded that:

1. They need a plant designed for specified quantities and varieties of mixes, etc.
2. They will need \$X thousand of which they have \$Z thousand in savings, among the three people who will put the company together and get it started.
3. The constraints will be:
   The available sources of sand, gravel, and cement, the prevailing market prices and conditions, the volume of sales within transport distance, the land sites available and reasonable for the plant, etc.
4. The criteria for success will be:
   Will the prospective income from sales be sufficient to meet all operating costs, pay off the initial plant investment, and earn a profit sufficient to warrant the risk involved in starting this project (compared to other potential investments)?

This is a very broad statement of the long-range goals for designing a transmit-mix plant for certain economic end results.

Smaller-sized problems issue from this large one, such as the design of the aggregate feed-conveyor. Here it might be:

1. Needs:	Convey sand and gravel at 100 tons/hr from 15 feet below ground to 40 feet above ground, on as steep a slope as possible.
2. Resources:	Use standard commercial equipment; lowest possible cost; manual operation to begin with, automatic later.
3. Constraints:	List grades of sand, gravel, water-content ranges; determine maximum loading, per cent on-time, operating temperatures, power supplies, etc.
4. Criterion:	Will the prospective return be sufficient to warrant the additional risk, according to current investment standards?

Student Project Exercises

To the degree appropriate in each case, answer the following questions in your needs analysis and validation for the selected project. If the question does not apply, go on to the next one.

1. What was the primitive or simplest statement of need?
2. What are the significant or special needs of:
   the end user (consumer), the distributor (sales), the manufacturer?
3. Is there a real conflict among any of these? State any such conflict as clearly as you can.
4. What is your projection about future needs?
5. What is your initial statement of the broad goal?
6. What are the criteria of success? List in order of importance.

REFERENCES

1. Moroney, M. J.: "Facts from Figures," p. 109, Penguin Books, c/o D. C. Heath & Company, Boston, 1951.
2. Hall, A. D.: "A Methodology for Systems Engineering," chap. 6, Needs Research, D. Van Nostrand Co., Inc., Princeton, N.J., 1962.
3. Asimow, M.: "Introduction to Design," chap. 8, Analysis of Needs and Activity, Prentice-Hall, Inc., Englewood Cliffs, N.J., 1962.

The Feasibility Study: Identification of the Problems

Four

"Life is the art of drawing sufficient conclusions from insufficient premises."
Samuel Butler

Is It Feasible?

The word *feasible* comes from the French *faire*, which means to make or do. Thus, in questioning feasibility, we ask about the eventual achievability of the proposal. Can it be devised or produced as expected? The investor, whether private or corporate, must be persuaded that his money will be usefully and prudently spent. Part of the prudence consists of limiting the spending to controllable stages, and part consists of demanding favorable early evidence.

Cursory examination of a proposal may be sandwiched into other activities, without costing much time or effort. As a proposal reaches the stage of serious consideration, however, the informal activity must be replaced by an organized examination to see if there are a variety of acceptable alternate solutions. Looking for these "acceptable solutions" is the purpose of the *Feasibility Study*. In normal practice such a study costs thousands or tens of thousands of dollars; but it must be done before more money is committed in further pursuit of the project. It is much wiser, if the project is shown for any reason to be unfeasible, to stop the work, accept the loss, and look for another prospect. This, incidentally, is done regularly in industrial research organizations where only a few original projects "live through" to completion. To pursue an ill-omened, unpromising project is to "continue forcing vitamins and medicines into a dead horse." Face facts, give up; and simply arrange for an honest burial!

Thus, the *Feasibility Study* has the purpose of both initiating the design and establishing the direction of its thinking, as well as providing convincing evidence to those who are supporting the project. The thinking on feasibility must, of course, include the technical-social-economic-legal feasibility, rather than only the technical fraction, with an insufficiently considered risk of other types of failure.

Now that the painful word "failure" has been used, let us observe something about success and failure for the engineer. In the medical profession, success is relative; it is transitory; it is never final. There comes a time when the whole hospital staff of doctors is powerless, and failure prevails. In the legal arena, one lawyer always prevails over another, no matter how brilliantly and forcefully the loser's case is argued. In the ministry or education, strictly professional success or failure is rarely discussed. In engineering, however, the situation is entirely different. One dam failure, one fallen bridge or building, one spacecraft abortion, or one chronic shaft failure in an auto—any of these, regardless of the reasons, has a direct and serious impact on the public reputation of the engineer responsible. His care on this particular task, his total record, and his reputation among his peers may have been beyond reproach, but now an indelible question mark will have been stamped on his professional image. We shall have much more to say about this in the chapter on Checking. However, we here re-emphasize the necessity for extreme care in the prediction of project success during the feasibility inquiry.

Elements of a Feasibility Study

Whether the feasibility is investigated by an old established firm of consulting engineers or by an individual working for himself, it is advisable to look carefully at a number of questions. Of course, the immediate aspects of the situation have much to tell about the depth and importance of these questions; but following are the feasibility questions suggested by the first column of Figure 2.1. The answers to the questions on this and the next page should lead to a set of "feasible" or useful solution alternatives.

Are the needs validated and the project goals identified?

What are the varieties of systems that can be involved (electrical, chemical—economic, political—code authorities, etc.)?

What is the initial technical statement of the problem?
What are some suggested solutions?
Can the solutions be constructed, and can they be expected to operate?
Which of the solutions are economically attainable?
Can each of the solutions be financed?
If funds must be borrowed, what will be the rate of interest and the schedule of payments? Does the finance plan provide for contingency?

In the previous chapter, we looked closely at the needs and the goals. After examining the above-listed elements of a feasibility study, we shall devote the rest of this chapter to determining "the initial technical statement of the problem," and the explanations that accompany it.

Varieties of systems involved. Looking at all the subjects, all the fields and systems permits a perspective view of the problems to be faced as well as a glimpse of all the possible avenues for their solutions. The broadest coverage and greatest generality should be initially sought, even though the possibilities are severely cut back upon further evaluation.

Initial technical statement of the problem. The goals are formulated by words which are carefully phrased not to *solve* the problem, but to state it in its most general technical form. The word *initial* is included because further work always refines and even changes the problem and adds and substitutes other problems.

Suggested solutions. Solutions to engineering problems are practically always synthesized, or put together from previously known parts. After trial combinations are assembled in the mind, they are analyzed to see if they meet the original need. Some of the analysis goes on in the originator's mind quickly, but much of it awaits the careful examination by mathematical, graphic, experimental, and logical methods. Several solutions are always sought at this stage.

Physical realizability. This is the first of three tests applied to those solutions synthesized above. Can the product or system be made? Are materials and manufacturing methods known? Will it perform as needed?

Economic feasibility. Each solution must be studied, and estimates must be made of the income that will probably result, the operating expenses that will occur, and the initial investment that will be needed. In each of these estimates there will probably be recognizable risks. The prospective earnings should be compared with the prospective risks in order to estimate the economic feasibility. Are there intangible benefits or risks that should be considered? Does the entrepreneur (venturer) have enough money available to finance the venture? Is he satisfied that the prospective returns will warrant his financing the venture? If the owner needs to borrow money to help finance the venture, what rate of interest must he pay and how soon must he repay the lender? Are the additional risks of borrowing justifiable?

Tools to Aid in Clarifying Goals, Stating the Problem, Formulating Criteria

Whereas they are not a cure-all nor fully useful in all cases we will describe here a few tools and concepts which have been found helpful to many in this stage of designing.

a. Input-output analysis. In operating an automobile, the gasoline, oil, water, air, etc., are expendables which comprise material inputs to the

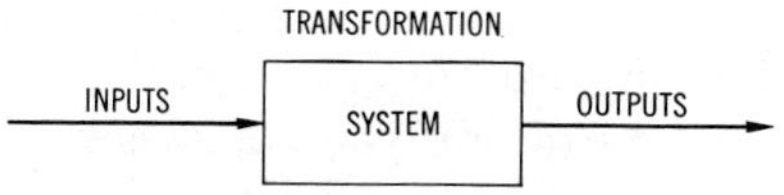

Fig. 4.1 Transformation by a system.

system. The major outputs are transportation of humans and materials, protection from the elements, and we suppose pleasure. The auto itself is the system, which together with the driver transforms the inputs to outputs. One quickly recognizes that there are additional inputs, different classes of inputs, in the complete "auto-highway-person" system. In general, the

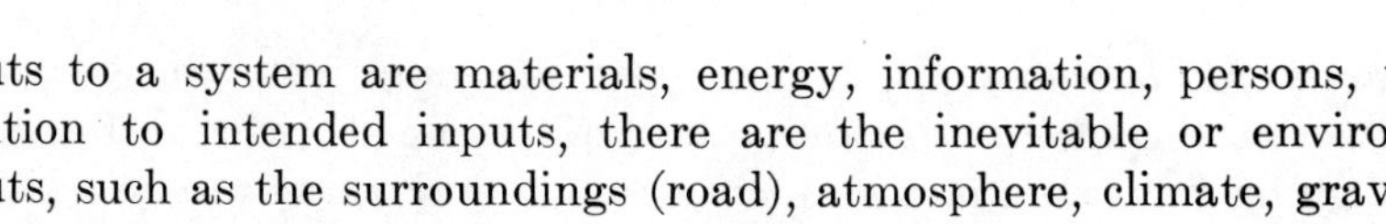

Fig. 4.2 Input-output analysis. Diagram to aid in setting boundaries or constraints on possible design solutions.

inputs to a system are materials, energy, information, persons, time. In addition to intended inputs, there are the inevitable or environmental inputs, such as the surroundings (road), atmosphere, climate, gravity, and so on.

The outputs are not only the desirable results we want, but also some usually inescapable undesired concomitants. In this case there is noise, smoke, dust, vibration, danger, etc. Thus, the basic diagram of Figure 4.1 can meaningfully expand into Figure 4.2.

Notice that the headings list Inputs as Limitations and Outputs as Features to Optimize. To be useful as a design tool, the columns ought to be filled out by the designer under these respective headings in as general and inclusive a way as possible. In certain cases; the problem situation is too simple or too obvious, so this diagram serves no need. In the more complex cases, particularly new ones, this diagram forces the engineer's thinking along more penetrating lines, and the exercise is valuable to him. Chapter 13 presents two specific illustrations.

One can think of the diagram alternatively as describing the design project, rather than the final desired system. Care must be exercised to avoid confusing these two. The design engineer uses the diagram for the *system;* the project manager uses the diagram for the *project.*

Consider the Inputs as "Limitations on the System." This means that one may not propose a system which violates any of the input conditions. Likewise, the outputs are "Features to be Optimized," because the design must progress by trading off desired vs. undesired outputs. The particular advantages and disadvantages of each feature are compared against the goals of the project and a balance is chosen to obtain the maximum gain with the minimum penalty.

Normally the goals of the project comprise the Desired Outputs. This is the place to start the listing. Intended Inputs and then Environmental Inputs follow next. Only as some relatively specific kinds of system are conceived, can the undesired outputs be imagined. A caution, however: The system should religiously be considered a "black-box," an undecided array of hardware, up to the last possible moments. Only in this way can one avoid the premature acceptance of some conventional, familiar, but less-than-best "solution." The later in the process the system is chosen, the better. This delay permits maximum review of alternatives, a psychological aid to keeping the inventor's mind open as long as possible.

The purpose of the Input-Output Analysis is to more sharply define the boundary conditions of the project and lead up to the technical statement of the problem. If this Analysis is complete, with all the factors and numbers included, one can point to it almost as the initial statement of the problem.

b. Design vs. Research. Design consists of going from the generally nebulous need to the specific concrete solution. On the other hand, Research consists of going from unorganized facts and nebulous questions to a generalization, a new understanding of certain natural phenomena.

Notice the significant difference in the comparison of Design and Research "as systems." The transformation is initially unknown in Design, whereas the output is the unknown in Research. See Figure 4.3. The transformation must be found in the case of Design, but Data in the case of

Research. When the transformation is found, it comprises the solution, the desired end result of the design activity.

This comparison is useful primarily because it emphasizes that a system must be synthesized. To date, the only means known for synthesizing systems is the human mind. Although much work is proceeding, neither mathematical nor logical methods nor computers have seriously competed with the mind in the area of creativity. We shall further discuss creativity

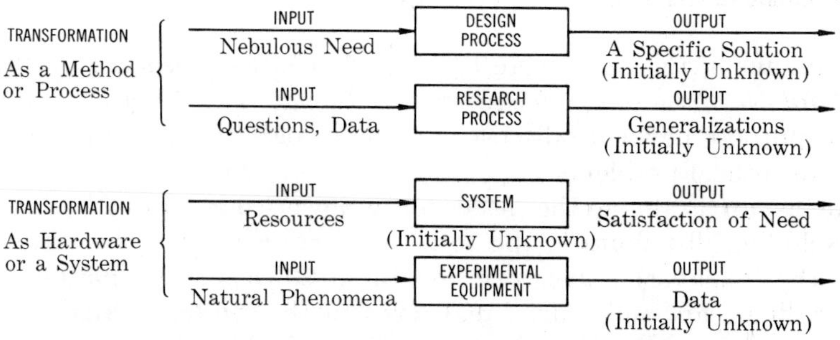

Fig. 4.3 Input-output comparison between design and research. In design the system is unknown, in research the output is unknown.

in Chapter 6. The object of a methodology of design, then, is to help the mind synthesize the best system possible, within the constraints.

c. Component disciplines within the design morphology. A discipline is a set of coherent theories, rules, laws and methods which presumably apply usefully in a given subject area. Further observations of the processes of Design have convinced some researchers that a common set of these disciplines is discernible in all large projects and that these form a basic set of engineering tools. We have discussed some of them before and we will enlarge on certain ones later. This group* has been identified as

Organization of information	Evaluation, Decision
Identification of problems	Design through experiment
Synthesis	Test, Proof of performance
Analysis	Prediction
Formulation of criteria	Communication
Optimization	Implementation

* These correspond to the "Anatomy of Design." See A. B. Rosenstein, Graduate Design Philosophy, *Proceedings of Conference on Engineering Design Education,* UCLA, Engineering Department, Los Angeles, California, September, 1962.

These are not all of the same depth or sophistication, but they form a backbone of skills and competences of which the engineer must be aware and would do well to develop. Each individual, naturally, will favor certain of these tools, according to his own special interests. However, somewhere in each major design, and to some degree of repeated attention, all of these instruments must be employed.

From Goals to the Engineering Problem

In Chapter 3, the goals were broadly defined as consisting of *Refined Needs, Available Resources, Constraints and Boundaries, Criteria for Acceptance*. Our purpose now is to discuss how these goals may be converted into technical (or other) problems.

Take as an example the need for a lower-cost automobile license plate. Excluding distribution, the total production cost to the state (and hence to the taxpayer) is between $0.60 and $0.75 for two plates. There were 80 million cars and trucks registered in the United States in 1962; and we may assume the license plate will continue to be replaced every five years, using the tab or sticker for the yearly identification. This method is an improvement on the annual issue of plates, in practice not so long ago. But a saving of even ten cents on each plate would total a million dollars annually for the whole country, or a hundred thousand dollars in each of the three largest states. In addition to the savings, it is certainly desirable to increase corrosion resistance, flexibility, and legibility. This much comes from common knowledge.

We shall itemize the first list of the goals below and demonstrate how that list moves into the initial statement of the technical problem. The whole project undoubtedly could justify an extensive research into many pertinent questions such as 1) all possible attachment places and methods, 2) all problems of truck, trailer, special vehicle license legibility, 3) present legal restrictions in all states, 4) importance of reflective night lighting, fire resistance, color change, 5) available applicable materials, etc. For simplicity, however, we shall proceed without benefit of such research; we shall continue on the current basis of numbered plates fastened to the front and rear of the car. The goals might thus be stated:

Needs	*Resources*	*Constraints*	*Criteria*
Lower cost.	Current production materials and methods.	Must fit all cars and trucks.	Cost.
Greater legibility.	Spend one-year's savings on new tools and equipment.	Normal person must see at 100 ft.	Legibility better than present type.

Needs	*Resources*	*Constraints*	*Criteria*
Longer life (corrosion resistance, flexibility, color permanence).	Fact: Steel: 10¢/lb Aluminum: 40¢/lb Plastic: 30–90¢/lb	Legal size (12 in. × 6 in.). Six letters or digits. Must not swing. Must be mounted between 12 in. and 60 in. off the ground. Must be fire-resistant, tamperproof, transferable; must use annual tabs.	Adaptability to mounting. Life longer than present type.

This is a contraction of the Input-Output analysis of Section *a* on page 59, but it is being used to allow concentration on a relatively small problem. As we think about the functions involved, let us take just the "number plate" as our starting point (temporarily disregarding the attachment and the tab). We make a chart by writing down all the reasonable alternate ways of doing this under the first word: "Number plate."

This chart, Figure 4.4, is made up by thinking of the alternate functions or constructions possible under a given heading. Each word represents a question as it is viewed in the higher position, and an answer in the lower position. On the right-hand side, the question is, "How can I make a composite plate?" The answer beneath it is "Add symbols to a base." To the question, "How can I make symbols?", we answer "Use paint, dye, or another layer." The "layer" can be laminated of aluminum foil or another plastic. This simply represents an organized first approximation, going through a variety of materials and processes to get some new ideas to explore. Other ways are possible, but this is shown particularly to demonstrate how an initial problem (i.e., number plate) immediately splits into sub-problems, each in turn subdividing all the way down to the final detail descriptive answers. This is the nature of all design problems. Only as the last, lowest layer of problems are solved is the design finally worked through. If the solution has been listed among these alternatives, some zig-zag line can be traced from the original question on top to the final answer at the bottom.

The goals entered intangibly, as each set of alternatives was mentally brought up. Some possibilities seem acceptable, Ⓐ, and others likely, Ⓑ, etc. Ⓐ is conceived as a thin, fiberglas-reinforced high-temperature plastic sheet with the numbers perforated into aluminum foil and protected by a transparent mylar cover-layer. If we have judged this correctly, the resulting first technical problem Ⓐ would probably be,

> "Can an adequate laminate of reinforced resin, aluminum foil and mylar be made in large quantities competitively?"

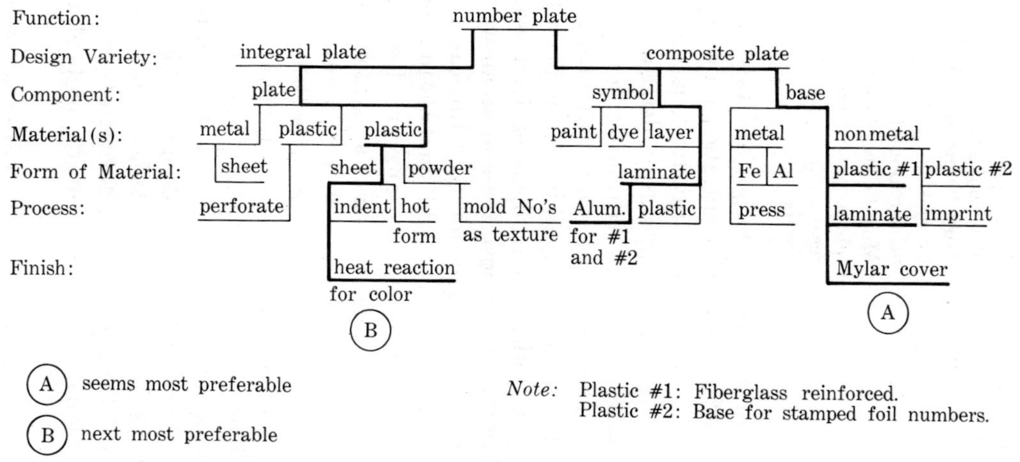

Fig. 4.4 License plate problem development, considering only new fabrication methods.

A second problem could be the search for a high temperature plastic, whose color could be changed by hot-imprinting (as Ⓑ).

Only further investigation would determine whether either or neither of these was worth time and effort. If they both proved blind alleys, the next likely problem statements would be sought, and then pursued. In general, as the problem is being uncovered and recognized, it is best to phrase all questions as broadly and generally as possible. Leave prospective solutions until as late as possible in the deliberations; concentrate on the end results, not on the way of reaching them.

BOOK PROJECT EXAMPLE

We turn now to our hypothetical transit-mix concrete business. Another need that would enter our transit-mix plant design would be the sand and gravel storage hopper. Let us give an example of converting those needs into a problem statement.

Need:	An array of storage hoppers arranged to weigh and discharge directly into the mixer truck two grades of sand and three grades of gravel, plus water and cement.
Resources:	Standard materials and special constructions, taking into account the supply and delivery systems connected to the hopper.
Constraints:	Store a minimum of two hours' usage of sand and gravel in separate bins sized to accommodate the respective average proportions of each grade. Assume 10 trucks, capacity 7 cubic yards each, 1 trip per hour. Design must suit the sand and gravel feed equipment as well as the mixer trucks.
Criterion:	Lowest cost per year per ton of concrete delivered to the truck, including any unusual costs due to loading and unloading this hopper.

The problem can now almost be written by putting these four statements together serially. Note that it is still broadly-enough described to allow the hoppers to be separate or together, made of any material, oriented in any way, gravity- or pressure-fed or unloaded, and designed through any geometry. Even though the final result is safely assured, a great freedom of choice in specific direction is open to the designer.

Student Project Exercises

The goals for the several suggested projects depend upon the assumptions. These assumptions can be made with one of two approaches.

a. Do research into the actual present situation, form some conclusions, and *then* write a set of assumptions.

b. On certain projects with many possible degrees of solution (ranging from amateur to professional), select a plausible set of assumptions in collaboration with your instructor.

Then, with this background,

1. Write out your project goals.
2. Deduce and write down the initial statement of the engineering problem.

REFERENCES

1. Asimow, M.: "Introduction to Design," chaps. 4, 7, 8, Prentice Hall, Inc., Englewood Cliffs, N.J., 1962.
2. Buhl, H. R.: "Creative Engineering Design," chaps. 2, 3, Iowa State University Press, Ames, Iowa, 1960.
3. Mantell, M. I., M. E. Reeder, J. A. Stevens: "Orientation in Engineering," chap. 8, Prentice Hall, Inc., Englewood Cliffs, N.J., 1963.

Information Sources; Laws and Effects

Five

"So it is in travelling,
a man must carry knowledge with him,
if he would bring home knowledge."
Samuel Johnson

Where Can I Find Out?

As the engineer learns more about his project, its particular technology and specializations, he usually realizes there are a great many things he doesn't know. He needs to learn about a range of subjects including operating principles, commercial equipment on the market, and data that have been published. He appreciates again that the engineer's life is one of continual self-education. He never knows in advance all he must know on a project. Incidentally, he can avoid a great deal of personal frustration if he accepts this perennial activity gracefully, and keeps up with his own "retooling."

So the engineer continually looks around for sources for the wide variety of information which he must collect and assimilate. Over the full period of a single project, the number of necessary sources can be amazing, covering the country and sometimes the world. To get a feeling for this breadth, let us look at the information-collecting period in the design of a sand conveying system, taken out of the Econocrete project.

The final design consists of carefully organized information—a very particular collection of old and new information. The old information concerns standard products and components which will be used. The new information describes how much, how many, and how to put them together. New information also includes new systems and structures never before built, new operating results, new costs, new relationships between people, etc.

BOOK PROJECT EXAMPLE

The standard components needed for a new system are described in the handbooks, catalogs, specification sheets and drawings known to the trade or easily found. What would these standard components and data be for our sand conveying system? Probably such a heterogeneous collection as:

Motors, wiring, controls, gear drives, signal systems.
Structural steel (plates, bars, angles), fasteners, welding facilities.
Conveyor sections, belts, hangers, guides, chutes, idlers.
Going labor rates, job classifications.
Proportions of various aggregates in standard concrete mixes.
Cost of money, amortization periods, available capital sources, etc.

Whereas, this conventional information is directly available upon inquiry, the specialized and new information must be developed bit by bit as a result of selection from a wide variety of sources.

Many Sources

The wide variety of sources the engineer must be familiar with is illustrated in our present case by listing these many broad classes:

Vendors of components, materials and supplies.
Regulating agencies of the government.
Trade associations in the field of the project.
Testing organizations.
Professional specialists.
Financial, insurance, tax rate sources.
General information sources.
His own company management.
And, of course, the client or customer.

This chapter deals with the problem of engineering information sources, both the standard types and special types. As we said in Chapter 3 most design time is spent in *finding* and *collecting* information, no matter what the field. When the information is at the engineer's fingertips, the design work moves quickly.

Economic Basis

As we have said, information costs time and money and certain types cost much time and much money. Thus, the engineer is continually balancing the benefits of hoped-for specific information against the time and money cost of obtaining it. He must keep a running progress report in his mind or on paper concerning the chosen minimum amount versus the likelihood of his securing it. As time moves, the types and amounts of "minimum" information change. If his confidence of success in a certain problem area goes up, the need of added information for use there can be cancelled. In our sand conveyor case, if a vendor will guarantee a belt's life, the engineer need not continue to collect information on varieties and performance of belts. Each decision comes as a balance between the risk of proceeding versus the cost of verifying. All engineering information thus has an economic base. If the risk is small, the information is relatively not important. Each engineer must decide* whether the risk will be large or small.

Types of Sources

There are many, many more sources of information than the engineer normally realizes he has access to. He starts out on a project feeling very much alone. The sooner he gets the many available people working for him, the faster his progress and the less his frustration.

The people available to provide information are 1) those paid to do so by government or appropriate philanthropic or private agencies, 2) those who have a financial interest in so doing, and 3) those who do it because of professional responsibility, friendship, or simply altruism as one person helping another.

These various sources cluster in the classes of "Public" or "Private" organizations. We list below the general varieties that are of interest to

* One development of probability theory calls information *"that which reduces uncertainty."* We mention this here because it leads to a quantification of the value of information, and because it stresses the condition of imperfect knowledge surrounding engineering decisions. Briefly, it is concerned with the probability of a hypothesis being true. This probability ranges from 0 to 1, from *impossible* to *absolutely certain*. Additional information permits computation of the change in probability, and hence a better decision under the hypothesis.

the engineer. This list is expanded into a more adequate reference list and certain key addresses are given in Appendix C.

A. *Public*

1. *Departments in the federal government:*
   Agriculture, Commerce, Health, Education and Welfare, Interior, Labor, etc.
2. *Agencies in state and local government:*
   Buildings,* Safety,* Standards,* Banking* and Insurance,* etc.
3. *Libraries:*
   General, university, special technical, (varieties of abstracts, indices, periodicals, etc.).
4. *Universities, research institutions, museums, expositions, fairs:*
   Wide variety of public and private institutions, collections and displays.
5. *Foreign governments, agencies and companies:*
   Embassies, consulates, attaches, and representatives.

B. *Private*

1. *Non-profit, semi-public organizations and sources:*
   Professional societies:
   Engineering, science, business, medicine, law, etc.
   Trade and labor associations:
   Commerce and manufacturing, agriculture, guilds and crafts, builders, utilities, etc.
   Magazines and newspapers:
   Research and public information activities and departments.
   Membership organizations:
   Consumers, motorists, farmers, veterans, hot-rodders.
   Non-profit laboratories and associations:
   Underwriters Laboratories (for electrical and gas apparatus fire insurance).
   American Standards Association (professional society approvals).
   American Gas Association (gas apparatus safety).
   American Society for Testing Materials (standard procedures).
2. *Profit-oriented organizations:*
   Vendors:
   All types of information: Catalogs, samples, tests, quotations; operation, construction; materials, costs, delivery, servicing, etc. Vendors include manufacturers, suppliers, financiers, etc.
   Current manufacturers, businessmen, competitors:
   Occasional conversations and possible plant visits.

* Those agencies marked with an asterisk are primarily *regulatory;* that is, within statutory limits, they establish and enforce the standards of use, construction and practice within their field of jurisdiction. For example, the plumbing department fixed the pipe sizes, the number of attached fixtures, the drain line slope, materials, etc.

Consultants:
Technical specialists in every field who sell their services. Information ranging from simple figures and recommendations to complete plant or system designs, ready to construct.

3. *Persons:*
Direct dialogue or correspondence.
Personal friends, acquaintances, associates, "friends of friends."
Faculty of suitable schools, colleges, institutions.
Philanthropic or service groups.
Public officials.

4. *Private philanthropic foundations:*
Carnegie, Ford, Guggenheim, Rockefeller, etc. Various specifically financed foundation-sponsored reports are available upon request.

In conclusion, the young engineer can obtain information from all of these. Only the profit-oriented sources are restricted at all. With the right approach much can be secured even from them. This leads us to a discussion about the accessibility of the various sources.

Suggested Procedures to Obtain Information

It is probably conceded by the reader that the many above-listed representative sources are literal treasure-houses of information for most imagined engineering projects. The questions are: Can I get the information? How do I go about it? How long does it take? Are there any secret methods? Will the people throw me out?

Without minimizing the problem of summoning the nerve to telephone or write to a stranger, we shall suggest a direct, common-sense approach.

Some information will be easy to furnish, some will be relatively time-consuming or expensive. Some will be unimportant to the supplier; some may be vital, and hence divulged with selectivity and care. Since the supplier's situation will not usually be fully known to the inquirer, the approach must be made tactfully and diplomatically, giving the supplier an escape by not pressing the demand too hard. The questioner must be ready to accept all he has asked for, much less, or possibly no information at all.

Specific Situations

Here are some specific suggestions for specific situations:

1. *Shall I tell what I know before asking my question?*
Set the stage by a brief background, so you don't waste the time of the respondent.

2. *Shall I ask for more than I need to get enough of what I want?*
   Generally ask for only that which you specifically need.
3. *Do I ask for information if I am not sure they have it?*
   Be reasonably sure you are not wasting their time; or else ask only for further references or sources.
4. *What if I ask for confidential information?*
   Don't worry, they won't tell you if it's really confidential. However, you can get marginal information by telephone or face-to-face conversation that you will never get in writing.
5. *How can I find the right person in a large organization?*
   Simply phrase the question so it gets passed quickly to the right person to answer it.
6. *What if the source is in a distant city?*
   Write a letter or look in the phone book for a local representative.
7. *Should I call or write?*
   If you can afford it, always call. A dialogue is vastly superior to the written word. Sometimes you should also write, to confirm the results of the call.
8. *How much should I expect a library or professional society to do?*
   The library will verify at no charge, if they have a copy; or they will make a copy for a charge. The professional society will only sell you a copy of any paper or book they have sponsored.

Recommendation. In summary, you can see the best strategy is:

a. Open the conversation by identifying yourself, your affiliation, your project and its purpose (whether by letter or by phone).
b. Ask your questions directly and with frankness and simplicity.
c. Stick to your problem. Don't be nebulous.
d. Keep the questions logically related to your project.
e. Make the timing realistic. Don't ask for immediate answers if you can wait a month.

Laws and Effects

Information must be readily accessible if the engineer is to move his project along actively. General information can come from handbooks and other broad reference books. Specific information must be traced down to whatever source available, but "scanning" information is seldom found. By scanning information we mean easily surveyed compact arrays of encapsulated ideas and principles, so the reader can glance at and encompass whole fields of phenomena or methods. Thus the slow, pencil-chewing method of idea search can presumably be replaced by a faster, more complete review of new approaches. Since such information summaries are rare, we have extracted[1] certain laws and effects and have attached them as Table 5.1. These items are solely for the purpose of starting trains of thought, and they include a wide range of subjects for the analogies that may be drawn. Detailed expositions of the various items can be found in standard textbooks or in summary form in Reference 1. A further list, Analogies from Nature, is given as Table 5.2.

We expect these lists to be used in the early phases of a project, when whole classes of phenomena are being considered as possible avenues of exploration.

A few examples of scanning information are also available in the form of encyclopedias of mechanical structures and motions, collections of fasteners and spring forms; handbooks of electric communication or control circuits, chemical process summaries, heat transfer devices, etc. See a reference librarian for help in locating appropriate references.[2,5,6]

Cause and Effect Chart (Input/Output)

We add a further tabulation for scanning purposes. Table 5.3 is constructed by listing physical variables as X and Y coordinates, and then placing the law (or one of the laws) that relates those two variables together in the box at a given intersection. For example, acceleration as an input can produce current as an output, using the piezoelectric effect. This chart is merely a starting point, because many of the intersections accommodate several alternative phenomena, making multiple layers out of the chart, in effect. For instance, current as an input (box 15), can produce electromagnetic radiation as an output, by means of:

1. Incandescent solid (lamp filament).
2. Excited gas (fluorescent lamp).
3. High velocity impact of electrons on a target (X-rays).
4. Radiation from a radio antenna (transmitter), etc.

To completely catalog in chart form all the known laws and effects as inputs and outputs would be very helpful, but very massive. The engineer is interested in biotechnical, psychological and other nonphysical relationships as well as the physical-chemical items listed. The reader is encouraged to compile an input-output chart in his own field of interest.

Keep in mind the fields of technology or science listed here and on the next page:

Electric Power	Thermostatics	Mechanics
Electronics	Heat Flow	Kinematics

Magnetics	Fluid flow	Dynamics
Optics	Acoustics	Elasticity
Luminescence	Aerodynamics	Plasticity
Organic chemistry	Nucleonics	Inorganic chemistry
Astronautics	Crystallography	Mining
Metallurgy	Biotechnology	Ceramics
Cyrogenics	Physiology	Botany

BOOK PROJECT EXAMPLE

Taking the sand conveyor of the first section of the chapter as our illustration, we could look to the following sources for the necessary information.

1. *Motors, wiring, controls, gear drives:*
   Several different company catalogs.
2. *Structural steel, fasteners, welding facilities:*
   Structural steel would be found in handbooks and warehouse catalogs.
   Standard fasteners: Nut and bolt company catalogs and handbooks.
   Welding facilities: Telephone book or trade reference file.
3. *Conveyor sections, belts, etc.:*
   Company catalogs.
4. *Labor rates, etc.:*
   Local union hall's business office.
5. *Proportions of mixes*
   Concrete-cement handbook, plus trade association data, plus architects' handbooks.
6. *Financing*
   Local investment banker; commercial bank; securities broker, financial consultants; etc.

Student Project Exercise

In cooperation with your instructor and the library and any available design reference facilities, list carefully for your project the information you now need and the sources you should follow up. Obtain the names and addresses of specific sources.

Table 5.1 Laws and Effects Reference Table.

Adsorption	Adsorption of gas on surface of a solid.
Ampere's law	Force on current-carrying conductor in magnetic field.
Archimedes' principle	Buoyancy of floating object due to displaced fluid.
Avogadro's law	6×10^{23} Molecules per mole of gas at standard temperature and pressure.
Bernoulli's theorem	Conservation of energy in fluids (change of normal pressure with axial velocity).
Biot-Savart law	Magnetic field produced by electric current.
Biot's law	Linear conductivity of heat.
Blagdeno law	Depression of freezing point due to impurity.
Boyle's law	Pressure $\times$ volume = constant, at constant temperature.
Brewster's law	Reflection of polarized light.
Brownian movement	Random motion of solids in collodial solution.
Cavitation	Collapse of bubbles in liquid under pressure change.
Charles' law	Pressure $\times$ temperature = constant, at constant volume.
Christiansen effect	Refraction of powders in a liquid.
Color mixing	Any color may be obtained by combining the primary colors (red, green, and blue).
Conservation of Energy	Energy cannot be created or destroyed.*
Conservation of Momentum	Vector sum of momenta is constant.
Conservation of Angular Momentum	Angular momentum of a closed system is constant.
Conservation of Mass	Mass of bodies in a closed system is constant.*
Corbino effect	Current flow in rotating metallic disc in magnetic field.
Coulomb's law	Force between electric charges, magnetic poles, or masses.
Coulomb friction (see friction)	Constant coefficient of dry sliding friction.
Curie-Weiss law	Temperature for transition from ferromagnetic to paramagnetic.
Dalton, partial-pressures	In gas mixture, total pressure is sum of partial pressures.
Debye frequency effect	Increase of conductance of electrolyte with frequency.

* Except in nuclear or astronomic systems.

Table 5.1 Laws and Effects Reference Table. (*Continued*)

Diffraction	Spreading of beam of energy behind an obstacle.
Doppler effect	Shift of received frequency due to motion of sound or radiation source.
Edison effect	Thermionic emission from heated metal in vacuum.
Elastic limit	Maximum stress without permanent deformation.
Electric wind	Motion of air away from point of corona discharge.
Electro-capillarity	Change of meniscus pressure with current flow between two liquids.
Electrodeposition	Electro-plating; (see electro-osmosis).
Electroluminescence	Production of light by alternating current through thin fluorescent films.
Electromagnetic radiation	The total spectrum that includes radio waves, infrared, visible light, ultraviolet, X rays, etc.
Electroösmosis	Movement of fluid through a porous wall by electric current.
Electrostriction	Self-compression or tension of electrified crystals.
Eötvos effect (Coriolis)	Change in body weight with east-west motion at surface of earth.
Expansion, thermal	Volume change of materials with temperature.
Faraday electrolysis	Chemical change proportional to electricity flow across an interface.
Faraday induction	Generation of electromotive force by change of magnetic linkages.
Fission	High-energy splitting of atomic nucleus under bombardment by nuclear particles.
Fluid flow: small sphere	Limiting velocity of falling sphere (see Stokes).
Fluid Flow: Mach	Shock wave effects as fluid velocity approaches the speed of sound in the fluid.
Fluid Flow: Reynolds	Change of flow and friction, from laminar to turbulent.
Friction: Newton	Shearing stress in gas or liquid.
Friction: rolling	Similar to sliding friction but having a smaller coefficient.
Gas laws, ideal	Relations of pressure, volume, temperature, and gas constant.
Gauss' law	Lines of force from a charged body.

Table 5.1 Laws and Effects Reference Table.
(*Continued*)

Gauss effect	Increase of resistance of magnetized conductors.
Geiger-Nuttal rule	Range of alpha particles according to lifetimes of radioactive elements.
Gibbs phase rule	Equilibrium conditions in mixed systems (temperature, pressure, concentration).
Gladstone-Dale law	Change of index of refraction with density.
Graham's law	Osmotic flow of gases and solutions through a porous wall.
Hall effect	Voltage is generated perpendicular to current in a magnetic field.
Henry's law	Behavior of dissolved gas in a liquid.
Hertz effect	Effect of ultraviolet light on spark discharge.
Hooke's law	Stress is proportional to strain below the elastic limit.
Hysteresis loss	Losses due to internal friction in electrical, magnetic, mechanical devices.
Impedance, frequency	Relation of electric impedance to frequency for capacitors and inductances.
Interference, optical	Light and dark bands due to optical cancellation or reinforcement.
Johnsen-Rahbek (Winslow) effect	Increase of friction or viscosity with applied voltage across interfaces between conductor, semi-conductor and conductor.
Joule's law	Heat produced in an electrical resistor by current flow and time.
Kelvin's principle of fluid drops	Minimum size of drops in supersaturated vapor.
Kepler's laws	Planetary orbits are ellipses; equal areas are swept by radius vector in equal time increments.
Kerr effect	Optical polarization due to electric field applied to a substance.
Kirchoff's laws	Sum of currents into a circuit point is zero; algebraic sum of voltages around a circuit is zero.
Kohlrausch's law	Limiting conductance of an electrolyte.
Lambert's cosine law	Illumination of a surface varies with cosine of angle of incidence.
Lambert's law	Absorption of radiation traversing thin layer.

Table 5.1 Laws and Effects Reference Table.
(Continued)

Le Chatelier's law	Equilibrium adjustments of a system under stress.
Lenz's law	Current and force induced so as to oppose motion of conductor in a magnetic field.
Leidenfrost phenomenon	Non-wetting of a hot surface by a dancing liquid drop.
Luminescence	Light produced through effects of heat, electromagnetic or a-c fields, particle excitation, mechanical stress, chemical reactions, etc.
Magnetic behavior	Permanent magnets, nonmagnets, etc.
Magnetic effects	Mechanical, optic, fluorescent, resistive; chemical effect of magnetic fields.
Magnetostriction	Change in length of a ferromagnet in a magnetic field.
Nernst effect	Electromotive force generated by heat flow across force lines of a magnetic field.
Newton's laws	Acceleration only by and proportional to a force; actions are opposed by equal reactions.
Ohm's laws	Current through impedance or resistance is proportional to voltage drop.
Organ-pipe resonance	Vibrations in a gas as function of pipe length.
Paschen's law	Sparking voltage of gas between electrodes.
Peltier effect	Absorption or liberation of heat by electric current flowing between different materials.
Photoelectric effects	Photo-emissive, photo-voltaic, photo-conductive.
Piezoelectric effect	Dimension change of electrified crystals; generation of electricity by stressing crystals.
Photosynthesis	Building up of organic chemical compounds in a living cell with the aid of light.
Photoelastic effect	Formation of color bands by polarized light through certain materials under stress.
Pinch effect	Reduction of cross section of liquid conductor carrying electric current.
Poisson's coefficient	Ratio of lateral change to longitudinal change of an elastic material under stress.

Table 5.1 Laws and Effects Reference Table.
(*Continued*)

Proximity effect	Crowding of the current in a conductor toward the side of a neighboring current-carrying conductor.
Pyroelectric effect	Charges developed between opposite crystal faces when the temperature is changed.
Purkinje effect	In the human eye: Color sensitivity change under dim light.
Radioactivity	Emission of nuclear particles through disintegration of atomic nucleus.
Raoult's dimension effect	Change of electrical resistance with change of dimension (strain gauge).
Resistance-pressure effect	Change of electrical resistance with change of hydrostatic pressure.
Resistance-temperature effect	Change of electrical resistance with change of temperature.
Seebeck effect (see Peltier effect)	Generation of electromotive force in circuit of different metals with junctions at different temperatures (thermocouples).
Shot noise	Random voltages in vacuum tubes due to variations in thermal electron emission.
Skin effect	Crowding of electric current to surface of conductor at high frequencies.
Snells' law	Relations between incident, refracted and reflected rays at an optic interface.
Stark effect	Split of spectral lines when source is in strong electric field.
State, change of	The succession of solid, liquid, gas phases of matter. Triple point is the temperature and pressure where all three phases exist simultaneously.
Stefan-Boltzmann law	Heat radiated from a black body varies as fourth power of temperature.
Stokes' law (fluorescence)	Wavelength of light emitted by fluorescent material is always longer than that absorbed.
Stress-strain effects	Elastic limit; creep; elastic fatigue; work hardening; stress corrosion.
Superposition theorem	In a fluid, the intensity of sound or electric waves at a point in space is the sum of individual component waves.
Surface tension	Effect of cohesive forces of fluids, manifest mostly at surfaces or interfaces.

Table 5.1 Laws and Effects Reference Table.
(Continued)

Thermodynamics, laws of	1. Energy is constant when heat is transformed into work or work into heat. 2. Heat cannot flow from lower to higher temperatures. 3. No finite number of operations can reduce any system to absolute zero temperature.
Total reflection angle	Critical angle of incidence for total reflection of a light beam at an interface.
Tribo-electricity	Electromotive force produced at the interface of two dissimilar metals by relative motion and friction.
Weber-Fechner law	In psychology, the minimum change of stimulus necessary for a response is proportional to stimulus already existing (e.g., sound, decibels).
Weidemenn-Franz law	The ratio of thermal to electrical conductivity is proportional to absolute temperature.
Wien effect	The increase in the conductance of an electrolyte with applied voltage.
Wien's displacement law	As the temperature of a body is increased through incandescence, the color of its emitted spectrum shifts toward the blue end.

Table 5.2 Analogies from Nature.*

Tubular structural members	Reeds, bamboo, bones
Levers, fulcrums	Muscle attachment to bone
Doors and hinges	Trap-door-spiders' nests
Lock bolt	Trigger fish fins
Traps and triggers	Venus fly-trap
Toggle	Knee
Safe packaging	Cranium
Great wind resistance	Radial fibers on seed carrier
Lubrication	Joints between bones
Cooling and control	Skin surface evaporation
Squeeze pump	Heart
Roller pump	Intestinal peristalsis
Strainer	Whale-bone baleen
Improved zipper	Brambles and heather webs
Automatic clasping device	Chicken leg and claws
Streamlining	Fish; birds
Retractable landing gear, flying	Birds' legs
Anti-wetting agents	Ducks' feather oil
Camouflage	Animal colorings
Infrared homing devices	Rattlesnake nasal "pits"
Sonar-Radar	Bat and dolphin navigation
Electric pulse generator	Heart pacer; electric eel
Ultrasonic communication	Bat and dog hearing
Illumination	Firefly; phosphorescent fish
Heat insulation	Hair covering
Level control	Inner ear canal
Jet propulsion	Squid siphon
Optimum plumbing	Veins of circulation system
Regenerated tools	Beaver teeth
Extruded products	Insect webs
Light sensitive motion	Morning-glory flower

* With help from Alger and Hays, Reference 4.

Table 5.3 Array of Laws and Effects.

Legend: INPUT to a specific box produces OUTPUT above.

OUTPUTS / INPUTS	*Acceleration*	*Current*	*Displacement*	*Electro-magnetic Radiation*	*Force*	*Frequency*	*Nuclear Radiation*	*Pressure*	*Temperature*	*Velocity*	*Voltage*
Acceleration	1	2 Piezo-electric effect	3 Newton's laws	4	5 Newton's laws	6	7	8	9	10 Newton's laws	11 Resistance/dimension
Current	12	13	14 Ampere's law Magneto-striction	15 Luminescence	16 Ampere's law	17 Ampere's law	18	19	20 Joule's law	21 Ampere's law	22 Ohm's law
Displacement	23	24	25	26 Diffraction	27 Hooke's law	28 Organ pipe resonance	29	30 Boyle's law	31	32	33 Piezoelectric effect
Electro-magnetic radiation	34	35 Photoelectric effects	36 Doppler effect	37 Luminescence	38	39 Diffraction	40	41 Radiation pressure	42	43	44 Photoelectric effects
Force	45 Newton's laws	46	47 Hooke's laws	48	49	50	51	52 Ideal gas law	53 Friction effects	54 Newton's laws; Stokes' law	55 Piezoelectric effect
Frequency	56	57 Ohm's law	58	59	60	61	62	63 Ampere's law	64	65	66 Impedance/frequency
Nuclear radiation	67	68	69	70 X-rays Gamma rays	71	72	73	74	75	76	77
Pressure	78	79 Piezoelectric effect	80 Boyle's law	81	82	83	84	85	86 Charles' law	87	88 Piezoelectric effect
Temperature	89	90 Thermo-electric effect	91 Ideal gas law	92 Stefan-Boltzmann law	93	94 Wien's dis-placement law	95	96 Charles' law	97	98	99 Pyroelectric effect
Velocity	100	101	102 Bernoulli's theorem	103 Geiger effect	104 Stokes' law	105 Doppler effect	106	107 Bernoulli's theorem	108	109	110 Electro-kinetic phenomena
Voltage	111	112 Ohm's law	113 Electro-capillarity	114	115 Ampere's law	116	117	118	119 Joule's law	120	121

REFERENCES

1. Extracted from C. F. Hix and R. P. Alley: "Laws and Effects," John Wiley & Sons, Inc., New York, 1958.
2. Greenwood, D. C.: "Engineering Data for Product Design," McGraw-Hill Book Company, New York, 1961.
3. Materials in Design Engineering, Annual October *Materials Selector* Issue; for instance, October 1965.
4. Alger, J. R. M. and C. V. Hays: "Creative Synthesis in Design," p. 42, Prentice-Hall, Inc., Englewood Cliffs, N.J., 1964.
5. Herkimer, Herbert: "The Engineer's Illustrated Thesaurus." Chemical Publishing Co., Inc. New York, 1952. (8000 designs and drawings, covering many engineering fields.)
6. Beggs, Joseph S.: "Mechanism," chap. 13, McGraw-Hill Book Company, New York, 1955.
7. "The Way Things Work," *An Illustrated Encyclopedia of Technology,* Simon & Schuster, Inc., New York, 1967.

Synthesis of Alternatives: Creativity

Six

"The sublime and the ridiculous are often so nearly related that it is difficult to class them separately. One step above the sublime makes the ridiculous, one more step makes the sublime again."
Thomas Paine

Re-checking the Problem Starting Point

The typical engineer in our assumed problem situation is undoubtedly champing at the bit now, impatient to be off. These delays at examining the needs, re-phrasing the goals, and coming up with a "problem statement" can be agreed to intellectually, but they certainly do not satisfy our emotional sense of accomplishment. The real drive in problem solving has always been for release of the emotions that accompany personal involvement. Easton[1] says:

"The creative thinker must have ambition; for a clear and powerful urge to accomplish something notable is the mainspring of all creative endeavor.

"When he meets with difficulties which inevitably arise in the course of his work, he must persist in struggling with them until he has overcome them, no matter how long it may take . . . .

"He must have enthusiasm, devotions, passions; creative thinking is not a purely intellectual process; on the contrary, the thinker is dominated by his emotions from the start to the finish of his work."

Thus the rewards of self-fulfillment, of the emotional gratification in "breaking the puzzle" and in communicating the solution to an appreciative audience must either be in hand or in sight before the problem-solver's internal pressure can be relieved.

The recommended preparatory activities in setting up the problem only further increase the frustration, instead of producing any clear progress. However, such preparation is vital, because an arrow aimed at the shadow isn't likely to hit the deer. To carry our metaphor a bit further, we must be sure it is deer meat we went hunting for, rather than possibly a bear skin.

To re-focus his wandering eye on the target, we suggest the engineer apply a check on his problem statement. This check consists of going through all the limiting words in his "needs statement" to see whether they cover as large a class of alternatives as acceptable. For example, let us say he wants to design a new pen. Statements of the problem might have been written like any one of these:

1. "A small manual liquid-dispensing device to draw lines."
2. "A manual fluid-type line drawer."
3. "A manual device to produce permanent linear records on paper."
4. "A paper imprinter."
5. "Means for recording geometric patterns in a plane."

Although one can write many word groupings that will describe the objective relatively well, note some observations about the above examples.

In No. 1, there are other than "liquid" writing agents; for example, the grease in a ball-point pen. Also "small" is unnecessary; "manual" covers it.

In No. 2, how about a powder, or heat?

We believe No. 3 describes the intent of the need fairly well. Numbers 4 and 5 are both too general, No. 4 permitting a large machine, and No. 5 going so far as to allow even a tape recorder!

The review of the actual engineering objective gives a fresh perspective, combs out some of the interference and prepares the mind for the trial solutions it must synthesize.

Synthesis

Now we have to build something up, something that we believe will do the job we see before us.

Experience tells us that we can never fully predict the outcome of a new assembly of parts. We cannot know the characteristics of the pieces well enough in advance to prophesy all the results when they react with each other. An obvious (although overstated) illustration is that of persons. Individually, they present one picture; together they may be a cooperative audience or a riotous mob. We are not thinking of predictable interactions or combinations, as with dice or symbols; we are thinking of fringe area interactions between components and parts. Will there be any unsuspected misbehavior between this relay and its controlling amplifier? Between this bearing and that camshaft? This pipe and that chemical? This shaft seal and that piston? We cannot always predict. This does not lead us to disregard science and mathematics in our synthesis. Quite the contrary: We use them for analysis as far as they are useful, and then recognize that we are on our own beyond that point.

Synthesis is combinatorial. This means we choose from the most likely classes of building blocks and mentally fit them together, searching for the various pieces that will yield that outcome which we sense, but

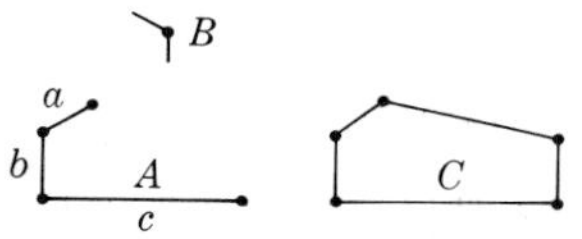

Fig. 6.1 Diagram of conceptual synthesis. Separate concepts A and B are mentally joined to produce the desired result C. (Adapted from Haefele)

cannot be sure of. This "fitting" process goes on in the mind, as with a jig-saw puzzle, searching for and testing piece after piece. Many times several pieces fit together but they turn out to be for "another puzzle" and must be discarded.

Haefele[2] offers an analogy that helps us understand the synthesizing process, symbolized by

$$A + B \rightarrow C$$

This reads: A and B are concepts established and available in the mind. In some way they come together to merge and yield the desired result C. The concept A may have been built of a number of carefully thought-out elements abc, but the sought-for whole result C awaited the recognition of B and its place in the organization (Figure 6.1). Result C was not exactly known; it was only suspected, until B was found to confirm it.

If synthesis is a combination, it implies multitudes of alternatives. Mathematically, the number of combinations of 10 different elements in groups of 10 is factorial 10 (i.e., 10 combined by 10's = 10! = 3,628,800). The mind deals readily with 10 variables, but it doesn't make 3,600,000 tests. It eliminates the unlikely prospects from consideration and looks only at the good prospects. But the mind still must fit sets together to see whether or not they produce the hoped-for result.

Sometimes we feel history portrays the geniuses of the past as receiving their great ideas and important discoveries as strokes from the blue, without forewarning or effort. We ordinary human beings react to this phenomenon with a shrug, never expecting that such a thing could happen to us. It has been established however, through many personal testimonies,[2] that insight does come after intensive preparation; and although the climactic synthesis is sudden, the path to it is already well worn by many unsuccessful probings, and many re-tracings. Ordinary persons *have* creative insights.

Creativity

This flash of insight, this "instantaneous" production of something new and important, has been the subject of much speculation, research, and discussion. It has reached the point where books have now been written[3] summarizing what conferences and other books have reported on creativity.

The synthesis of alternatives in design is closely enough related to creativity to justify a brief resume of some of the conclusions. The consensus of the studies made over the last two decades is about as follows:

1. The experiences of creative persons are very similar, according to the accounts of those in widely differing fields such as music, poetry, mathematics, and business.
2. Problem-solving and creating can be similarly described (Table 2.1).
3. Certain personal characteristics seem to be shared by all creative people.
4. There are recognized aids and handicaps to creativity.
5. Certain working conditions can help creativity in most persons.

We can begin by acknowledging that each of us has hereditary and environmental limitations to his creativity. But we also know that awareness of the situation can lead to improvement. We generally know what kind of results we want; what we need to add is some understanding of ways to achieve them. Here are some well-known characteristics of the engineering excellence which we all seek: *Elegant simplicity, fewest moving parts, new and unique function, and most efficient use of materials and parts.* Familiar examples come easily to mind: the transistor, the zipper fastener, the Polaroid camera, Nylon, the jet engine, the food waste disposer, the heat-sealed plastic bag, fiberglas textiles, the immersible temperature-controlled skillet, extruded wire insulation, catalytic-cracked gasoline, plywood, the parachute, and so on.

Studies[3] of the "creative person" have found that beyond a minimum IQ of approximately 110 (about that needed for college) and under 150 or 160, the value of the IQ has little correlation with creativity. But the creative person does seem to possess the personality traits that set him off from the general run of people. The creative person is

Highly self-sufficient, self-assertive, adventurous;
More perceptive than judgmental in observation;
Emotionally quite sensitive; more aesthetic than social in his values;
Willing to take greater risks in hope of greater gains;
More reliant on intuitive perception than sense perception.

Creative persons are reported[4] to have the attributes of self confidence, constructive discontent, positive expectations, open-mindedness, and the courage of their own convictions.

The experiences of the creative person have most usually been described in a sequence such as this:

1. Preparation: Collection of pertinent information.
2. Concentration: Digestion of all aspects of the problem situation.
3. Incubation: Relief or relaxation away from problem, for some hours or days.
4. Illumination: A rush of insight, relief and understanding about a solution.
5. Verification: Test and completion of the details.

It would be mistaken to assume that this sequence operates as a simple 1, 2, 3. The discovery or illumination phase is notoriously unpredictable. Hours, weeks, months of preparation and concentration may yield nothing, or only fractions of the expected outcome. These activities are all carried on as long as necessary or until the search is given up.

One final comment here: The insight or illumination usually arrives unbidden and inauspiciously timed. The following times and places have been reported for a number of important discoveries or creations[5]:

Time	*Place*
Day	Chair
Night	Bed
Day	On foot
4 A.M.	Pullman car
Night	Before a fire
Evening	Royal dinner
Day	Church, etc.

Most inventors will confirm this picture of the nature of their experience. As a result of circumstances, it is believed that the subconscious mind is the primary agent in the synthesis, and the person's preparation has provided both the materials and the many trial inter-associations. This of course leads us to stress again the importance of preparation and concentration, even though, in the end, we might have to wait for an inspiration to flash us an answer.

Hindrances and Aids to Creativity

"Set" or "fixity" are the words usually applied to the greatest obstacle in the path of invention or creativity. Habit is usually a friend because it saves time and effort; but to the engineer seeking significant new ideas, habitual, familiar forms and methods of solution must be consciously and strenuously avoided.

We are not suggesting you "forget all you have learned," but we do say that there are many more solutions than those you now happen to know. "Breaking set," as it is called, consists of jumping the old fences and roaming in entirely new and unfamiliar (but probably fruitful) fields. Other hindrances have been found to consist of unsympathetic or hostile audiences, tension or anxiety in the inventor, fear of ridicule, and quite naturally, inadequate preparation. Also set-breaking requires a time of relaxation.

The aids for innovative thinking, whether by individuals or groups, can be described under the following summarizing headings, with their accompanying explanations.

1. *Develop Your Backlog of Ideas*
   Make a continuing effort through curiosity, observation, and collection, to build your own storehouse of information.
2. *Review Historic Information*
   Go over applicable laws and effects;[6] check the literature; discuss with the experts, if available, to reduce the problem areas.
3. *Break Mental Set*
   Imagine a completely new set of fundamental conditions such as operating on another planet. Also, use inversion, where the driver and the driven are interchanged, or the nouns and verbs, in the statement, etc.
4. *Produce a Flood of Possible Ideas* (brainstorming)
   Noncritically list all remotely related ideas, in a freewheeling permissive atmosphere.
5. *Use Analogy**
   Direct analogy—similar problem with known answers.
   Personal analogy—uninhibitedly play the role (become a "little piece of air," a part of the machine, etc.).
   Symbolic analogy—use fictional, poetic, nontechnical literature for idea suggestions.

* J. R. Dixon, Swarthmore University (private communication).

6. *Analyze Morphologically* (Reference 3, page 256)
   Set up a matrix along the dimensions of the major variable components in a solution. For the example of a student lamp:
   1. Types of light source: Incandescent, fluorescent, round, straight, bulb.
   2. Types of structure: Gooseneck, pantograph, drop cord, pole, bracket).
   3. Types of control: On-off, dimming, one-two-three filament, automatic.

   Cross check each cell of the matrix for feasibility (i.e., incandescent, gooseneck, on-off; etc.).

Patents and the Inventor[7]

A final aspect of creativity is the recording and protection of new ideas by the innovator. This is important because new ideas not recorded

Fig. 6.2 A sketch that changed automobile ignition. (Courtesy General Motors Corp.)

have been lost and never recovered; and new ideas publicly known but not protected can be freely appropriated by others and may similarly be lost to the inventor. An example of an important recorded sketch is shown in Figure 6.2, C. F. Kettering's idea for automobile ignition, still in use today.

A patent is usually important either to the inventor or to his company. The Constitution of the United States recognized this and has provided, in Article I, Section 8:

> "The Congress shall have power . . . to promote the progress of science and useful arts, by securing for limited times to authors and inventors the exclusive right to their respective writings and discoveries."

Hall[8] has summarized the situation well:

"A patent establishes the right of the patentee to exclude others from making, using, or selling the patented article for a period of 17 years. It does not bestow upon the inventor the right to make, use, or sell the invention himself, only the right to exclude others. (The inventor already has this right under common law unless excluded by the patents of others.) In exchange for this right, the inventor makes a full and complete disclosure of his invention to the public. These facts, established firmly in the patent law explain our present concern for patents . . ."

Patentable objects fall into one of several categories.[9] The present situation has evolved from the basic Patent Act of 1836, from a great number of interpretations by the courts, from some recent congressional modifications, and most recently from the Act of 1953 (#35 USC). The six categories are:

1. *Process*
   A process is utilization of the forces or elements of nature to effect useful changes in materials, environment or information.
2. *Machines*
   A machine is an assemblage of coacting mechanical parts which receive energy and transmit it to an object, or otherwise produce a useful result.
3. *Manufactures*
   A manufacture is any useful article made by man (as distinguished from those found in nature) whether made by hand or automatic machine.
4. *Compositions of Matter*
   A composition of matter is any useful mixture or chemical compound.
5. *Designs*
   Ornamental appearance
6. *Hybrid Plants*
   Asexually reproducible; other than tuber propagated.

Excluded as nonpatentable are *products of nature*, discoveries of new *natural laws* or principles, *new business methods*, *ideas*, and *data*. Furthermore, an *aggregation* of elements operating independently is not an invention; nor is the *substitution* of one material for another if no new result occurs beyond the effects of the material's own properties. Finding a *new use* for an existing material or machine does not qualify as invention. Finally, the *substitution of equivalent elements* to perform the same task doesn't comprise invention (e.g., levers for pulleys or wedges).

To be patentable an invention must be new, that is to say, not published or publicly known in practice prior to the inventor's disclosure. The inventor may test or try out an invention before applying for a patent; but if it has been in public use for more than a year, before application is filed in the patent office, it is not patentable.* To meet the test of patentability, the invention must be more than an obvious advance to a person having ordinary skill in the art at the time the invention was made.

* Certain foreign patents would be invalidated, however, if the invention had been in public use even for a day. Patent laws vary among the different countries of the world.

The inventor may be required to prove inventorship by suitable evidence in one or more of these ways: 1) written, dated, and witnessed disclosure fully describing the invention, understandable to anyone skilled in the particular art, 2) dated and identified documents leading up to the invention and possibly developments after it, 3) a formal patent application, 4) a witnessed operating model.

Original Idea Disclosure or Design Record

The immediate concern of this section is the proper recording of an idea, both for later possible patent application and for present use on the project. Several well-established do's and don'ts have come out of practical experience and patent law. Generally, the rule is: Keep a bound, serially numbered notebook; write ideas down clearly and completely; and have the important ideas read, dated, and witnessed as soon as convenient after entry.

Do's	*Don'ts*
1. Describe (in writing, in ink, and by sketch) good ideas, and their embodiment, operation and advantages, as soon as conceived. Sign your name and the date.	1. Don't omit the date, or put down a date other than the date signed.
2. Have an associate (or preferably two) familiar with the field but not a co-inventor, read and witness the disclosure with his name and date of signing: "Read and understood by: ________________ date ______ ________________ date ______"	2. Don't write the disclosures on loose pieces of paper.
3. Make the record permanent, through the use of ink or equivalent, and then by storage in a safe, familiar place.	3. Don't omit any of the description or features from the disclosure.
4. Keep all the records of your development chronologically in the notebook, using every page, later crossing out the material you don't want.	4. Don't wait for "world-beater" ideas to start recording. Put down even those which are fairly good. But don't put down *every* idea you think of.
5. Included in your day-to-day record all calculations, sketches, data and curves that pertain to your idea.	5. If you are really interested in patentability, don't allow any significant or unexplained delays in physically "reducing the invention to practice."

An example of a disclosure page from an invention that was patented by one company is shown in Figure 6.3. This is simply an illustration of a sketch disclosure prepared at the time of conception. In practice, this is followed by more formal and more complete descriptions of the occasion and reason for invention, the advantages over competition, the detail construction and operation as planned for final use.

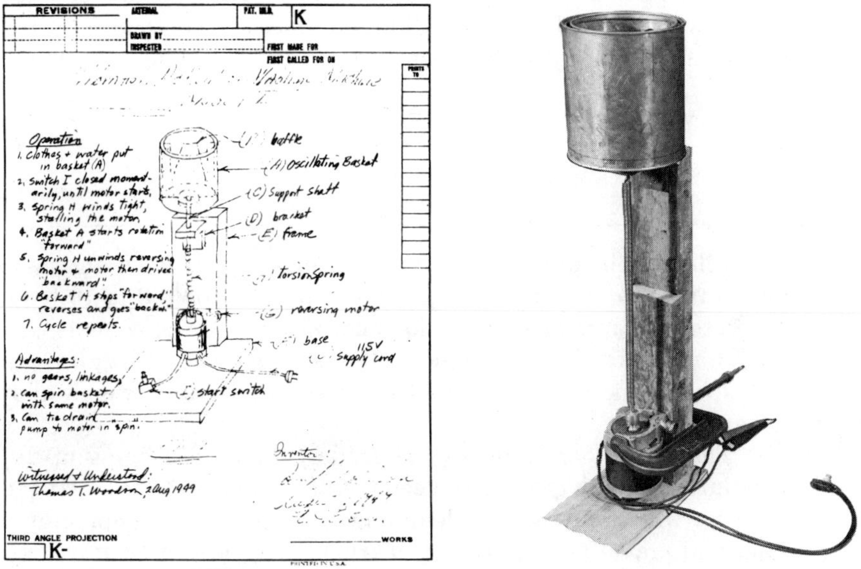

Fig. 6.3 Example of a written patent disclosure and a photograph of the experimental embodiment.

More reference will be made to the engineering design records in the chapter on written communication (Chapter 17), and in Appendix F.

SUMMARY on Creativeness

The engineer starting to put together an array of solutions must first re-check to be sure he is working on the right problem.

Synthesis of alternative solutions consists of manipulating memory to bring together already available ideas in a new form to yield desired new results.

The desired creative "inspiration" has been reported to follow preparatory work, concentration, and incubation; it also must be verified and the details completed.

Hindrances to innovation consist of mental fixity or "set," personal anxiety, hostile surroundings, too little knowledge of the field, and excessive busyness. Aids are available, in the form of techniques to free the thinking process from rigidity, and to supply quantities of fresh ideas to be evaluated. Patents can be granted to an inventor as a right to exclude others from manufacture, use or sale of his invention. Patentability includes only processes, machines, manufactured objects, compositions of matter, designs for appearance, and certain plants.

Adequate recording, reporting and witnessing in an engineer's notebook are necessary to support an application for a patent. These practices (except witnessing) should be followed even if there is no present serious intent of patenting.

BOOK PROJECT EXAMPLE

We will take our problem to be the loading of weighed amounts of sand and gravel into mixing trucks. These trucks can hold 8 to 10 tons of the concrete mixture. The mix ratio is normally 1:2:3 of cement, sand and gravel, respectively by weight, with a normal 5 gal. of water per sack of cement (1 sack = 1 cu ft = 94 lb). Sand and gravel weigh about 100 lb/cu ft.

The gravel comes in several grades, the largest being up to $1\frac{1}{2}$ in. in diameter. Sand, gravel, and cement are delivered to the plant in bulk.

We will start the problem simply as one of transfer, how to get sand and gravel from piles into a truck. Here are some ways:

1. Screw conveyor.
2. Airborne, through pipe.
3. Waterborne, through pipe.
4. Gravity drop, out of bin.
5. Belt conveyor.
6. Bag conveyor.
7. Bag carried by hook.
8. Bag slid down ramp.
9. Dissolvable bag.
10. Freezing sand into block, to melt later.
11. Catapult the bags, or the sand.
12. Load by crane or power shovel.
13. Load by hand shovel.
14. Explosive charge for propulsion.
15. Truck scoop-up while running, as a steam locomotive takes water from a track tank.
16. Electrify the particles of sand.
17. Take trucks to sand-gravel pit.
18. Dump directly out of bulk-delivery truck.
19. Tunnel a mixer truck underground, through the pile.
20. Bulldoze sand off the edge of a bank.
21. Use vibrating conveyor to lift the sand.
22. Animal power to lift the sand.
23. Sand to be lifted by falling on high speed wheel.
24. Blast the sand up by steam jets.
25. Truck loads itself from piles of sand, etc.

Without going through an actual evaluation, let us say we choose item 4, gravity drop from a bin. We will list in Figure 6.4 a variety of geometric shapes in which the bin or storage hoppers could be made, for loading into the top and out of the bottom. These can be either six separate discrete bins (two sands, four gravels) or all six assembled in one cluster. The latter option allows more variety, so we shall show them assembled.

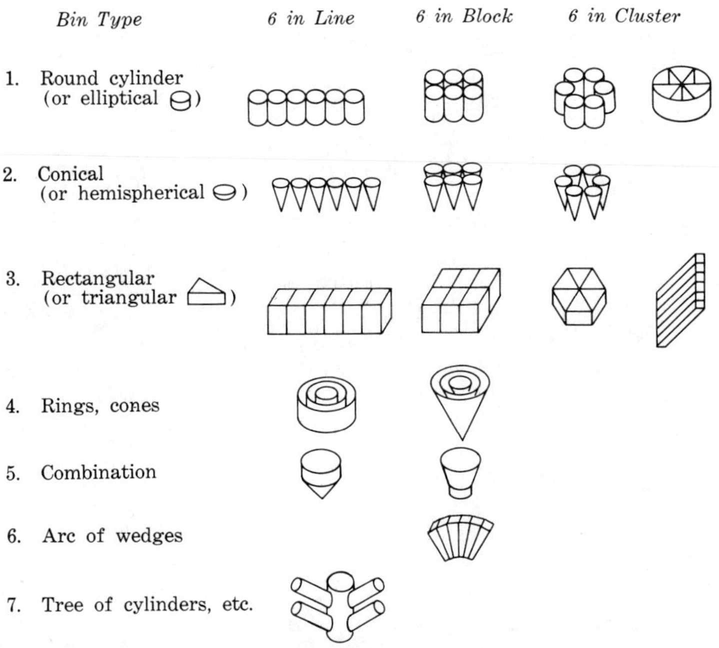

Fig. 6.4 Assortment of possible sand-hopper geometries.

Other shapes can be thought of, but this suggests a starting assortment to consider.

REFERENCES

1. Easton, W. H.: Creative Thinking and How to Develop It, American Society of Mechanical Engineers, New York, circa 1955.

2. Haefele, J. W.: "Creativity and Innovation," chap. 4, Reinhold Publishing Corp., New York, 1962.
3. Parnes, S. J. and H. F. Harding: "Source Book for Creative Thinking," Charles Scribner's Sons, New York, 1962.
4. Alger, J. R. M. and C. V. Hays: "Creative Synthesis in Design," chap. 3, Prentice Hall, Inc., Englewood Cliffs, N.J., 1964.
5. Haefele, J. W.: p. 88. See Reference 2.
6. Hix, C. F. and R. P. Alley: "Laws and Effects," John Wiley & Sons, New York, 1958.
7. Woodling, G. B.: "Inventions and Their Protection," 2nd ed., Clark Boardman, New York, 1954.
8. Hall, A. D.: "A Methodology for Systems Engineering," p. 136, D. Van Nostrand Co., Inc., Princeton, N.J., 1962.
9. Personal communication from L. R. Kempton, General Electric Co., Bridgeport, Conn.

Seven

Estimation and Order-of-Magnitude Analysis

"There is nothing more frightful than ignorance in action."
Goethe

Confidence from Experience

The newcomer to any game plays cautiously until he learns the rules well and can measure the risks. The newcomer to engineering is in a sort of game: He must spend years learning the rules, and the stakes are very high. During the learning period, his efforts are plagued by mistakes, by errors of omission and commission, by errors of ignorance and carelessness, and even by occasional errors on the part of the book or the instructor. The average student scores 20 to 40 per cent error continually, depending upon the grading scheme. Furthermore, this "game" of education is not a pleasurable divertissement; it is the student's major activity and preoccupation. The critical nature of the game is further affirmed by the high risk. The stakes include not only several years of the student's life, several thousand hard-to-come-by dollars, success in his expected vocation, but also—and most importantly—his own acceptable self-image.

Small wonder it is, then, that during his training years the student's confidence level—*his total confidence level*—takes a plunge, even though only his technical performance is directly involved.

At this moment our purposes are to reassure the reader by reminding him of the other, the successful side of his experience and to urge him to put that side to work. Incidentally, let us recall that the engineer in practice must adapt to failure, because while he is perfecting his final plan, his developmental steps must be revised repeatedly. This is even more true of the engineer as an inventor. Charles F. Kettering, Vice President of Research of General Motors Corporation, remarked that "The school child is taught it is very dangerous to fail. The inventor fails 1000 times, and if he succeeds once, he is in."

An engineer needs confidence to estimate in advance, to give general predictions of the size, cost, performance, or schedule of a proposed job. He uses his accumulated knowledge to accomplish these things. The accumulated common knowledge of an engineering student is surprisingly inclusive, and if thoughtfully applied will serve well for many estimations.

But what is this useful information that we claim most engineering students have? In exploring the more sophisticated edge of knowledge, we are likely to overlook the solid foundation of experience, of facts, lore, and probabilities that the modern person "knows he knows." Even though disclaiming any expertise on materials, the normal college person knows generally where concrete, or synthetic plastics, or iron or copper are used. He knows sizes and weights generally; he knows poisons, nutrients, fuels, solvents. He knows arithmetic well. (And perhaps 75 to 90 per cent of the math used by an engineer is plain arithmetic.) The engineering student also knows algebra, trigonometry, geometry, graphics, plus some physics and chemistry. He can search out from handbooks and other sources the prices he doesn't have at his fingertips.

We can hear, however, a protesting clamor: "This information is enough for engineering estimation?" No it isn't enough, but it is the most important part; it is the foundation for the common-sense evaluation of validity. Educated common sense, with all its weaknesses, is still the basis of final judgments. We recognize, of course, the Achilles heel represented by such things as optical illusions, fallacious reasoning, conditioned reflexes; but, as George Bernard Shaw is quoted, "Common sense is instinct; enough of it is genius." Common sense helps us distinguish what isn't common sense.

Estimation

Estimation is a projection based upon limited certainties. It is thus distinguished from "guessing," which usually is an emotionally directed speculation. It is the difference between going out on a diving board whose support you have examined, versus diving into an unknown pool. Estimation, the engineer does frequently; guessing, he should never do. When challenged to guess, the appropriate response is, "I don't know" or "Give me more time." When the individual makes an estimate he relies upon introspection, upon careful weighing and comparing of his own several related experiences. This estimate finally comes out to the public ear and eye as an affirmation by an internally secure person, who is willing to risk the declaration, in spite of "what others may think."

Estimation is the creative (or synthesizing) twin to "order-of-magnitude analysis," which is the critical partner of the team. They both are helpful in "cutting the problem down to size."

Order-of-Magnitude Analysis

Numerical answers to engineering problems vary over a wide range, most of them lying between some values, say 10^{-6} and 10^{9}. But, looked at as orders of magnitude, this range is much less imposing: Only 15 options await, since the exponents lie between -6 and $+9$. Notice we are using the word "order" in its mathematical sense, as the degree of exponentiation. One of the objects of order-of-magnitude analysis is to find out which one of these 15 exponents is closest to any anticipated answer. We may possibly be interested in one further decimal point (i.e., which one out of 150), depending upon the circumstances; but this is not usual.

In addition to indicating probable answers, such analysis is important in checking answers already found directly by the straightforward use of physical principles and mathematics. (Checking will be treated in detail in Chapter 14.) In this connection, we realize that to provide an adequate engineering check, a different but parallel method must be used, such as checking by the conservation of energy where the problem had been worked out using Newton's laws. Doing the problem over again the same way checks only the arithmetic, not the thinking, and generally checking the thinking is more important and certainly more difficult.

Estimation and order-of-magnitude analysis are very similar. For our present purpose, we distinguish the former by the fact that some system or method of solution must be conceived; whereas with the latter, the solution system is given, and only numbers must be supplied. In most cases, we use our past experience for the information sources, particularly if we cannot immediately locate the required fact in a book. Some simple illustrations follow.

Estimation: Examples

Parking autos, I. For a simple beginning example, let us estimate how many cars per 10,000 sq ft (100×100 ft) can be parked in an ordinary rectangular parking lot. As a first rough estimate, we can draw upon memory and assume an area 9 × 18 ft (≅ 160 sq ft) for each parking space itself.* We will allow an equal amount for aisles, waste corners, and entrance space, since we are including parking at any angle. These assumptions then lead to the following numbers, suitably rounded off.

$$\text{Ratio} = \frac{10{,}000 \text{ sq ft/unit area}}{320 \text{ sq ft/car}} \cong 30 \text{ cars/unit area}$$

The assumption could have been different, such as 20 ft instead of 18 ft, or 75 per cent of car space for aisles, versus 100 per cent, etc.; but these would have only changed the final figures to 28 or 36 cars/10,000 sq ft, respectively. Notice that this is done very simply and even without the use of graphics. At this stage, the uncertainties do not warrant much detailed consideration.

Parking autos, II. We can extend the method to a more complex case. Assume a suburban triangular corner lot 300 × 260 × 150 ft, as shown in Figure 7.1, and estimate the number of cars that can be parked there. This

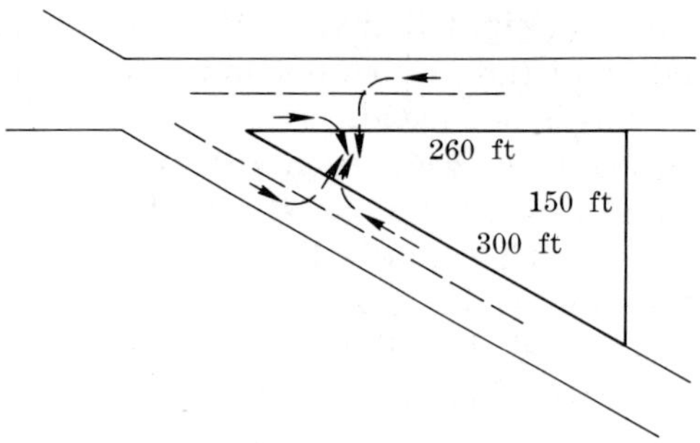

Fig. 7.1 Parking lot corner location and access.

time we need some further assumptions and some graphics. We will assume angular parking and sketch an approximate picture to find the dimensions of aisle and stall spaces. See Figure 7.2. From this sketch, we measure the aisle width to be 19 ft, the stall box width as 18 ft, and the unit stall length as 9 ft. We have estimated the minimum turning radius closely enough from past experience, simply by imagining a large car turning around in the street. We have assumed this to be a 20-ft inside radius. The 60° parking angle seems logical because of the angles between the sides of the lot.

* If one's memory is poor, the size of cars can be readily paced off. Large cars are approximately $6\frac{1}{2}$ ft wide by 17 ft long. Small cars don't change the picture, normally, except they can be used to occupy otherwise unusable patches and corners in a parking lot.

Now our problem is to sketch the aisles inside the given triangular lot to yield the best use of the area, under reasonable car-handling conditions. We will not consider "backing-out" of the aisle; thus we need a loop aisle so the cars can continue around to the exit. Now we spend a little time

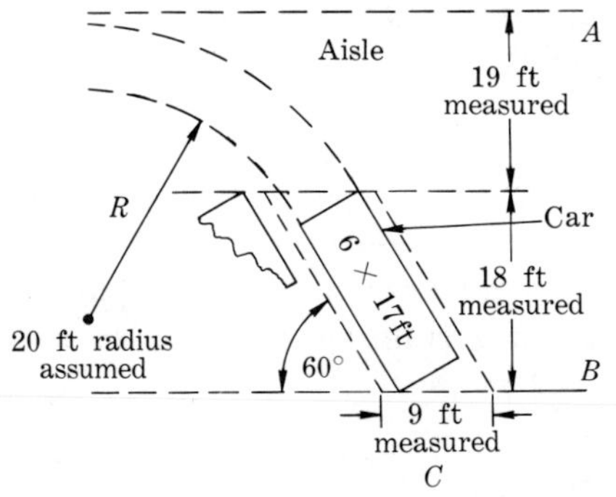

Fig. 7.2 Space and aisle estimates for auto parking.

making a number of alternate sketches, each time visualizing how the driver would maneuver his car into and out of the various possible parking places; so we stop and count the stalls we have created. In this particular case (see Figure 7.3), we obtained 64. The parking lot is not perfect, but, on the

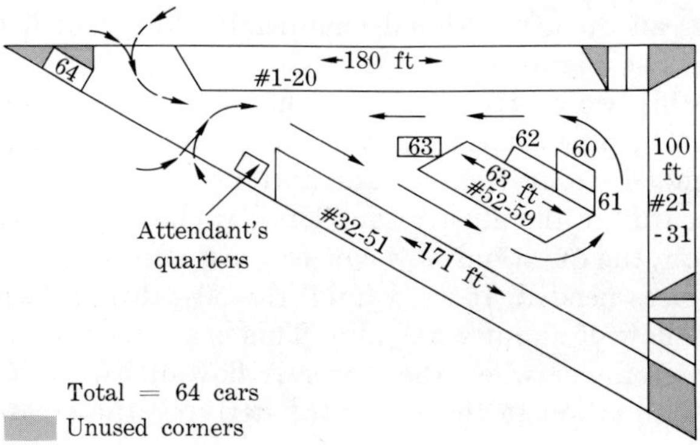

Fig. 7.3 Estimated parking stall allocations, using the dimensions of space and aisle from Figure 7.2.

other hand, the space is fairly well used. The ratio (cars/area) per 10,000 sq ft for this particular lot is

$$\frac{64 \text{ cars}}{130 \times 150 \text{ sq ft}} = 33 \text{ cars}/10{,}000 \text{ sq ft}$$

This is close to our first rough estimate of 30 cars/10,000 sq ft. The second example represented a more detailed examination of a given space, but we can't be sure at this stage that even this is correct. We do know, however, that it is neither 10 nor 100 cars per 10,000 sq ft. Hence our "order-of-magnitude" certainty. For comparison we are including here the standard dimensions* for parking, currently accepted by highway officials. Referring to Figure 7.2, the standards are (60° angle):

A, Aisle: 18 ft	C, Length: 9.2 ft
B, Stall: 18 ft	R, Radius: 19.3 ft

Water supply for mountain camp. For another simple example, let us estimate how we would bring water from a running stream into a tank

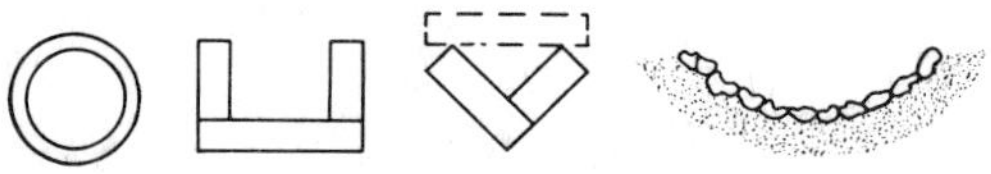

Fig. 7.4 Possible water channel sections. Water supply project for mountain cabin.

(let's say a 50-gallon gravity tank) to supply water for a vacation cabin in the woods. A natural supply point is 100 ft away upstream, guaranteeing among other things a clean, continuous water supply. Our problem is transport. Shall we use pipe, an open rock-lined channel in the ground, or a wooden flume or trough? See Figure 7.4.

As we think about this, we discard the open channel in the ground as too easily contaminated. The pipe could be laid on the ground; and the wood flumes could be suspended from tree trunks and possibly covered as shown by the dashed "board" in the illustration. Thus, we have two reasonable ways of doing this job; the questions now concern cost and convenience.

Next we check the sizes needed. If we wanted the 50-gal tank filled in 15 min, we would need a flow of about 4 gal/min. This is a stream of water about as big as a person's finger when the water is flowing two feet per second, as shown by the equation in the footnote† deriving the cross-sectional area, A, of the stream.

* American Association of State Highway Officials. *Policy on Design*, 1954, quoted in Reference 3.

† $A = Q/V$ (area = flow/velocity)

where

$$Q = 4 \text{ gal/min} \cong 1000 \text{ cu in./min} \quad (1 \text{ gal} = 231 \text{ cu in.})$$

$$V = 2 \text{ ft/sec} = 24 \text{ in./}(\tfrac{1}{60} \text{ min}) = 1500 \text{ in./min}$$

$$A = \frac{1000 \text{ cu in./min}}{1500 \text{ in./min}} = \frac{2}{3} \text{ sq in.}$$

This area would require a pipe one inch in diameter. If we were using the wood "vee" channels, we would need two boards each about three inches wide to avoid splashing over, or one-half board-foot per running foot of channel (per foot of channel length). (A board-foot is one square foot of wood, one inch or less thick.)

Now we need to arrive at costs. The most convenient reference is a broad-coverage catalog (such as that of Sears, Roebuck), in which wood, metal, and other supplies can be found listed at retail prices. Of course, one can also phone the retail plumbing or lumber suppliers. In any case, we find

Wood: 15¢–20¢/ board-foot in the sizes we need
1 in. iron pipe: 30¢/ft
$\frac{3}{4}$ in. copper pipe: 58¢/ft (One size smaller than iron pipe for the same flow rate.)

Assuming one-half board-foot for each running foot of wood channel, the comparisons are

Wood: 10¢/running foot, materials only
Iron: 30¢/running foot, materials only
Copper: 60¢/running foot, materials only

Since we would do the work ourselves, the cost of labor is disregarded, and it seems that wood should be our choice. On one final check through, we ask whether these are all the choices. Someone suggests plastic pipe, so we look that up: It is corrosion-resistant, flexible, easily connected, sanitary; it has a smooth interior; it could be in one piece and simply laid on the ground. It seems to be a natural choice. The price of $\frac{3}{4}$-in. diameter plastic pipe is 10¢/ft; 1-in. diameter is 16¢/ft. Considering the labor needed with wood or iron pipe, or the cost of copper tubing, and the plastic's sanitary advantages, the plastic pipe (high-density polyethylene) is certainly the preferred choice.

Thus our final estimate is the use of this plastic tubing, probably $\frac{3}{4}$-in. diameter at 10¢/ft.

These examples constitute simple engineering designs because they involve some technology. As far as they go, they present specific recommended solutions. Yet, common information was the basis for the solution in each case.

Order-of-Magnitude Analysis: Examples

Water heater. As a simple example of order-of-magnitude analysis, let us estimate the heating recovery time for a 30-gal. electric water heater.

We assume:

1. All electric energy goes into heating the water (no losses through the tank wall during the heat-up period).
2. Water temperature starts at 60° F and finished at 160° F.
3. Heating power is the conventional 1500 w minimum.

We go to handbooks to look up the Btu and watt-hour equivalents. We know that power multiplied by time yields the quantity of heat delivered to the water:

$$Pt = H \qquad \text{or} \qquad t = \frac{H}{P}$$

Also the heat H delivered to the water is the product of the weight of water and the change in temperature. Assuming a temperature rise of 100° F,

$$H = \text{volume} \times \text{density} \times \text{change in temperature}$$
$$H = (30 \text{ gal})(8.3 \text{ lb/gal})(100°\text{F}) = 25{,}000 \text{ Btu} \qquad \text{(See Appendix A for reference data)}$$

The power is converted from kilowatts to Btu/hr by the relation:

$$P = 1.5 \text{ kw}\left(\frac{3414 \text{ Btu}}{\text{kw-hr}}\right) = 5000 \text{ Btu/hr} \qquad \text{(See Appendix A)}$$

Thus, the time is

$$t = \frac{H}{P} = \frac{25{,}000 \text{ Btu}}{5{,}000 \text{ Btu/hr}} = 5 \text{ hr}$$

(*Note:* The *maximum* electric input is 4500 w usually; thus the time could be as short as two hours. The heat input for gas is usually 30,000 Btu/hr, so the time would be less than one hour for a gas water heater.)

Motorized knife sharpener. How much power should it take to run the grindstone of a kitchen knife sharpener? We shall decide what we need and then assume the necessary quantities.

$$\text{Power} = \text{force} \times \text{velocity:} \quad P = F_t V_t$$

where

F_t = tangential friction force on the grinding wheel

V_t = tangential wheel velocity

The friction force F_t is the pressure on the knife blade F_n, multiplied by the coefficient of friction f; i.e., $F_t = F_n f$. Thus we need to estimate three quantities: F_n, f, and V_t.

Assume:

F_n: Pressure on knife blade is 1 lb
f: Coefficient of friction is 0.5
V_t: Surface speed of grinding wheel is 15 ft/sec (e.g., a 3-in. wheel at 1500 rpm)

Then, recalling that 1 hp = 550 ft lb/sec, the power is

$$P = fF_nV_t = (0.5)(1 \text{ lb})(15 \text{ ft/sec}) \frac{1}{550 \text{ ft-lb/sec/hp}}$$

or

$$P \cong \tfrac{1}{100} \text{ hp}$$

Notice that the effects of pressure and speed are proportional. If we had estimated high or low, the order-of-magnitude would still have been close. Twice (or half) the pressure or speed would have yielded twice (or half) the horsepower: $\frac{1}{50}$hp (or $\frac{1}{200}$hp) but not $\frac{1}{10}$ hp (or $\frac{1}{1000}$ hp).

Strength of chain. As another example, suppose you need to hoist the engine out of a sports car to put in a replacement. You have some steel chain, whose links you measure and find made of $\frac{1}{8}$-in. diameter wire. Is this chain strong enough to lift out a 300-lb engine? What about nicks in the chain? What about the engine getting caught askew during hoisting (and lifting the car)?

If the chain is in pure tension, each side of the link carries half the load. We learn the safe tensile stress from an engineering handbook*

$$S_s = \text{safe tensile stress} = 20{,}000 \text{ lb/sq in.}$$

The load capacity of the chain is, therefore,

$$L_s = A \cdot S_s \quad (\text{safe load} = \text{area of wire} \times \text{safe stress})$$

where

$$A = \frac{2\pi d^2}{4} = \frac{\pi}{2}(0.125 \text{ in.})^2 = .02 \text{ sq in.}$$

Thus,

$$L_s = (.02 \text{ sq in.})(20{,}000 \text{ lb/sq in.})$$

or

$$L_s = 400 \text{ lb}$$

This gives you a safety excess of only 100 lb over the 300-lb engine weight. The rest of the safety factor is in the S_s value. Since the ultimate tensile breaking strength is probably at least 60,000 lb/sq in., we have a

* References 3, 4 at the end of Chapter and Appendix A.

total factor of safety to breakage of four. Even with this safety factor, the chance of the engine getting caught, and the likelihood of nicks in the chain would probably motivate you to double or triple the chain carrying the engine to the hook of the hoist.

Sun-powered flying satellite. The engineer uses order-of magnitude analysis routinely to separate technically feasible new ideas from the unfeasible. Someone has suggested the new idea of a solar-powered plane which will continuously circle the earth high within the atmosphere in order to provide a synchronous communication satellite. Let us find out whether it appears feasible.

The top surfaces of the plane's wings and body would continuously receive solar energy via solar cells connected to suitable electric motors, these motors in turn would drive the propellors and maintain the 1000 mph speed necessary to keep pace with the sun. We shall consider only the steady-state problem of maintaining the speed, not the occasional problem of getting the airplane up there from the ground. First, we obtain the solar radiation constant, which is 2.00 cal/sq cm/min or 442 Btu/sq ft/hr, measured at the earth, on perpendicular incident planes. Second, we realize that the aeronautical wing-airfoil index, known as the "lift-drag ratio," L/D, is the key to estimating the required power to keep aloft. This ratio represents the weight that can be carried per unit of propulsive force. Looking up the above index for high flying, fast planes[5] we find it ranges from 10 to 20; so we choose a value midway between.

Let us assume $L/D = 15$ or $D = L/15$. This ratio informs us that the drag, D, is $\frac{1}{15}$ the lift, L, the weight of the airplane. The power required is the product of force and velocity; so, to keep our airplane aloft at a velocity, V, of 1000 mph, overcoming a drag force D, we need the following power per pound of airplane weight:

$$\text{Power (per lb)} = DV = \tfrac{1}{15}\text{ lb} \times 1000\text{ mi/hr} \times 5280\text{ ft/mi} \times \frac{1}{778\text{ ft-lb/Btu}}$$

$$= 450\text{ btu/hr (per lb)}$$

Since the sun is supplying only 442 Btu/hr/sq ft and the best solar-cell and electric motor conversion efficiency is less than 10 per cent, we can expect an effective driving power of less than 44 Btu/hr/sq ft. For convenience in comparison, let us convert this driving power of 44 Btu/hr/sq ft into 440 Btu/hr/10 sq ft, which approximates the power available.

Thus the power required to push the plane would equal the available driving power only if 10 sq ft $\cong$ 1 lb. This is a *very* light weight section of a wing. Even if such a wing were made out of the lightest metals (aluminum or magnesium), the weight limit of 1 lb would permit a single metal thickness of only 0.01 in., the thickness of three sheets of paper! Also, the airplane must carry dense and heavy equipment such as motors, controls, fuselage, etc.

Thus we are forced to conclude that either the available energy must increase to the theoretical limit, or densities of structural materials must decrease by factors of at least 10 (or both) before such an idea as the solar-powered flying satellite becomes feasible.

SUMMARY

We see that surprisingly quick and accurate (presumably) estimates of engineering situations can be obtained in a few minutes. This is a practice all engineers develop and use. If fact, as someone has said, estimation and order-of-magnitude analysis are the hallmarks of the engineer. Even in complex situations, simple equivalents are found and estimated as preliminary information guides or as checks on other computations.

Estimates are made by:

1. *Imaginary measurements.* Figuratively applying in the mind yardsticks, forces, or motions to the object under scrutiny.
2. *Rough calculations from simple equivalents.* Applying a physical law to a simplified form of the situation.
3. *Graphics.* Sketching to establish forms, relations, quantities.
4. *Intuition.* From memory, and only partly reasoned, correlating with past experiences.

In conclusion, let us note that numerical answers are carefully shown only to a *significant* number of significant figures. Answers must be rounded off ruthlessly to avoid giving an impression of higher-than-justified accuracy. Use 0.3 for $\frac{1}{3}$, rather than 0.333, when no more than the first 3 can be relied upon.

Student Exercises

7.1 Estimate the size and cost of a tank to use as a reservoir supplying water to a U.S. city of 5000 persons.

7.2 How much would a wood fence cost to go around the rear of a residential lot whose eastern, western, and southern fenced sections are 60, 40, and 75 ft respectively? The object is to keep the dog in, retain some privacy, and yet allow breezes to penetrate.

7.3 How thick is one coat of outside paint on a house if the label says that a gallon will cover 300 sq ft?

7.4 How many turns of cotton twine should you put around a 100-lb sack of sugar to carry it safely by hand?

7.5 Estimate by detail-tabulated items the cost of a student (yourself) to live for a school year in a school dormitory, versus the cost in either an apartment or a rooming house (or at home if commuting is possible).

REFERENCES

1. Johnson, L. H.: "Engineering Principles and Problems," chap. 2, McGraw-Hill Book Company, New York, 1960.
2. Smith, R. J.: "Engineering as a Career," 2nd ed., chap. 16, McGraw-Hill Book Company, New York, 1962.
3. Perry, R. H., Editor: "Engineering Manual," pp. 6–19, 6–31, 6–72, McGraw-Hill Book Company, New York, 1959.
4. "Marks Mechanical Engineer's Handbook," 6th ed., pp. 6–37, McGraw-Hill Book Company, New York, 1964.
5. "Marks Mechanical Engineer's Handbook," Reference 4, pp. 11–75.

Engineering and Money*

Eight

"The use of money is all
the advantage there is in having it."
Benjamin Franklin

Why Money?

Money is not listed as a course in the engineering curriculum, so why should there be a chapter on the subject? Why not let engineering students concentrate on the necessary courses in engineering, mathematics, physics, chemistry, and humanities? Why not leave the study of "unrelated" subjects such as money to the student of business?

In responding to these objections, let us ask what the engineering student can accomplish *without* money. The student, or the engineer, can *think* without money, but no one is called an engineer just for thinking!

The engineer is a *doer*, and he soon learns that he cannot get much done in a vacuum. Money is the "life blood" of business and government, and it is safe to say that the engineer who understands the money side of engineering is most likely to succeed in getting his job done.

To perform an engineering task you must understand technology and be creative, but you must also find *someone* who is so impressed with your engineering plans that he will invest his "savings" to get the job started.

* This chapter written and copyrighted by William D. McIlvaine, Associate Professor, Engineering, California State College at Long Beach.

Someone has to plan to spend the money: Hire the needed talents, organize the office, provide the supplementary services, such as typing the letters and specifications, reproduce the drawings, plan and perhaps build test facilities, plan manufacturing procedures and systems. Someone has to be sure that the customer buys the products that the engineer has designed. Statistics will indicate that the time spent on solving technical problems and the time spent on organizing and administering the work are about equal and that, even in highly technical work, the administrative costs are often greater than the technical costs.

The typical engineer in a responsible position needs to answer money questions perhaps more frequently than he needs to answer technical questions. Such an engineer must know what it is likely to cost to get his job done. There will also be questions such as: What price will the customer pay for the product or service? What will our competitors charge for their designs? How much money will be required to set up a manufacturing system? How much money will be required to manufacture each item? Will the cost per unit produced remain constant or will it go down as the number produced goes up? If the cost per unit varies drastically with the number produced, what risks and what prospective advantages are involved?

The young technical man in his first job will probably not have to answer many money questions such as those asked above. But each time an engineer is promoted to a more responsible assignment, it is likely that the number of management (and therefore money) questions that he has to answer will increase. Engineers do get promotions for answering more complex questions that are purely technical, but statistics on salaries strongly suggest that employers place a high premium on an employee's willingness and success in accepting responsibility for *planning* and *organizing* engineering work. And the engineer finds that every time he accepts more responsibility for planning and organizing, he spends more time answering money questions.

The important point at this stage of your study is that you need to be aware of money problems in engineering. The next several pages will describe some of the important relationships between engineering and money. An understanding of these relationships will help you complete design work that is feasible, practical and likely to be rewarding, in both the recognition and the remuneration that you receive.

Long before you become responsible for any segment of technical work, you will be judged on the *practicality* of your ideas, in terms of money. Without some elementary understanding of the language of money, your assignments will have to be stated in technical language only. The more complicated assignments require engineers who understand more than one language. Complicated assignments require engineers who can work with supervisors who fluently use the language of the technical work and the language of the financial world, engineers who are truly bilingual. Without an awareness and understanding of money in its relationship to engineering, you will always need an interpreter.

We believe that the language of the financial world is best learned by actually working with money problems rather than by reading about them. Experience in part-time, year-round or summer employment may provide some of this background. Working your way through college as a "cooperative or work-study" student, who alternately works and studies, is another constructive way of becoming familiar with economic aspects of engineering. If you are seriously interested in the engineering aspects of the technical world, in contrast with the purely scientific aspects of the technical world, you will find that work experience is as important as formal education. You will also understand more fully the role of money in engineering.

Fundamental Relationships between Money and Engineering

You have already been introduced to the idea that engineering work requires "inputs" and produces "outputs". Suppose we use the input-output

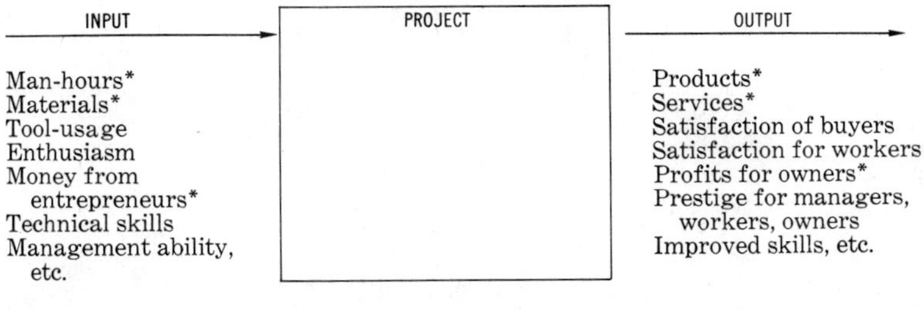

Fig. 8.1 Inputs and outputs for an engineered product.

method for a gross analysis of a typical engineering project. We first used the input-output analysis on the product in Chapter 4, Figure 4.2. Our observations would show that when considered as a project, it would probably shape itself schematically as shown in Figure 8.1.

The inputs and outputs marked with an asterisk (*) in Figure 8.1 can readily be measured in dollars. In the case of tool-usage it is easy to measure in dollars if the tools are rented; but it is not so easy when the tools are already owned. However, intangible inputs and outputs such as enthusiam, management ability, technical know-how, personal satisfaction and prestige may be related to money; but it is difficult if not impossible to measure these attributes in terms of money. Therefore, if we are to follow our analysis logically, it is essential to keep in mind that measuring the inputs and outputs in terms of money *is not a complete reporting* of all of the observable events. In other words, we can be sure that conclusions about an engineering project are incomplete if they are stated solely in terms of money.

In spite of this limitation, we can learn a great deal from observing engineering project activities, and studying those elements of an engineering project which can be directly measured in dollars. It may help us to be objective in our observations if we visualize an engineering project as a black box until we have systematically observed and recorded the dollar inputs and outputs.

Gross Analysis of the Engineering Project

The first action observed in all engineering projects is that money is placed in the black box by someone who is sufficiently enthusiastic about

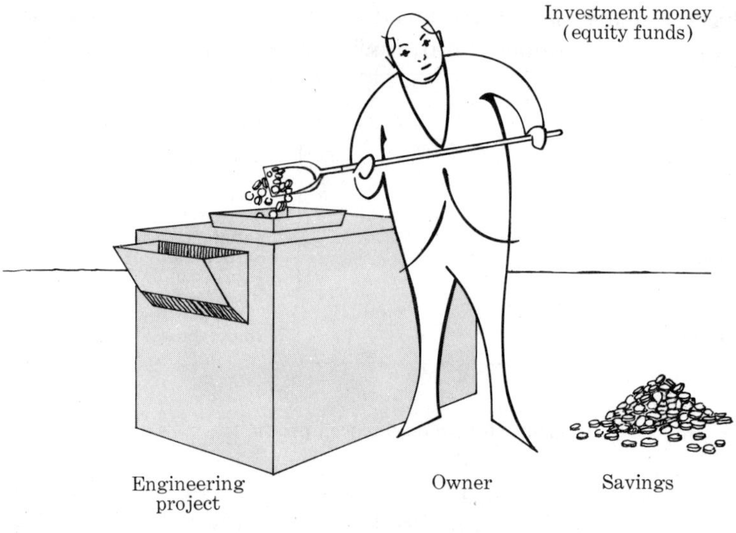

Fig. 8.2 Investment to start an engineering project.

the goals to warrant investing his savings in the project. The man who shovels the first money into the black box is always an owner (as in Figure 8.2). Modern business is generally organized in the form of a corporation, and the owner will be found to be a large number of persons, the stockholders of the corporation. An owner is usually referred to as an entrepreneur when he starts or has a major control in a business himself. But, whatever label is used to describe the owner, the money he is putting into the black box is *money he has saved from other possible uses.* The money going into the box at this stage is referred to as *equity funds* or *owner investment money.*

The typical uses of the investment money are to purchase or rent machines that are required to get the job done. The most important observation at this stage is that each shovelful of money is money that somebody has decided to risk on the probable success of this engineering venture. In each case, the owner of the money was sufficiently enthusiastic about the venture to put his money into it rather than use the money to purchase some item or service that would bring more immediate satisfaction.

The next step that will be observed in the operation of this black box (Figure 8.3) simulating our engineering project is that customers will put in money to purchase products or services from the venture. Note the outlets where money flows to those who keep the project operating: the workers, the suppliers of materials, the owner of the building or the rented tools, the suppliers of electricity or fuel, the government unit that runs the community in which the business operates, and so on. A detailed list of these outlets would be almost endless in a complex project. We should also note that these flows are frequent and that the total of these outward flows is related to the total money flow in from customers.

In this diagram (Figure 8.3) the owner or stockholder appears to be anxious. He has invested most of his savings and, up to this time, he has received no reward from his investment. His anxiety is also justified by the fact that organized businesses fail at the rate of more than a thousand per month, even in the so-called "good" business years. Since most businesses are collections of engineering projects, all owners have good reason to be intensely interested in each individual engineering project.

If we continue our economic observations of engineering projects, we will find that about half of them receive additional money from *lenders* who looked like owners, but who put money into the black box in exchange for a promise that they would be repaid with interest. In many cases the promise to repay is backed up with a legal document mortgaging company property to lenders.

We have not included the lender graphically in the diagram, but we should report that the black boxes which received loan money also had a time-controlled door that opened every three or six months in order to pay interest to the lenders. If we observed a black box over its entire lifetime, we would also observe that the original amount borrowed (the principal) was also returned to the lender, and that when all loans were repaid, the time-controlled door no longer opened.

However, in our studies of these black boxes over a period of time, we observed that the owners were very pleased with the results of many of the ventures. Usually, in these cases, the black box had begun to bulge as though it needed to expand its capacity. In all successful cases, we noted that the box now held more money than had been deposited by both the owners and lenders. There was now an excess of inputs over outputs. The money flows at this stage were so complicated that we had to query owners,

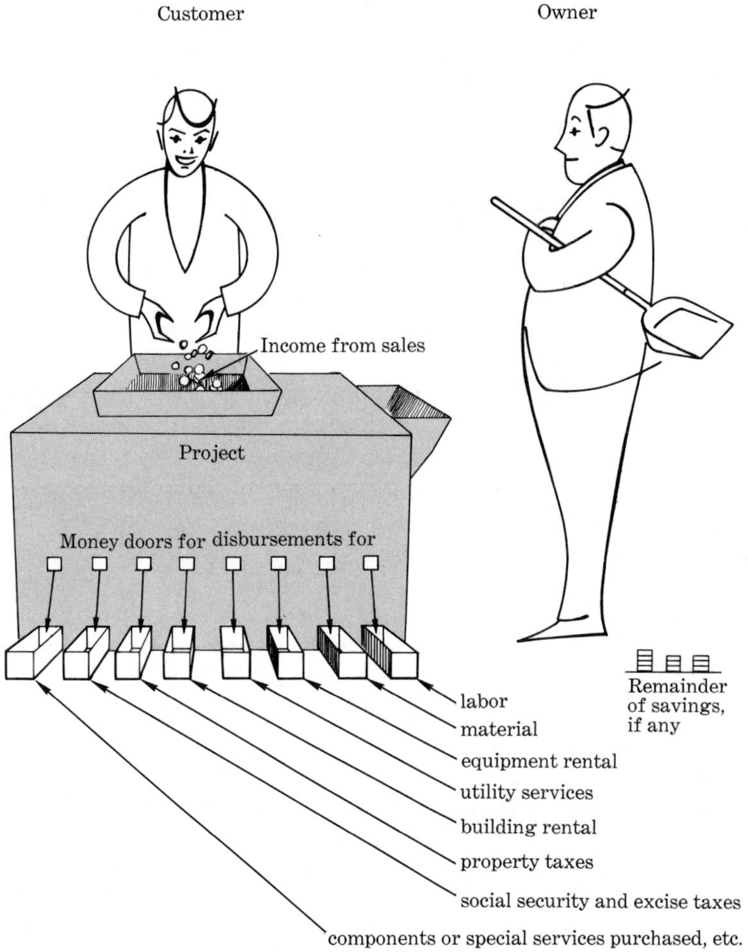

Fig. 8.3 Situation in early stage of operation of an engineering project.

accountants, lawyers, and United States Treasury tax experts before we could clearly identify any system that would apply to all of the black boxes observed. As a result of our studies and consultations, we found how the more important money flows can be identified. These are indicated in the diagram of Figure 8.4. The quantity of money in each of the depicted flows is the quantity for a given period such as a year, a quarter, or a month. The financial statement that reports these quantities is usually labeled a statement of income.

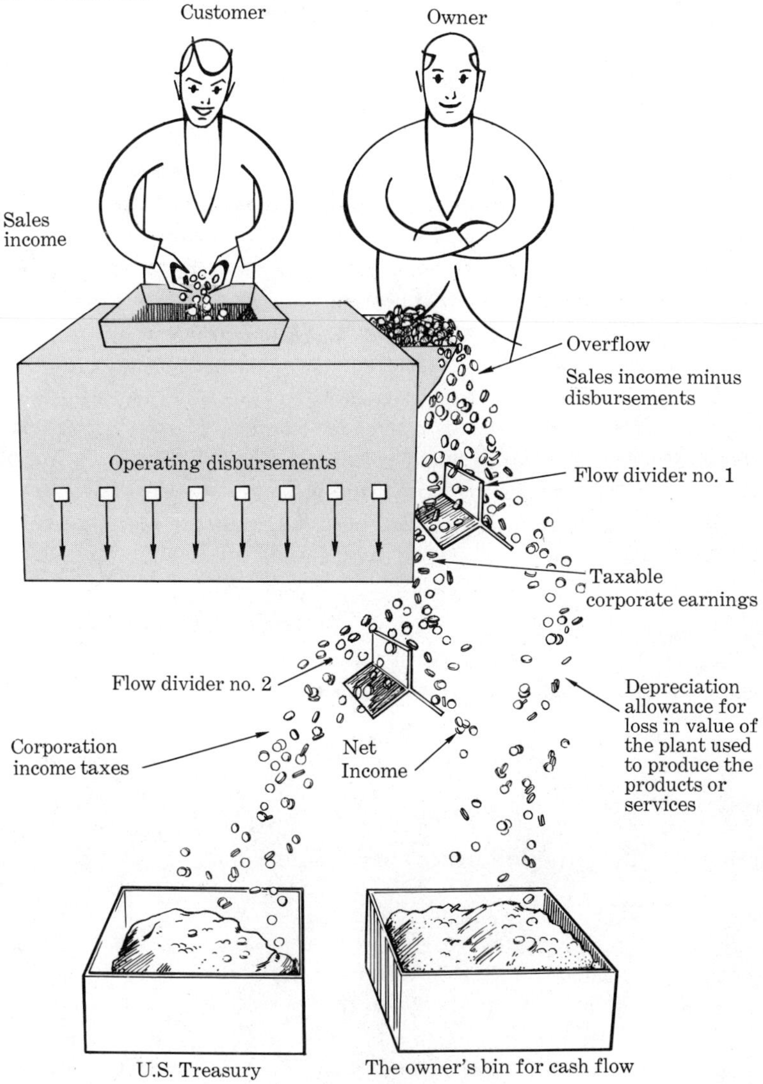

Fig. 8.4 Cash flow in a successful engineering project operation.

The flow diagram would be too complicated if it attempted to explain all about money flows in a business. In particular, we should explain that the position of Flow Divider No. 1 can be adjusted a little to the left or right depending on the judgment and experience of the owner concerning the rate at which his plant decreases in value. Also, the position of Flow Divider No. 1 affects the amount of Taxable Corporate Earnings, so the government has issued a booklet describing permissible depreciation allowances. This sets a limit to the amount Flow Divider No. 1 can divert overflow into the owner's bin.

The position of Flow Divider No. 2 is fixed by law, but the detailed provisions of the law are much too complicated to describe here. The approximate percentages paid to the federal government in 1965 were 22 per cent of the first $25,000 of corporate taxable income plus 48 per cent of all added taxable income above $25,000. However, the task of drawing up statements that were acceptable to the Treasury and fair to the owners must have been difficult, because accountants, lawyers, managers, owners, and government men spent many hours compiling and reviewing these reports.

Our observations would be incomplete if we failed to record that the owners seemed very pleased with the cash flow. We should also report that this cash flow was used mainly in two ways: (a) some was used to pay dividends to the owners and (b) some was "retained" by the owner to finance other projects or to improve the corporation's ability to compete. We also found that the black box analogy could help us analyze an engineering project *only* when we insisted that *each* investment of cash flow should be studied as a "new" black box.

Conclusions about the Economics of Engineering Projects

What conclusions can be drawn from our observations of engineering projects as black boxes? Did we observe any fundamental phenomena that will help engineers to understand the relationships between money and engineering? Let us draw the following conclusions:

1. Money flows, such as the income from sales, operating disbursements, income taxes, etc., are vital elements of engineering projects, and any detailed study of an engineering project must examine these flows carefully, since the inputs from and outputs to the owners are closely related to these basic money flows.

2. The relationships between the money inputs and outputs of the owners are regarded by the owners as the principal and most tangible measure of the "success" of an engineering project.

3. The specific information for fully describing the "money success" of an engineering project is a list of the amounts and dates of each transaction that is one of the following quantities: (a) an investment, (b) a net income, or (c) a return of investment to the owner, via the cash flow of depreciation allowances.

Remember there were some factors excluded from the foregoing analysis, because we decided at the beginning to exclude those factors which could not readily be measured in dollars. Some of the "other factors" that are regarded as additional or partial measures of success are: (1) the volume of sales of the product compared with the sales of similar products of competitors, (2) increased understanding of business or technology that may result in future successful engineering projects, (3) improved employee morale and (4) enhanced company reputation for manufacturing products that are durable and trouble-free. There are many "other factors" but those listed above should serve as a reminder that the analysis of money flows is only *one* way of measuring "success."

The Business Unit

Before we undertake any detailed study of prospective money flows in an engineering project, a word should be said about the money factors outside of, but connected with, our engineering projects. The most immediate set of "outside" factors is the complex set of money flows in the individual business unit. Our engineering project is only one of several flows, or perhaps only one among hundreds. The "word" that should be said about the individual business unit might be as short as a few paragraphs but it could also be as long as a whole text on the Theory of the Firm.

Figure 8.5 shows money flowing as if it were an incompressible fluid circulating through the most important of the many possible channels of a typical individual business unit. The diagram also traces some of the most important channels connecting the typical individual business unit with the national economy.

Now notice the similarity between Figures 8.4 and 8.5. The money flows in the Typical Business Unit (the left-hand side of Figure 8.5) are similar to the money flows in our engineering project, Figure 8.4, in that each flow is the measured total for a specific period of time, such as a month, a quarter, or a year. Money flows such as these are reported in a financial statement which is usually labeled "Income Statement," and each such statement always indicates the exact period for which the money flows were measured.

It should also be noted that the Operating Unit looks very much like the black box that was used as a model for the engineering project. This resemblance is intentional because the typical business unit is usually a *collection* of engineering projects. The following are the most important money flows in the operating unit (the numbers in parentheses refer to the numbered channels in Figure 8.5): income from sales (1), wages and salaries paid to production workers (2), materials purchased (3), miscellaneous

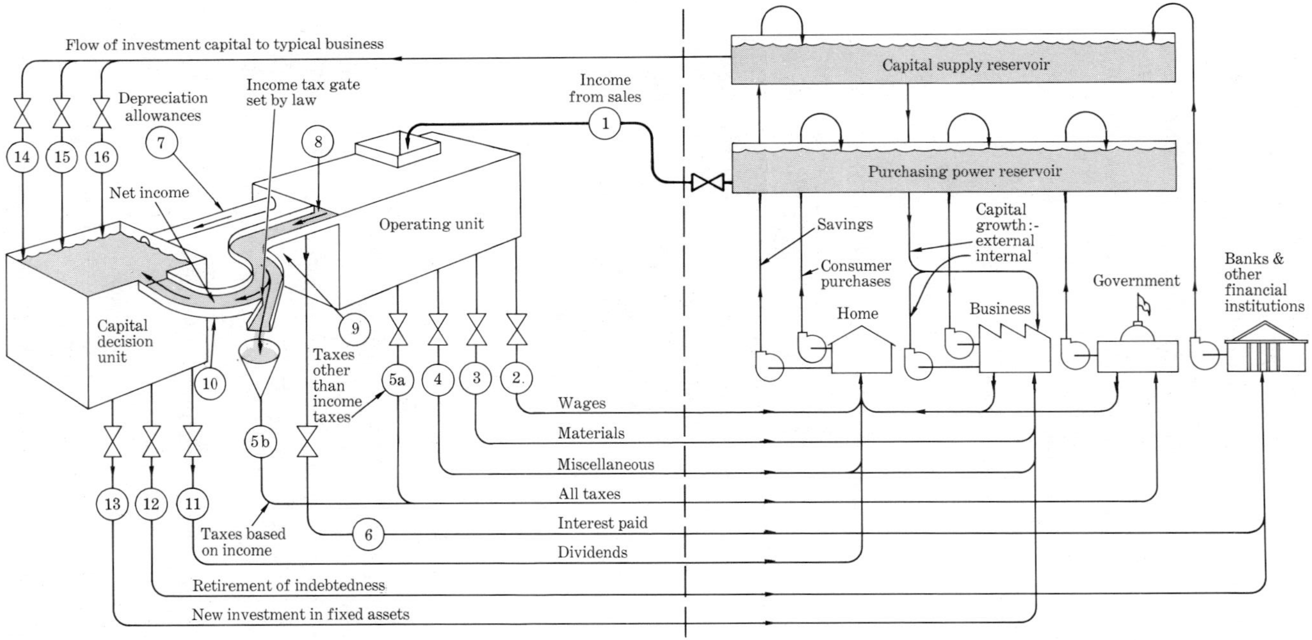

LEGEND AND CODE FOR NUMBERS

Flow control valve

Pumps that deliver variable volume and pressure

1. Income from sales
2. Wages and salaries
3. Materials purchased
4. Miscellaneous (services, etc.)
5a. Taxes other than income taxes
5b. Taxes based on income
6. Interest paid to banks and bondholders
7. Depreciation allowances
8. Profit from the operating unit
9. Taxable income (8) minus (6)
10. Net income (9) minus (5b)
11. Dividends to stockholders
12. Debt retirement to banks, etc.
13. Added investment in fixed assets
14. Short - term loans from banks, etc.
15. Long - term loans
16. Equity funds: money paid to the business unit for *new* shares of stock

Fig. 8.5 Money flow between the business unit and the national economy.

services and salaries (4); and property taxes, sales taxes, and all other taxes *not* based on the earnings of the business unit (5a). These are the essential money flows that must be stimulated (if income) and controlled (if outgo) for "success" in the Operating Unit.

It should also be noted that there is a second set of Typical Business Unit money flows (Flows 5b, 6, 7, 8, 9, and 10) which occur only when the Operating Unit is "successful." The "success" of the Operating Unit can be described as the happy situation that exists when, as the result of operations, money accumulates and must be distributed to those who may have some claim to the accumulation of money. Usually the highest priority claim is that of the "lenders," and this claim is paid off periodically through Channel 6.

Before discussing the other claims on the accumulation of money, it is necessary to describe "success" a little more carefully. Technically, "success" *does not begin* until the output flow of the Operating Unit is sufficient to "fill" Channel 7, the depreciation allowance channel. A "full" flow in this channel occurs when the rate of flow of depreciation allowance money is a reasonable approximation of the rate of decrease in the value of the fixed assets used in the business. Graphically, Channel 7 is shown at a lower elevation than Channel 8 to suggest that Channel 7 must be filled before there is any flow in 8.

Any output of the Operating Unit beyond the amount required to "fill" Channel 7 is discharged through Channel 8 and is described as Operating Profit. Next the Operating Profit is reduced by the amount of the Interest Paid Channel 6 and the remainder is labeled Income before Taxes, and is shown in Channel 9. Channel 5b removes the Corporate Income Taxes leaving the Net Income to flow through Channel 10 into the Capital Decision Unit.

It is appropriate to point out that the addition of this Capital Decision Unit constitutes the main difference between the money flow diagram for an engineering *project* and the money flow diagram for a typical *individual business unit.* The flows from the Net Income and Depreciation Allowances are the "end money products" of an engineering project. The value of our "money model" of an *engineering project* was that it enabled us to identify and measure these end results. Since a business unit is *more* than a single engineering project, we need to identify and measure the "end money products" of a collection of engineering projects. This identification and measurement is complicated because each individual output of *one* engineering project is likely to be used as the investment in one or more *other* engineering projects in the same business unit. The Capital Decision Unit is our model for describing the mixing of the outputs from all engineering projects and the re-allocation of these flows to new engineering projects within the Business Unit. A little later we will discuss the re-allocation of at least a portion of these flows to those who have invested in our Typical Business Unit. This discussion will be deferred to the next section, the Business Unit and the National Economy.

The sum of Flows 7 (Depreciation Allowance) and 10 (Net Income) is the Cash Flow, and it is this money flow about which the owners regularly (and of course enthusiastically) make decisions. Perhaps the easiest decision is to refrain from deciding to re-allocate the accumulating money. In this case the amount of money in the Capital Decision Unit will increase with every periodic success of the Operating Unit; and, in terms of financial statement analysis, the Working Capital of the company will increase. Since Working Capital is the *status* of an accumulation rather than a *flow* of money, it can be measured only by information shown on the Balance Sheet rather than by information shown on the Income Statement.

The Business Unit and the National Economy

Owners are typically enthusiastic about a successful business, and while it is possible that *all* cash flow will be retained in a successful business, it is traditional for the owners to elect to withdraw some of the cash flow from the business. Such transactions are shown in Figure 8.5 as flows from the Typical Business Unit (the left-hand portion of Figure 8.5) into the National Economy (the right-hand portion of Figure 8.5). If the money withdrawn from the business is from the depreciation allowance, the owner is merely removing some of his investment from the business, and such a transaction would be an upward flow in Channel 16. If the source of the money withdrawn from the business is the net income, then the withdrawal is labeled "dividends to stockholders" and its flow to the National Economy is shown in Channel 11. The balance of the net income *not* withdrawn is kept in the business and this quantity is a part of the Balance Sheet with the traditional label, "retained earnings."

The function of the Capital Decisions Unit that is most important to engineers is the decision to use cash flow from depreciation allowances and retained earnings for *new* or expanded projects of the business unit. It is this flow and this type of a decision that supports the design of new products and the design and purchase of new or improved production facilities. The *amount* of engineering work performed by a company is usually proportional to the Cash Flow.

If the owners believe that the business should be expanded very rapidly, they may wish to obtain more money than the amount generated internally by the cash flow, which consists of the depreciation allowances plus retained earnings. In such a situation, the business unit must look for external sources of money, and in Figure 8.5 these sources are shown collectively as the Capital Supply Reservoir in the National Economy. This "reservoir" is symbolic only, but it does represent the savings of individuals flowing either directly to the business unit or indirectly through a financial institution such as a bank or an insurance company. A thorough understanding of economics is necessary if one is to understand fully the functioning of the National Economy and the symbolic Capital Supply Reservoir. As a minimum, engineering students should know that the money flows from the reservoir to the Typical Business Unit are the *initial* transactions on which the creation of a business depends. These initial transactions are the investment of equity funds in the form of ownership, and without such a transaction no privately financed engineering project could be started. In publicly financed engineering projects, such as the design and construction of a harbor and port facilities, the initial act is an appropriation of money that a government unit is willing to allocate to a long-term project.

The engineering student should also know that many engineering projects require many more dollars of investment than can possibly be obtained from the internally generated cash flow. In such situations, the business unit must seek investment funds from the external Capital Supply Reservoir, shown symbolically in Figure 8.5. The student should also become familiar with the most common types of investment transactions through which external funds are obtained. The most common types of transactions are: Short-term loans (Channel 14), long-term loans obtained by selling bonds or debentures (Channel 15), and the sale of *new* shares of stock (Channel 16).

There also may be occasions when the owners may wish to reduce the investment in the business, and, if this is the case, they are likely to give priority to retiring some of the indebtedness of the company. Such repayments of short-term loans or the retirement of bonds would be a flow of money through Channel 12 back into the National Economy.

The student should be warned that the actual money flows are by no means as simple as the diagram suggests, particularly in the National Economy. Also, the complexity of this network flow cannot be appreciated fully until subjects such as circuit theory and its supporting disciplines have been studied. In addition, the diagram of the national economy is simplified to such an extent that it represents only the *major* flows. It should also be pointed out that each element in the money flow circuit is not an unchanging element such as cast-iron valve or a concrete tank. The constituents of economic circuits act more like people whose characteristics and behavior may change with the seasons or the styles of the moment.

The analytical techniques of the engineer are being used increasingly by scientific and technical managers, but such techniques provide answers that have limited validity. The economic circuit includes many elements with variable and seemingly unpredictable characteristics, and the analysis of an *economic* circuit is more difficult than the analysis of most *physical* circuits. In the opinion of William Morris, one of the foremost students of scientific management, engineers and managers need to regard the management of a business enterprise as an *experimental* process in which the application of scientific and technological methodologies should be treated as experiments. Ability to make more reliable predictions will depend on more experimentation and on more careful study of the results of experiments in the management of a business enterprise.

SUMMARY

Engineering depends upon money. There must be persons who are so fully persuaded about a given engineering project that they are willing to invest money in the venture. Such investment involves risk, and even if the venture is "successful" the investor obtains a profit only after the elapse of time, often several years. Engineers must remember this owner's *expectance* of a future profit if their designs are to attract investment capital.

A flow model can be used to describe the money inputs and outputs of an engineering venture, and an important output that the designing engineer must try to help create is the cash flow, the sum of the depreciation allowances and the net income.

An introduction to the principal terms of accounting has been presented, and the engineer should be able to use these terms in discussing proposed ventures with owners or with prospective investors.

This chapter has discussed basic concepts and terms, but it is too short a chapter to provide an understanding of the principles and techniques of engineering project economic analysis. In engineering education the study of such principles and techniques is usually called "Engineering Economy," and in business education such problems may be studied in courses in accounting, business finance, or management of capital. A few Engineering Economy texts are listed among the references at the end of this chapter.

Engineers who make the important decisions such as "do" or "don't do" or "do it this way rather than that way" are likely to be those who are well educated in the field of Engineering Economy as well as in the physical and technological aspects of engineering.

REFERENCES for Accounting and General Business

Anthony, Robert N.: "Management Accounting," Richard D. Irwin, Inc., Homewood, Ill., 1960.

Bonneville, Dewey and H. Kelly: "Organizing and Financing Business," Prentice-Hall, Inc., Englewood Cliffs, N.J., 1959.

Chase, Stuart: *How to Read an Annual Report,* available from Standard Oil Company (New Jersey), Room 1626, Rockefeller Plaza, New York, N.Y.

Grant, Eugene L. and Lawrence F. Bell: "Basic Accounting and Cost Accounting," 2nd ed., McGraw-Hill Book Company, New York, 1963.

How to Read a Financial Report, available from most local offices of Merrill Lynch, Pierce, Fenner, and Smith.

Morris, William T.: "Management Science in Action," Richard D. Irwin, Inc., Homewood, Ill., 1963.

REFERENCES for Engineering Economy and Management Decisions

Barish, Norman N.: "Economic Analysis for Engineering and Managerial Decision-Making," McGraw-Hill Book Company, New York, 1962.

Grant, Eugene and William G. Ireson: "Principles of Engineering Economy," 4th ed. (revised), Ronald Press Co., New York, 1964.

McIlvaine, William D.: "Engineering and Economics—An Introduction to the Economic Analysis of Engineering Proposals," California State College at Long Beach, 1966.

Morris, William T.: "The Analysis of Management Decisions" (revised edition), Richard D. Irwin, Inc., Homewood, Ill., 1964.

Taylor, George A.: "Managerial and Engineering Economy," D. Van Nostrand Co., Inc., Princeton, N.J., 1964.

Preliminary Design

Nine

"If you can measure that of which
you speak and express it in numbers you
know something about your subject; but if you
cannot measure it, your knowledge is of a
very meager and unsatisfactory kind."
Lord Kelvin

Alternate Usages

In several industries, the term Preliminary Design means the exploratory work which precedes a contract proposal, using the phrase in almost the same way that we have been using Feasibility Study. At other times this preparatory work is called advance engineering or proposal engineering. We reserve Preliminary Design, however, to describe that mid-design activity by which we put the problem into symbolic form and examine the variables mathematically. This can be simple or complex. In some cases, we cannot find very satisfactory ways of expressing the variables mathematically, as in certain chemical corrosion problems, where we don't know enough, or in vacuum tube or transistor circuits, where the devices have very irregular characteristics. But by far most engineering problems yield to some form of mathematical analysis, and we help ourselves immeasurably by putting the situation into symbolic form.

Stages in the Preliminary Design

The several elements of the Preliminary Design activity which meet our present purpose and scope are the following, in the order of their use:

1. *Input* of several possible solutions from the preceding Feasibility Study.
2. *Selection* of the most promising concepts from the several suggested solutions.
3. *Formulation of a useful model* for this selected solution. The model preferably is mathematical, but, initially it may be graphic, verbal or analog (see next chapter).
4. *Sensitivity analysis,* by which we find which variables are most critical to the proper performance of the system, and which values of those variables are in the critical ranges.
5. *Compatibility analysis,* which is used to find possible interference or mismatch between various components or parts of the system.
6. *Formulation of a criterion,* in order to determine and measure which values we are attempting to maximize, or minimize.
7. *Optimization.* Ordinarily optimization can be done only late in the development of a system, after a full understanding has been reached. This step might better be called maximization (of profit or output) or minimization (of cost or penalty). The step, by any name, consists of determining the values of the design parameters so as to best meet the objectives, as expressed by the criterion. The criterion, then, is the basis for measuring the effectiveness of this step.
8. *Prediction of performance.* Given the values of the design parameter(s) that were computed by the preceding step, the system performance is now projected. How will it behave under the several imposed conditions? Future versus present? At the worst extreme of tolerances?

Again, as with "design-stages" described in previous chapters, these phases are not sacrosanct. Each design project will have its own "personality," its own special needs, its own program and set of objectives. Designers all differ in talent and experience. What would be an obvious procedure for one designer might never occur to the next. The degree of mathematical analysis depends not only on skills but also on time, inspiration, and alternatives. The eight stages above will be too few for one project, too many for another. The spotlight turns again to the designer. He must decide what the project needs, what outside talent he should call in and how he can best use his own time.

Illustration: Pole Support

All of these steps can be illustrated in the pole-bracing problem shown in the accompanying sketch, Figure 9.1.

In the example of the transit-mix cement plant, the electric power line ended at the plant site, but the last power pole needed a cable brace to resist the conductor-wire tension. The cable must be installed at least 15 ft above the ground, in order to clear the trucks in the service road. Furthermore, the innermost edge of a sidewalk lies 10 ft from the foot of the 8 ft-high boundary fence F, as shown. Although such bracing is the responsibility of the power company, the special conditions give us a reason to examine it and to recommend what we think is the best solution. We can experiment here with the proposed Preliminary Design procedure.

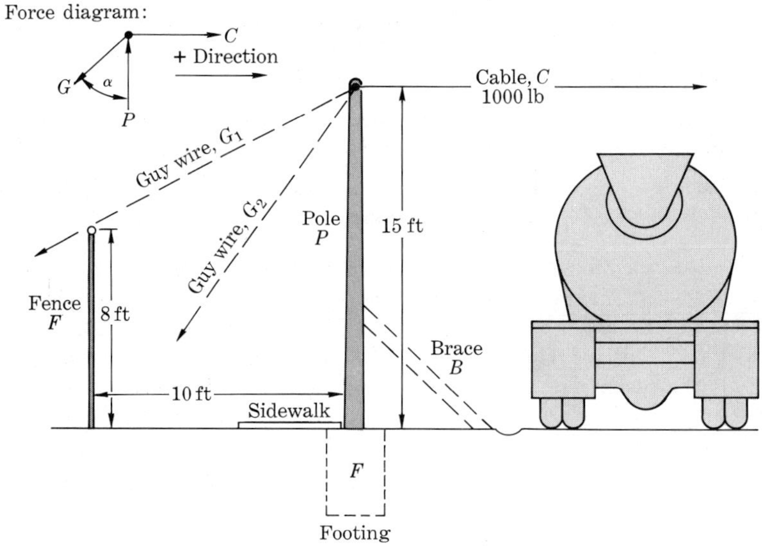

Fig. 9.1 Pole-bracing alternatives: Heavy footing, timber brace, or guy wire.

We shall propose the following alternative solutions; the reader will undoubtedly think of others.

1. Front brace B, of timber or pipe.
2. Heavy poured concrete footing F, assuming the pole P is strong enough.
3. Guy wire G_1 or G_2, going to the ground either outside or inside the fence F.

Without going into all the reasons,* we will select the guy-wire method as the most promising and make a mathematical model for it. This possibly overdignified phrase, "mathematical-model," is simply an equation written around the pertinent variables, representing in this case the forces acting

* *Note:* Tension members are practically always simpler and cheaper than compression (B) or bending (F) members performing the same function.

on the top of the pole. We need to find the best length, size and location of the guy wire. We shall substitute symbols for the realities, remembering that *the equation is not the pole and wires;* it is only an idealized sufficiently faithful equivalent of them. A little later, we shall show the shortcomings of the model, and, in the next chapter discuss models and modeling generally.

In the present case, we choose our criterion, our measure of goodness or success, to be simply the guy-wire weight. We want the lightest weight, of course. What we are really concerned about is the total cost of the whole support system, but the guy-wire weight seems to be a reasonable first approximation for the significant variable.

We start by finding the relation between forces and wire size and length. The small diagram of forces at the top of Figure 9.1 shows the guy-wire force G to the left, the cable force C to the right, and the pole force P up from below. The pole length is p; the wire length is $g = p/\cos \alpha$.

Making use of Newton's laws, we sum up these forces at the top of the pole in the horizontal direction, and equate them to zero. (The positive direction is to the right.)

$$\Sigma F_h = 0$$

Thus,

$$-G \sin \alpha + C = 0$$

or

$$G = \frac{C}{\sin \alpha} \tag{1}$$

A reasonable expression for the weight of a steel tension member* can be taken as the following:

$$\text{Weight} = \frac{\text{force (lb)} \times \text{length (ft)}}{6000} \text{ lb} \tag{2}$$

Using this expression (2) for weight, the weight of the guy wire is

$$\text{Weight} = W_g = \frac{1}{6000}\left(\frac{C}{\sin \alpha}\right)\left(\frac{p}{\cos \alpha}\right) \text{ lb}$$

Where: Force $= G = C/\sin\ \alpha$ and length $= p/\cos\ \alpha$, with $p = 15$ ft. So,

$$W_g = \frac{Cp}{6000}\left(\frac{1}{\sin \alpha \cos \alpha}\right) \text{ lb}$$

* For instance, a member 20 ft long carrying a load of 3000 lb would weigh 10 lb $\left(\text{i.e., } \frac{3000 \times 20}{6000}\right)$. This can be checked in a handbook, such as: R. H. Perry: "Engineering Manual," McGraw-Hill Book Company, New York, 1959. (Steel in Tension, p. 6-72.) (Density of Steel, p. 3-16.)

or

$$W_g = \frac{Cp}{3000 \sin 2\alpha} \text{ lb} \qquad (3)$$

Equation (3) is the mathematical model we were seeking. In this case the expression represents cost as well as weight, because they are nearly enough proportional, for our first iteration.

We suspect there will be a value of angle α which will minimize the weight (cost), because we see that as $\sin 2\alpha$ approaches zero, Equation (3) tells us that the weight becomes very large.

The differential calculus now helps us to find the minimum value for Equation (3). Set the derivative equal to zero, and solve for α.

$$\frac{d(W_g)}{d\alpha} = 2 \times \frac{Cp}{3000}\left(\frac{-\cos 2\alpha}{\sin^2 2\alpha}\right) = 0$$

Multiplying by $\sin^2 2a$, the derivative becomes:

$$\cos 2\alpha = 0$$

$$2\alpha = 90°, 270°$$

$$\alpha = 45°, \text{ (135° is not physically possible)} \qquad (4)$$

Thus, we find by Equation (4) that the minimum weight occurs at $\alpha = 45°$, and that the weight is

$$W_g = \frac{Cp}{3000}\left(\frac{1}{\sin 90°}\right) = \frac{1000 \times 15}{3000}\left(\frac{1}{1}\right) = 5 \text{ lb} \qquad (5)$$

Unfortunately, as we are about to fold up our papers and depart, we realize that at $\alpha = 45°$, this guy wire will go right through the fence F, and that we don't yet have the solution. Should we still try to use the guy wire? Should we go over the wall or down to the inside foot of the wall? The weights for the inside and outside locations are found to be respectively: $G_2 = 5.4$ lb; $G_1 = 5.3$ lb, as shown below and at the top of the next page.

$$\text{(Inside)} \qquad G_2 = \frac{Cp}{3000}\left(\frac{1}{\sin 2\alpha_2}\right)$$

where

$$\alpha_2 = \tan^{-1}\frac{10}{15} = 33°40' \quad \text{and} \quad \sin 2\alpha_2 = 0.92$$

or

$$G_2 = \frac{5}{0.92} = 5.4 \text{ lb}$$

$$\text{(Outside)} \qquad G_1 = \frac{5}{0.94} = 5.3 \text{ lb}$$

since

$$\alpha_1 = \tan^{-1} \frac{10}{7} = 55°00' \quad \text{and} \quad \sin 2\alpha_1 = 0.94$$

Since these two answers are so close to each other, we must depend on other limitations to decide which is better. Such things as room for the cable ground-anchor, access over the fence, and type of soil would lead toward the final determination. Other things being equal, we would choose G_2 because of simplicity and the proximity to the pole.

We have selected the optimum value of the variable α under the conditions of the problem. Let us check sensitivity. From Equation (3), we see that the weight changes in direct proportion to C and to p; but near

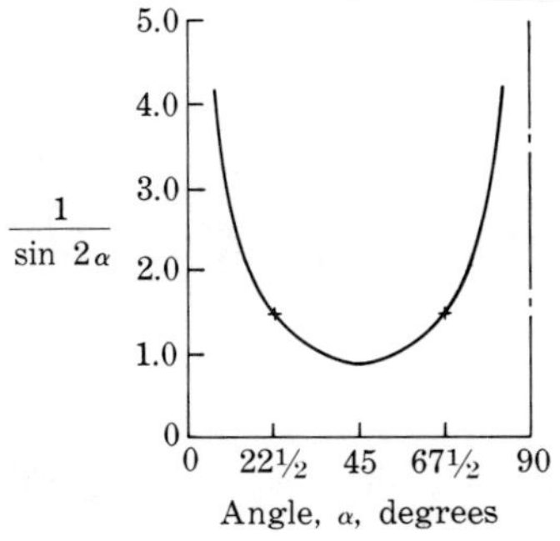

Fig. 9.2 Guy-wire minimization. Weight function $1/\sin 2\alpha$.

$\alpha = 45°$, the angle has very little effect. However, note that for $\alpha < 22°$, or $\alpha > 67°$, the weight increases very rapidly, simply because of the characteristic curve shape (i.e., $1/\sin 2\alpha$). See Figure 9.2. Thus we conclude the most sensitive variable (parameter) is α, but only at the higher or lower values. Between 40° and 50°, α is insensitive. C and p are not especially sensitive, since they enter in direct proportion. This knowledge of sensitivity guides the engineer in deciding upon the final values to select.

Looking at compatibility brings us face to face with the limitations of the math model. We do not know from the equation how to attach the guy wire to the pole or to the ground, and we have no idea of the cost of the labor. Thus we must search for answers, looking for standard parts or at least for usual commercial practices. The through-bolt strap and ground anchor are suggested solutions using conventional parts, as shown in Figure 9.3. If the pole or ground conditions had been otherwise, different terminations would be indicated. As it turned out, labor and termination costs were fairly constant, so we did not need them in our mathematical model.

The criterion we have been working with all along has been "lowest cost," plus practicality.

Thus, the final result of our "minimization" was 5.4 lb plus the cost of strap and anchor (which, incidentally, may well equal the cost of the guy wire itself).

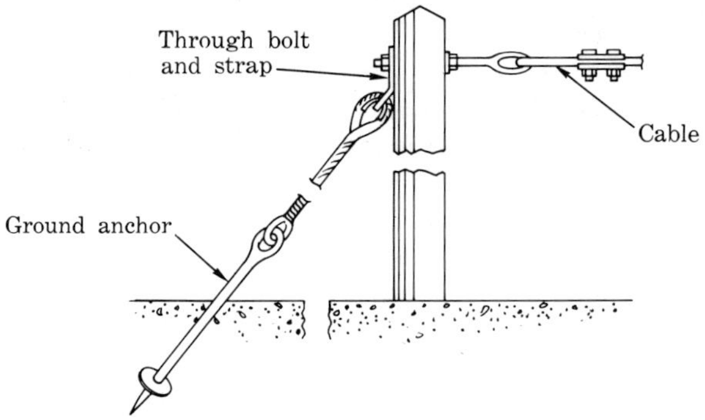

Fig. 9.3 Practical pole guy-wire detail.

This brings us to one of the important conclusions concerning a mathematical optimization study:

> The results of the analysis help the engineer to understand what is going on in his problem, but his decisions usually are over-ridden with considerations which are not or cannot be included in his equation. His judgment must prevail, rather than the dogma of the mathematical solution.

The final step, "prediction," in this case is unnecessary, because we know the weight already (5.4 lb) and no other aspect had been under question.

Thus our preliminary design in this case has brought us (at least as concerns the guy wire) almost to the end of the total design. But in retrospect, we see that it allowed us to examine easily many ways (various values of angle α) to support the pole. It helped us to measure the best way and then to find the needed numbers. With skill and similar intentions, most engineering problems can yield useful results from such optimizing analysis.

Illustration: Conveyor Belt

The first illustration, pole-support, was in the field of statics and the model was formed by the use of Newton's laws of forces. We now introduce a conveyor illustration in which the mathematical model is based upon an energy equation. That is, the energy supplied by the motor must equal the energy stored in the sand plus any energy lost. Referring to Figure 9.4,

$$E_i = (E_{sp} + E_{sk}) + E_f$$

where

E_i = energy in (from the motor)
E_{sp} = potential energy stored (lifting the sand)
E_{sk} = kinetic energy stored (in the moving sand)
E_f = energy lost in friction (sand sliding on the belt during acceleration at the point of sand impact)

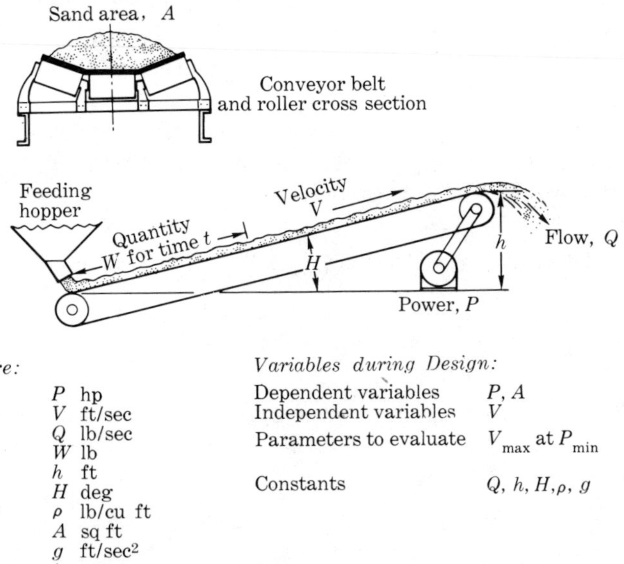

Nomenclature:

Power	P	hp
Belt velocity	V	ft/sec
Flow rate	Q	lb/sec
Amount	W	lb
Height	h	ft
Angle	H	deg
Sand density	ρ	lb/cu ft
Sand area	A	sq ft
Gravity	g	ft/sec^2
Time	t	sec

Variables during Design:

Dependent variables	P, A
Independent variables	V
Parameters to evaluate	V_{max} at P_{min}
Constants	Q, h, H, ρ, g

Variables in Operation:

Dependent variables	P, A
Independent variables	Q, V
Constants	h, H, ρ, g V_{max}

Assumed Trial Values:

Q = 50 lb/sec
h = 55 ft
A = 0.5 sq ft, initially
ρ = 100 lb/cu ft
g = 32 ft/sec^2

Fig. 9.4 Sand conveyor layout, showing variables, parameters, and constants.

Our object is to find whether there is an optimum velocity V for the belt that will minimize the needed motor power P, while delivering a given sand flow rate Q. Naturally, this value of power allows us to specify the cheapest motor. Different belt sizes and structures would be required depending on whether it were best to run the belt as fast as possible, as slow as possible, or at some preferred value in between. We assume there is no direct power recovery from the sand delivered off the top end of the belt.

We will proceed in the same way as before: (1) formulating an expression involving all the design variables and parameters, (2) finding the value(s) of velocity V for the minimum power P, by differentiating, (3) examining the sensitivity of the power P to various values of the parameters, and (4) concluding with recommendations. Our initial criterion is minimum power P, only. Later we would extend this into the total cost of the conveyor system if that became a concern.

The energy or work represented by the several terms above will now be expressed through use of the symbols. The reader is familiar with the form of the potential energy and the stored kinetic energy, but we have repeated them in the note below.* For the friction loss, it can be shown that the lost energy with constant driving velocity equals the stored energy of the driven member ($\frac{1}{2}MV^2$); hence the third term below is the same as the second term.

Thus, considering a weight of sand W on the conveyor, the total energy required to raise it to height h and velocity V becomes

Input = potential + kinetic + loss

$$E_i = Wh + \frac{1}{2}\left(\frac{W}{g}\right)V^2 + \frac{1}{2}\left(\frac{W}{g}\right)V^2$$

or

$$E_i = Wh + \frac{WV^2}{g}$$

Let us convert energy to power by dividing both sides of this equation by time t. The sand flow rate Q is W/t lb/sec. Thus the equation becomes

$$\frac{E_i}{t} = \left(\frac{W}{t}\right)h + \frac{W}{t}\frac{V^2}{g}$$

or

$$P = Qh + Q\frac{V^2}{g} = Q\left(h + \frac{V^2}{g}\right)\frac{\text{ft lb}}{\text{sec}} \tag{6}$$

When we differentiate P with respect to V, and equate to zero, we find that the minimum power occurs at $V = 0$. This is shown by

$$\frac{dP}{dV} = Q\left(0 + \frac{2V}{g}\right) = 0$$

or

$$\frac{2V}{g} = 0 \quad \text{thus} \quad V = 0 \tag{7}$$

* *Note:* Potential energy = force × distance acted through
Kinetic energy = $\frac{1}{2}$ mass × velocity²
Loss energy = Initial friction between belt and sand

Incidentally, we find it is a minimum (rather than maximum) by obtaining the second derivative and noting it is positive at the critical value of V.

$$\frac{d^2P}{dV^2} = \frac{2Q}{g} \quad \text{(positive)}$$

In the case of Equation (6), we could have observed by inspection that power P was minimum at velocity $V = 0$. The ensuing differentiations were only confirmations.

Equation (6) tells us that the slower the conveyer belt goes, the less power it will require. As yet, we don't know whether this means it must go 10 or 1.0 or 0.1 ft/sec. Substituting the actual numbers for our supposed

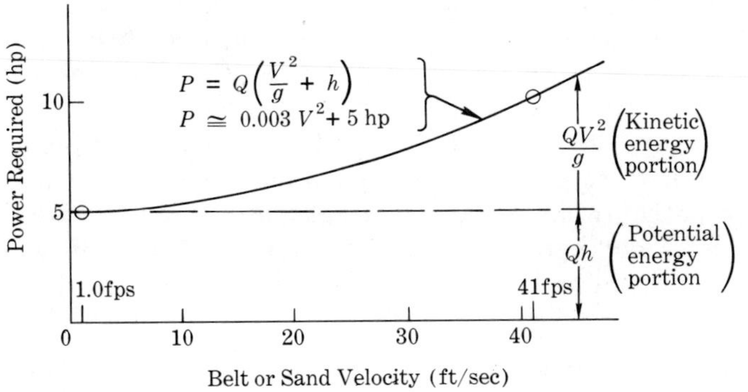

Fig. 9.5 Sand conveyor power characteristic. Belt velocity vs. driving power required.

situation into the equation will tell us this, however. Taking the numerical "trial" values from Figure 9.4, we find Equation (6) becomes

$$P \cong 0.003V^2 + 5 \text{ hp} \tag{8}$$

Thus $P = 5$ hp at $V \cong 0$. Naturally, no sand is moved if $V = 0$ in actuality. Also using $Q = AV\rho$, Equation (6) can be converted to the variable A.

$$P = \frac{Q}{g}\left(\frac{Q^2}{A^2\rho^2}\right) + Qh = \frac{Q^3}{g\rho^2A^2} + Qh \; \frac{\text{ft lb}}{\text{sec}} \tag{9}$$

Equation (8) is plotted in Figure 9.5, where the fixed-value potential energy portion (Qh) can be seen compared to the variable-value kinetic energy portion (QV^2/g). Using a flow rate of 100 tons/hr introduced in chapter 3 and assumed in Figure 9.4, a first approximation for velocity yields the value,

$$V = \frac{Q}{A\rho} = \frac{50 \text{ lb/sec}}{0.5 \text{ sq ft} \times 100 \text{ lb/cu ft}} = 1 \text{ ft/sec}$$

On the curve, $V = 1$ ft/sec is very close to $V = 0$, and hence contributes relatively no kinetic energy portion to the power requirement. In fact, the velocity must go to 41 ft/sec for the "kinetic" portion to equal the "potential" portion (as shown by intercepts in Figure 9.5).

Variables, Parameters, Constants

Notice that in Figure 9.4, we have listed the arrangements of the variables, etc., in two columns; one is Design, the other, Operating. These two columns are interpreted as follows.

In Design, we assume fixed constant values of flow Q, height h, density ρ, and angle H. Our object is to choose a value V which, when fixed in practice, would yield the lowest power P. Thus, the V_m at $P_{\min}$ becomes a fixed Design Parameter, the result of our optimizing operation.

However, in practice, the men who operate the system would probably vary V to be able to increase Q with a fixed maximum area A available on the belt. Particularly, they would do this since at reasonable flow rates the power per unit flow is not appreciably affected by V. The fixed values (constants) in operation will be the height h and angle H and the density.

Sensitivity

The power equations (6) and (9) and the curve in Figure 9.5 both allow us to check the sensitivity of the design variables and parameters.

From Equation (6),

(a) Power is insensitive to velocity up to $V \cong 15$ ft/sec.

(b) Power is proportional to Q, and it is proportional to h up to $V \cong 15$ ft/sec, where the power becomes influenced by V.

(c) Power varies as V^2, when $V^2/g \gg h$.

From Equation (9), the "kinetic" portion of the power $(Q^3/g\rho^2A^2)$ varies inversely as the square of the area A. Thus, the larger the area, the lower the kinetic power required.

Conclusions

In summary, the conveyor belt's width and speed should be chosen to balance the cost of a wide belt (to obtain large area A) against the cost of structure, motor and continuous power that would be required if the velocity component were appreciable. The following values seem reasonable: Let

$$V = 5 \text{ ft/sec}$$

Then

$$A = 0.1 \text{ sq ft} \qquad \text{and} \qquad P \cong 5 \text{ hp}$$

With the initially assumed A of 0.5 sq ft, the belt could be only 12 or 15 in. wide; and this reduction applies, of course, to rollers, supports, drums, structures, etc. The added sand-friction wear and increased flexing due to the higher speed would be taken into account, however, in arriving at the final choice.

General Discussion: Variables, Parameters, and Constants

Let us explain the difference between *variables* and *parameters* and identify them in the two preceding illustrations, the pole guy wire and the sand conveyor. *Variables* are quantities which may assume different values at any time, either during the design study or in actual operation. *Parameters* assume different values *only* during the manipulations of a design study, but are constant in the actual device or in operation. In the guy-wire case, the angle α was a parameter, allowed to vary while the cost was being minimized, and then fixed at $\alpha_2 = 33° 40'$. There were no variables in that case.

For the conveyor, the parameters were velocity V and sand area A; both were relatively fixed by the minimization of power $P(V = 1 \text{ to } 5 \text{ ft/sec}, A = 0.5 \text{ to } 0.1 \text{ sq ft})$. During operation the flow rate Q probably will vary, and hence the power P.

Constants are values that are unchanged or unchangeable during design or operations.

As a final illustration, suppose an airplane had been designed for best meeting coast-to-coast passenger load requirements. The wing area, the length, the engine power, and the maximum payload all would be design parameters, not determined until the design had been finished. When the plane flies, whether in real life, or on paper, the variables include its actual speed, altitude, thrust, passenger load, head wind, speed, losses, etc. You can see that the design must anticipate widely varying values of the variables to help freeze the values of the parameters at their optimum, if such variables have an impact.

In summary, the *variables* are the quantities in a system whose numerical value is not necessarily fixed either during a design or during actual system operation. The *parameters* are measures of a system, and their values change during the considerations of designing; but once designed, the values presumably remain fixed through the lifetime of the system. It is the choice of right values for the design parameters that measures the talent and skill of the designer.

The criterion should be an expression,* including the decisive parameters, of such a nature that the process of maximizing (minimizing) it would produce the most desirable design. To be useful, it must carry the symbols of the variables and parameters (at least of the parameters), so that the other constraining relations can be "substituted in" and as many symbols eliminated as possible.

This criterion can be concretely arrived at only when the equations are expressed explicitly. In our first illustration, the criterion is W_g, and the final parameter is α (a single variable).

$$W_g = \frac{Cp}{3000 \sin 2\alpha}$$

In the second illustration, the criterion is P, and the final parameter is V (a single variable).

$$P = Q\left(h + \frac{V^2}{g}\right)$$

In more complex cases, the criterion may be dependent on two or more design variables, generating curved surfaces. These naturally lead into the reasons for the use of advanced mathematics to determine the maxima and minima of our criterion expressions.

REFERENCES

Asimow, M.: "Introduction to Design," chap. 5, Prentice-Hall, Inc., Englewood Cliffs, N.J., 1962.

Johnson, L. H.: "Engineering Principles and Problems," chaps. 4, 7, McGraw-Hill Book Company, New York, 1960.

Krick, E. V.: "An Introduction to Engineering and Engineering Design," John Wiley & Sons, Inc., New York, 1965.

* It is quite an art to formulate criteria, particularly in complex designs, but some expression is necessary to measure design success. When the criteria include intangibles such as safety, appearance, life, maintainability, or reliability, all must be reduced to single variables such as dollars or "utility" or other units to be susceptible to analysis and explicit answers.

Engineering Problem Modeling

Ten

"The best carpenters make the fewest chips."
German proverb

Perfect Solutions?

Many types of analytical problems continually clamor for the attention of the practicing engineer. Each problem has some legitimate claim upon his time, but to be effective, the engineer must sort the problems, with more or less deliberation, into several types and priorities. He will delegate certain problems to other engineers, and disregard some; but he finally comes down to those problems for which *he* must assume responsibility:

(a) Problems with which he has some ready familiarity,

(b) Problems about which he knows relatively little, directly and specifically.

Only the latter stimulates any concern or agitation in him (or us). Contrary to common belief, the engineer frequently *does* find himself unfamiliar with problems and answers, even in his own special field. How does he, then, go about arriving at specific answers? We have talked generally about the problem-solving process in Chapter 2, but here we will come to grips with some of the analytical tools to put muscle on the bones of a skeleton procedure.

If he works diligently enough, immersing himself in all the facts of the problem situation and in all of its background, is it possible for the engineer to come to a perfect answer? Can he make a "true" analysis? Of course not. The "true" solution to any problem, taking into account all the conceivable relevant factors and related effects, is understandably quite impossible. First, no one can know *all* the relevant factors, or predict *all* the possible effects. Second, many influences are only slightly relevant and can be neglected. Only by extended experimentation can an engineer come to understand this relevancy and correctly choose which of these marginal influences to neglect. It is a truism that all the detailed macroscopic and microscopic phenomena of the natural world bring a staggering, overwhelming complexity into even the simplest problem, even to such a "transparent" question as: How does a paper clip bend? What is really going on in the metal crystals? Thus, we must always retreat from the real phenomenon; we must simplify by substituting one problem for another. We must reach practical solutions in a reasonable time.

After the engineer's first approach to his original problem, his successively deeper penetrations of its intricacies uncover other "sub-problems" which stand in the way of his final solution[1]. These smaller problems, as he recognizes from experience, usually fall into more familiar categories and remind him of similar problems he has solved or at least seen before. Because we can solve only familiar problems, we are led into the subject of models, the mental idealizations of physical situations. By the use of one or more of these models, we can uncover useful answers to the original questions asked of the engineer.

Values of a Model

There is a direct method of engineering problem-solving, and that consists of immediately constructing a likely embodiment of the device or system and subjecting it to use and observation. However, this method is so seldom used anymore that the profession has forgotten its possibility. We look instead for models because:

1. We find it much too expensive to keep constructing all the suggested alternative systems until we find a satisfactory one, and this method is even less practical to find the best system. Also, most systems of any complexity don't work the first time anyway, but require extended revision. In addition to being expensive or impractical, this direct process may be destructive or dangerous.
2. Many factors, attributes or qualities are found to be irrelevant; only a few important ones remain.

3. We can, with models, better define the problem, develop perspective about it, and break it into smaller problems.
4. Refinement of models improves the precision, permitting prediction to be made to further and further limits.
5. Finally, we can, with models, examine quickly the interactions of many variables, determining their sensitivities. With the digital or analog computer, vast numbers of combinations (even millions of them) can be screened quickly and economically for their applicability.

The abstraction of a problem from its specific singular situation into a class or group of problems brings it from a strange field to a familiar one. The engineer can then develop the confident understanding necessary for him to make a significant contribution to his client, and hopefully, to the art.

Limitations and Cautions

One should recognize early some of the limitations in describing nature by models and in predicting the behavior of any natural physical system.

1. There will always be error between the calculated predicted results and the measured results.
2. The degree of error depends upon the engineer's assumptions and in turn upon:
   (a) what he understands the problem to be and how much of the situation he decides to include,
   (b) what he needs to get out of the solution,
   (c) what models are available to him and what skill he employs in using them,
   (d) how much time and money he can spend on the solution,
   (e) and, upon test, where in the system the measurements are actually made and what the effects are of the measuring instruments upon the system.
3. Models are not unique. Different ones may be employed to compare results from different viewpoints. The different models may be employed on the same problem; or differing models may be advantageous for the different sub-problems.
4. All mathematical models are idealizations using the techniques of a constructed logic, not necessarily natural and certainly not complete. Thus, the resulting answers carry no inherent guarantee of accuracy. Unfortunately, the most meticulous care in carrying out the mathematical operations cannot correct fallacious, imprecise, or deficient assumptions. The uncertainties also carry through, and only the problem-solver can identify the number of credible, significant figures in the answer. To avoid misleading the reader, each engineer should present his numerical answers accompanied by an

implied* or stated estimate of accuracy. It hardly seems necessary to mention it, but no mathematical formulation can describe what is *actually going on*. Such formulations can only serve as guides for repeated experimentation, which in the end, persists as the ultimate arbiter.

Aspects of Solutions

When the typical practicing engineer learns of a problem, his background, his store of general information, and his experience usually lead him to block out some boundaries and make a relatively prompt order-of-magnitude estimate about the answer. How large must a certain beam be to carry the given load? Will the teeth of this gear break if the mechanism jams? How much will it cost to heat-treat these cutters? Will the electronic circuit oscillate if the coil-wire diameter is reduced 50 per cent? What will be the chemical reaction's yield if we double the contact surface area? An engineer practicing in the field would have an estimated answer to each of these problems.

In response to the question on beam size, the engineer would mentally consider the arithmetic on length, stresses, and area and be able to provide an answer: Say, a standard I-beam weighing 25–40 lb/ft. The engineer doesn't just guess or "pull a figure out of the air," but he does make rational, quick estimates. He uses familiar fundamentals, including his own "rules of thumb" and some of the methods referred to in Chapter 7 on Estimation. Naturally, the engineer then makes a conscious effort to suspend judgment, and keep his mind open to the unfolding logic that will take place as he proceeds to analyze carefully. He is not restricted or hampered by his first answer, but he remembers it.

There are other questions, however, that the engineer would hesitate to answer in this impromptu way. For example, what is the optimum takeoff angle function for a VTOL (Vertical Take-off and Landing) plane? Can a nuclear power plant be made safe enough to install in the middle of a large city? What is the value to be placed upon a human life in highway design? How much should be spent upon aesthetics in proportion to the functional cost of a product? Can silicon control rectifiers be made inexpensive enough to be used as motor speed regulators? These deeper questions bring up the nagging possibility that some questions have no answers, or that they may have different answers, depending on the questioner and the audience. Incidentally, the questions above are all engineering questions.

Most problems are expected to yield answers, but the illustrations just cited suggest that sometimes solution methods may be found, sometimes

* As was observed in Chapter 7, the implied accuracy of 0.3 is one significant figure, not 0.333 or even 0.33. The fraction $\frac{1}{3}$ may mean 0.3, 0.33, or 0.3333 depending upon the context. The safest statement is the explicit degree of uncertainty, computed by the engineer for his specific assumptions. For instance, $.333 \pm .001$ or $123{,}400 \pm 500$.

not. Sometimes answers can be checked and found "correct" or "incorrect" by submitting them to experiment; sometimes no answers can be found; sometimes the answer cannot be checked.

The following permutable alternatives* ought to be kept in the back of the mind as the engineer approaches sticky, baffling problems, whether the problems are new or have been already worked threadbare:

Affirmative	*Negative*
1. A method of solution is used.	No method is used (none found).
2. An answer is found.	No answer is found.
3. The answer is correct by test.	The answer is incorrect by later test.
4. The method is proved rigorously.	The method is not proved.

Illustrative examples under the *Affirmative* column would include conventional problems in algebra, geometry, mechanics, chemistry, statistics, and bio-technology. However, under the *Negative* column, we would find such problems as the geometric trisection of an angle (no method), the best way to stop urban smog (no method, no answer), how to build the Tacoma Narrows Bridge (answer incorrect by later test), how to predict the schedule and cost of a project (method not proven).

Naturally, the engineer's previous experience with problems of the given type, his training and skill, plus the state of the art, and the number of constraints, all combine to decide the outcome of the particular attempt.

In summarizing this section, we remember that the engineer's objective in solving problems, in getting answers, is to meet the client's or supervisor's needs, and to help himself to do so. He is searching for practical, timely answers to real problems, which, in the cases that concern us, he has never dealt with or solved before. These new problems, we suggest, should be converted by the act of modeling into familiar form and size, leading toward progressively more and more useful and realistic answers as the fidelity of the model iteratively improves in use.

Modeling serves many of the needs which arise on an engineering project. A summary of these served is listed below, digested from the more elaborate Table 10.2, presented at the end of the chapter. Project needs served by modeling are:

1. Self-communication, organization, record of development.
2. Clarification, understanding, visualization.
3. Analysis, synthesis.
4. Prediction.
5. Communication with and record of instructions to others.

* Developed from notes written by Dr. Paul Brock.

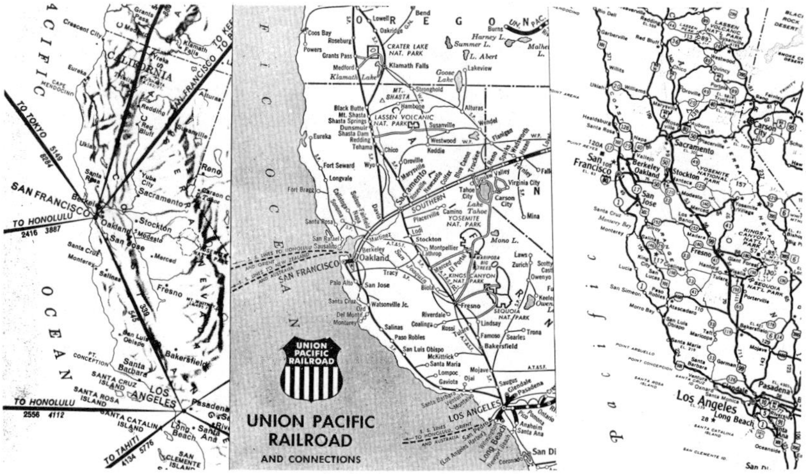

(a)

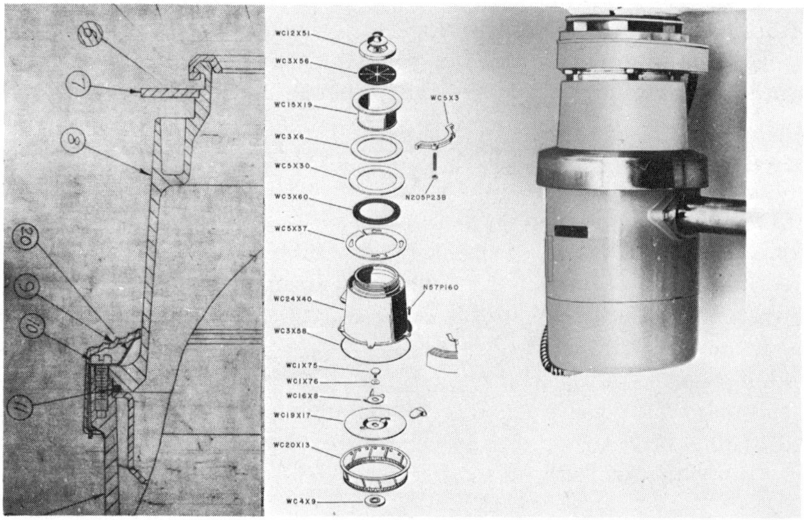

(b)

Fig. 10.1 Iconic models in two dimensions. (a) Transportation maps showing the differences between ship, rail, and auto markings; (b) Product delineation showing the differences between production drawing, service drawing (exploded), and photo. (UCLA photo)

Variety of Models

Models are generally classified under three names:[2]

Iconic: It specifically *looks like* the original.

Analog: It specifically *behaves like* the original (obeys the same laws of action).

Symbolic: It compactly and abstractly *represents the principles* of the original.

These descriptions are deceptively simple; because these three tools together comprise the bulk of the engineer's kit. Of course, he can fall back upon his other implements of intuitive creativity, assistance from others or trial and error, but the three varieties of models listed above are by far the most useful, versatile, and dependable tools the engineer can bring to bear.

We will amplify our description of these models only briefly here, deferring detailed treatment until the next chapters.

The Iconic Model

Iconic* models are representations that *look like* the reality. They are visual geometric equivalents, either as miniatures, enlargements, or duplicates made to the same scale. They usually show only part of the features

* Iconic: From the Greek, *eikon*, image, likeness.

Fig. 10.2 Iconic model applied. Process equipment layout integrated with complex electrical systems controls and indicators for the production of acetylene. (Courtesy Rohm and Haas, Inc.)

(a)

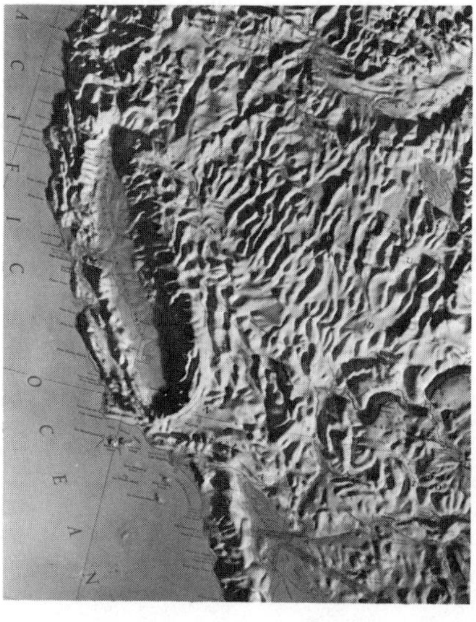

(b)

Fig. 10.3 Iconic models in three dimensions. (a) Molecule of barium titanate; (b) Relief map; (c) Pre-production plaster model of an electric range; (d) Plant layout for material flow and equipment placement. (Courtesy General Electric Co. and UCLA)

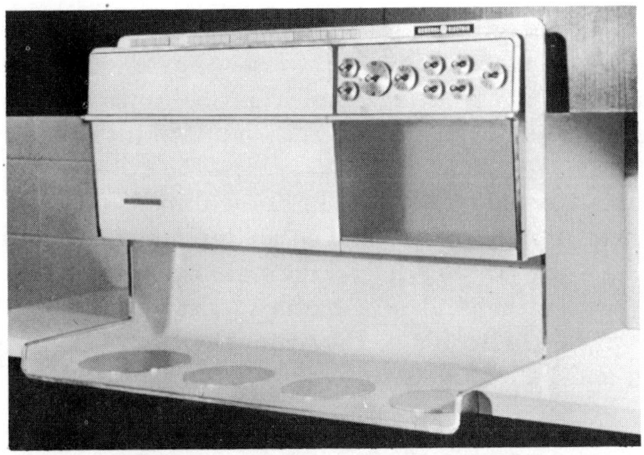

(c)

(d)

Fig. 10.3 (*Continued*)

or attributes, only those parts relevant to the immediate purpose. Thus, an automobile road map does not show any railroads, canals, or airline routes even though they have passenger traffic. For geometric replicas, iconic models are scaled in a ratio of linear dimensions (i.e., $\frac{1}{100}$ scale model of a 1000-ft dam is 10 ft long).

The first class of iconic models includes the pictorial representation, two-dimensioned simplifications. In this class belong all pictures, maps, drawings, charts, and diagrams which appear as visual equivalents. Certain charts and diagrams, such as a business organization charts and circuit wiring diagrams, are not iconic models because they functionally *behave* like the reality rather than look like it, and hence these belong to the analog models which will be described next. Illustrations of two-dimensional iconic models include road maps, manufacturing drawings, photographs, exploded drawings, special-purpose drawings such as rotated sections or phantoms, and so forth. See Figure 10.1a,b. A working application is shown in Figure 10.2, where a chemical process plant drawing is incorporated into the control boards.

A second class of iconic models comprises the three-dimensional physical replica, such as those shown in Figure 10.3. The scaling may be uniform, such as the reduction of the world to the size of a desk globe, or nonuniform, such as the exaggerated vertical heights in a highway drawing, or in the relief model of mountains as in Figure 10.3b. Illustrations of physical iconic models include the stick-and-ball models of molecules and crystals, the

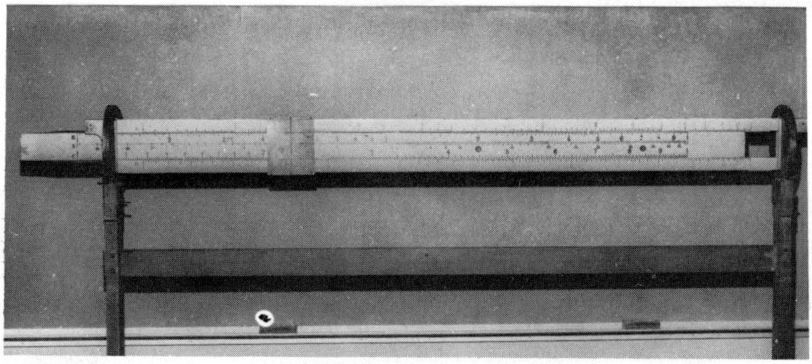

Fig. 10.4 Iconic and analog model combination. Classroom demonstration slide rule looks like a pocket slide rule and behaves like arithmetic operations on numbers. (UCLA photo)

miniature plant layout, the large-size demonstration slide rule, the three-dimensional model of a freeway interchange, the appearance design model (mock-up) of a new product, etc. Note that miniature models for ship-towing tanks and wind tunnels are not iconic models; they are built as analogs. Figure 10.4 shows a large slide rule, in this case both an iconic model and an analog. It is both a geometric replica (an operating equivalent physically) and an analog for the multiplication, etc. of numbers.

The Analog Model

Analog* models are convenient operating systems which *follow the same principles* as the original; and the model's only use is when it *is operat-*

* Analog: From the Greek, *analogos*, ratio (*ana* + *logos* = again + word, thought, principle).

ing. Except for one type, analogs are usually meaningless *visually*. The analog may be a partial subsystem or it may be substantially complete representation of the selected problem. For example, the ship-towing model and wind-tunnel airplane mentioned above may be complete shapes, but, of course, only the hydrodynamic and aerodynamic properties are being examined. See Figure 10.5. Such models would not be used to study structures, load capacity, weight or cost.

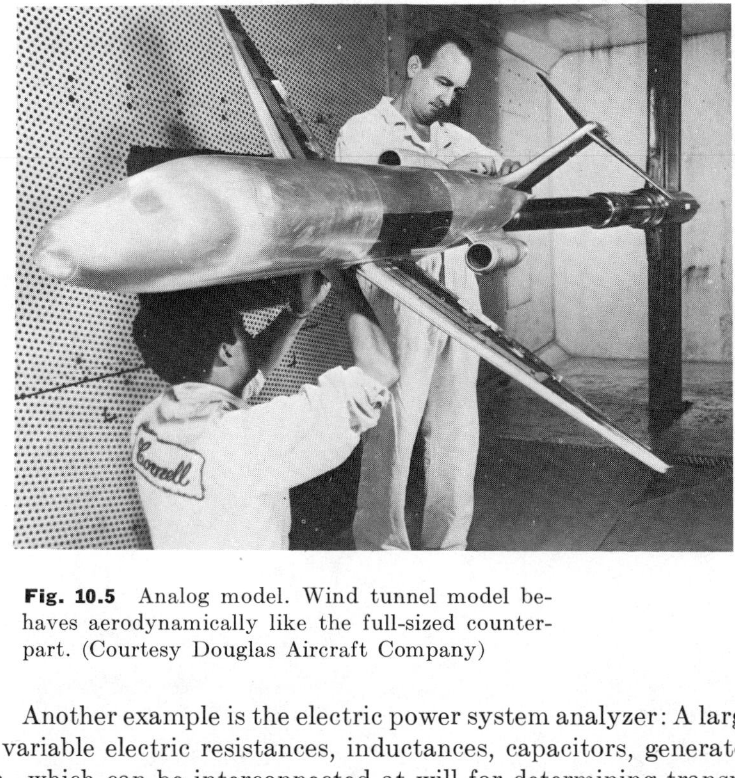

Fig. 10.5 Analog model. Wind tunnel model behaves aerodynamically like the full-sized counterpart. (Courtesy Douglas Aircraft Company)

Another example is the electric power system analyzer: A large assembly of variable electric resistances, inductances, capacitors, generators, meters, etc., which can be interconnected at will for determining transmission system capacity and stability before embarking on new construction programs. See Figure 10.6.

There are simple analogs as well as complex. The everyday slide rule is an analog that adds and subtracts logarithms to multiply and divide the corresponding numbers. A sine wave on paper is an analog for a rotating crank arm: The Y-axis or ordinate represents vertical displacement of the crank pin, and the X-axis or abscissa represents time. Note that, in each case, the model simply obeys the same laws or follows the same rules. Wheels (or gears or belts) can be used to make continuously variable adders or multi-

Fig. 10.6 Analog model. Electric power system analyzer showing the panels of adjustable system elements and the control console with read-out meters. (Courtesy Southern California Edison Company)

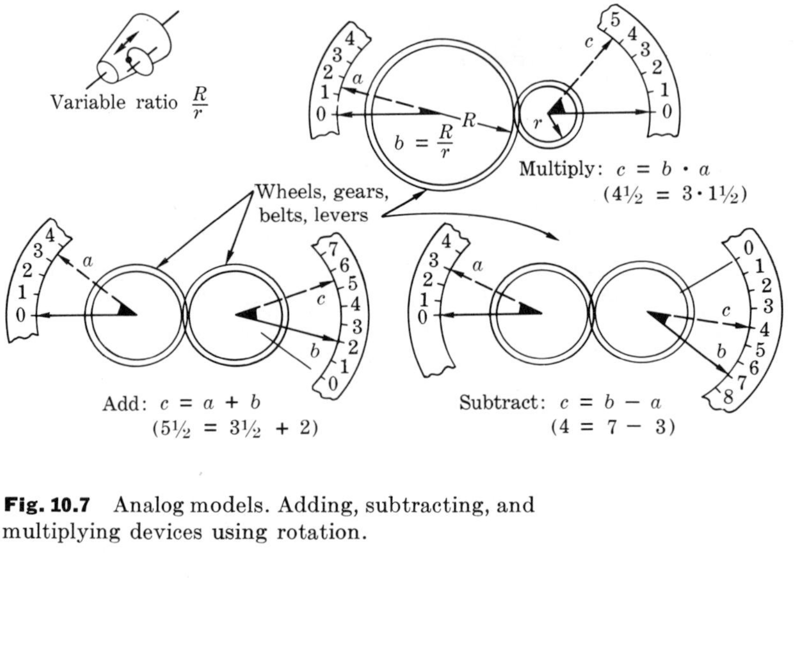

Fig. 10.7 Analog models. Adding, subtracting, and multiplying devices using rotation.

pliers, as shown in Figure 10.7. The abacus (Fig. 10.8a) is the adding analog that has been used for centuries, just as the hour glass and sun-dial have long been time analogs. The desk calculator and digital computer (Figure 10.8b,c are also numerical analogs.

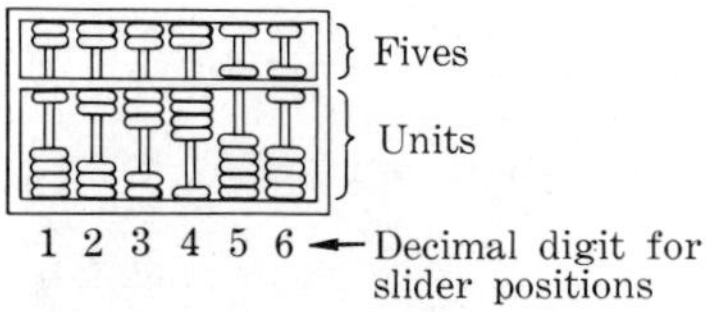

(a)

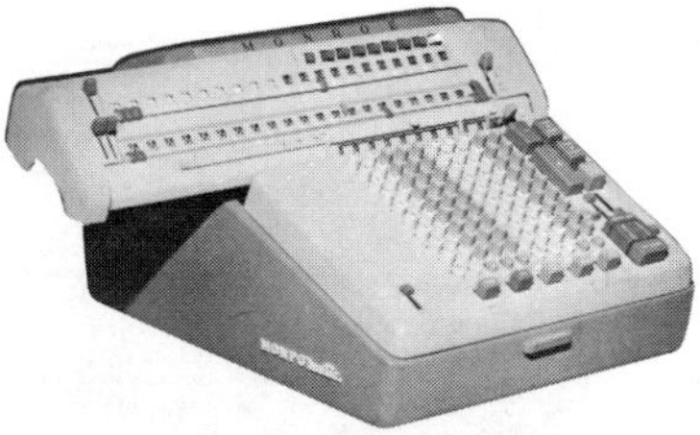

(b)

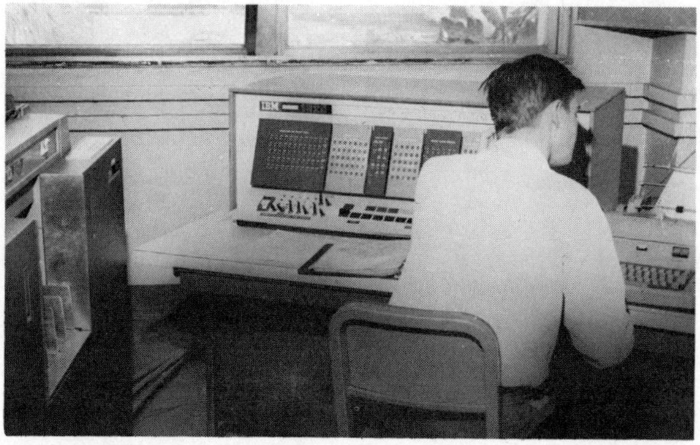

(c)

Fig. 10.8 Analog models. Numerical problem solvers: (a) Abacus; (b) Desk calculator; (c) Digital computer. (UCLA photo)

The reasons that analogs are used have been summarized at the end of this chapter, but briefly, they are easier to build and use than direct or scaled embodiments of the original. For example, water and electricity both follow Kirchoff's laws in closed circuits. Similarly, heat and electric current both flow in three-dimensional conductors according to the same equations. As an example of the latter, it is simple and economical to cut out and instrument a sheet of conducting paper for the electrical equivalent of heat flow into a door corner of a refrigerator. See Figure 10.9. Constructing and varying the analog is much easier than constructing and varying the door in the real case, using fiberglas, sheet metal, airspaces and plastics.

Other analog examples can be cited: Electric resistance strain gages to measure deflection; electric flip-flop or relay circuits for counting in computers; structural analysis miniatures; varnish coatings on tension-

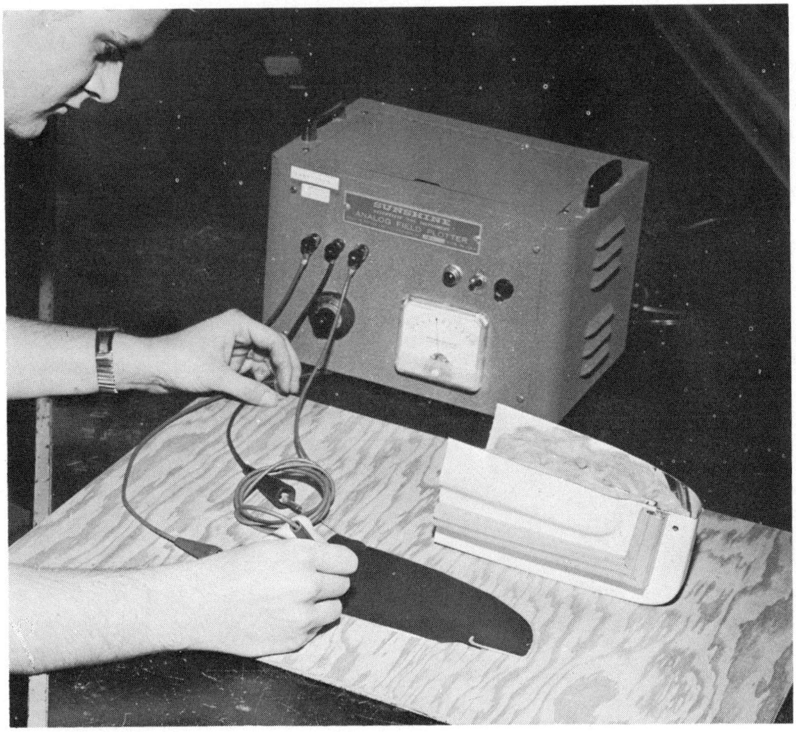

Fig. 10.9 Analog model. Electrical current stimulation of heat flow into corner of refrigerator door. Equipotential (equal temperature) lines are plotted and heat flow computed from current flowing through the black conducting paper between several boundaries. (UCLA photo)

stressed specimens; flow of a liquid simulating the flow of granules; polar planimeter for measuring areas, etc.* Many of these appear to be only measuring devices; but analogs are not necessarily measuring devices. However, since they actively translate one phenomenon into another, measurement is a useful application.

* This domain includes practically all transducers which convert one physical quantity into another. See Table 5.2.

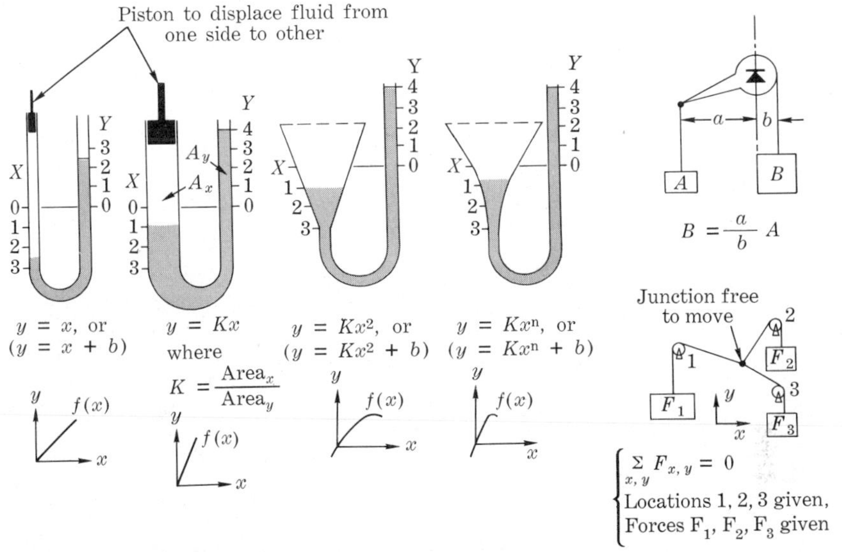

Fig. 10.10 Analog models. Devices to solve a variety of equations.

Fig. 10.11 Analog model. Flight motion simulator. Pilot trainer for Boeing 727 jet aircraft. (Courtesy United Air Lines, Inc.)

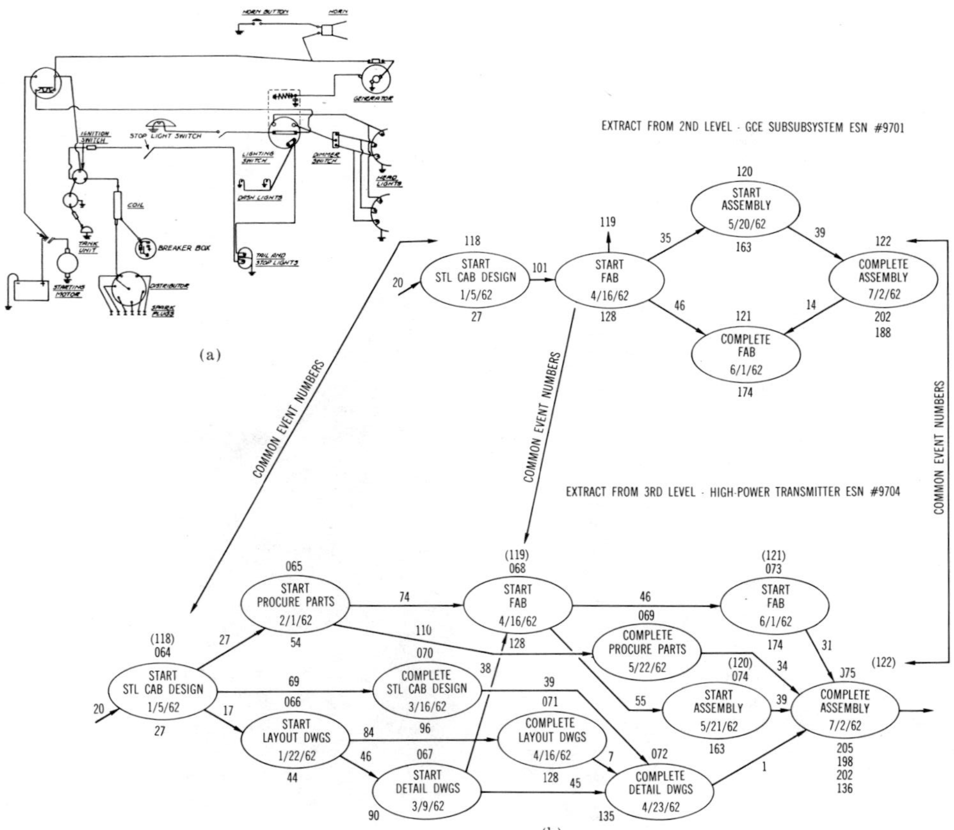

(b)

MODEL FA4N

Item	Dwg. Number	Description
1	524B118-G3	Motor Assembly
2	383D107	Housing
3	423C143	Shredder Element
4	383D119	Bottom Flange
5	423C137	Top Flange
6	423C138	Flywheel
7	5431650-P1	Bearing (Cone) (2)
8	5431650-P2	Bearing (Cup) (2)
9	727A533	Capacitor
10	524B142	Relay (Starting)
11	524B123-2	Seal Assembly
12	5431882	Cover Bottom Flange
13	727A515	Protector, Motor
14	383D115	Carton & Fillers
15	524B135	Stopper
16	524B143	Support Ring
17	524B130	Switch
18	5431717	Handle
19	5431731	Plate for Stopper
20	727A524	Impeller (2)
21	524B144	Clamping Ring
22	5431538	Elbow
23	5431651	Cup Insulator (2)
24	5431939	Cover
25	38-155	User Book
26	NP-176407	Monogram Metal Cal
27	524B145	Gasket
28	727A553	Seal Seat
29	524B138	Retainer Seal
30	524B129	Actuator Arm Assembly
31	5423655	Deflector
32	5417002	Gasket
33	5423657	Gasket
34	5431766	Elbow Bracket
35	727A527	Screw - Self Tapping - $\frac{1}{4} \times 20$
36	5431637	Retainer Assembly

(c)

Fig. 10.12 Analog and symbolic model combinations. Process and organization models that behave like (do not look like) the reality. (a) Electric circuit; (b) PERT scheduling chart; (c) Bill of material assembly list.

Some very simple analogs can be used to solve equations. Figure 10.10 shows liquid and string analogs for a few equations.

Another important member of the analog family is the equipment simulator, Figure 10.11 showing an example used for thorough, realistic training of aircraft pilots without leaving the ground or endangering their craft. Other simulators are used for automobile safe driving research, or for spacecraft high- or low-gravity tests.

The one analog which is significant visually is the circuit flow or relationship chart. Figure 10.12 shows examples where the lines, the flows, the organization and hierarchies are analogous to the reality; whereas the diagrams themselves don't necessarily portray a physical geometric relation.

Analogies between Technical Fields

The analogies between differing technical fields are very specific, so that a table of equivalents can be drawn up. As mentioned above, many mechanical, electrical, and thermal systems follow the same laws, In fact, the same equations describe the behavior, and only the units need conversion. Advanced engineering courses treat these so-called isomorphisms extensively. Because of the increasing use of these analogies in practice, many engineering educators now stress the similarities between fields rather than the differences. Table 10.1 is a comparison between mechanical, electrical, and thermal systems.*

The Analog Computer

The analog computer is a widely-used instrument by which solutions to many time-varying complex problems can readily be obtained. The original type was the mechanical analog, using levers, shafts, gears, etc. (See Figure 10.13.) A newer electrical type uses decimal-stepped electrical elements and permits the answers to be read on meters or moving charts (see Figure 10.14).† In both of these cases, the given problem is converted by the engineer to its analog statement; the instrument is set up and interconnected for the indicated gear ratios, dial settings, etc. The power is turned on; the answers are read off or recorded on tapes as the machine simulates the system. Then the numbers are converted back to the desired values and units, answering the original questions.

* Analogous relations exist in fluid flow, acoustics, electromagnetics, etc.

† E.A.S.E., Electrical Analogue Simulation Equipment (Beckman).

Table 10.1 Comparison of Analogous System-Variables and Parameters.

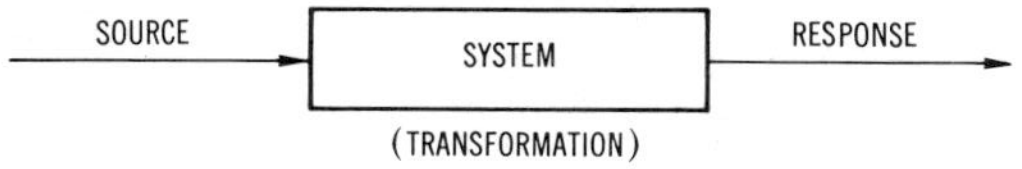

Description	*Mechanical*		*Thermal*		*Electrical*	
Conventional source* variable:	Force	f, lb	Temperature	θ, °F	Voltage	e, volt
Primary response variable and its derivatives:	Displacement,	x, ft	Heat,	H, Btu	Charge,	q, coul
	Velocity,	$v = \frac{dx}{dt}, \frac{\text{ft}}{\text{sec}}$	Heat flow,	$Q = \frac{dH}{dt}, \frac{\text{Btu}}{\text{sec}}$	Current,	$i = \frac{dq}{dt}$, amp
	Acceleration,	$a = \frac{d^2x}{dt^2}, \frac{\text{ft}}{\text{sec}^2}$	Heat flow change,	$\frac{dQ}{dt}, \frac{\text{Btu}}{\text{sec}^2}$	Current change,	$\frac{di}{dt} = \frac{d^2q}{dt^2}, \frac{\text{amp}}{\text{sec}}$
System parameters: (Physical elements or qualities)	Viscous friction,	$D, \frac{\text{lb sec}}{\text{ft}}$	Thermal resistance,†	$R, \frac{°\text{sec}}{\text{Btu}}$	Resistance,	R, ohm
	Inertia,	$M, \frac{\text{lb sec}^2}{\text{ft}}$	. . .		Inductance,	L, henry
	Elasticity, (Stiffness = K)	$\frac{1}{K}, \frac{\text{ft}}{\text{lb}}$	Thermal capacity,	$C, \frac{\text{Btu}}{°\text{F}}$	Capacity,	C, farad
Fundamental equation: Series circuit (one degree of freedom)	$M\frac{d^2x}{dt^2} + D\frac{dx}{dt} + Kx = f$		$R\frac{dH}{dt} + \frac{H}{C} = \theta$		$L\frac{d^2q}{dt^2} + R\frac{dq}{dt} + \frac{q}{C} = e$	
Applicable circuit for above equation:	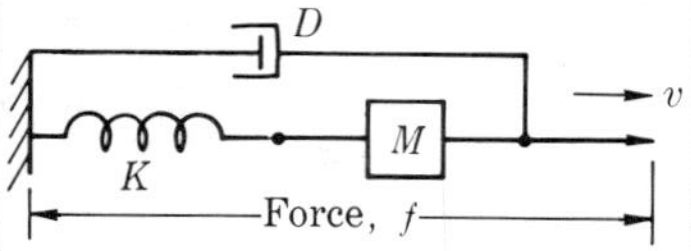		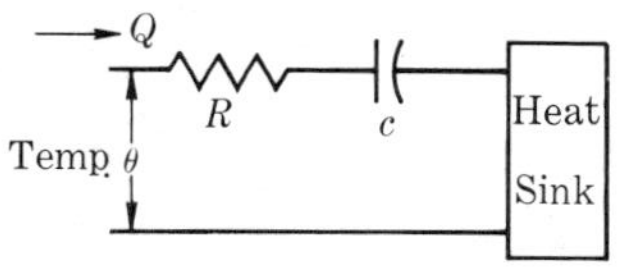		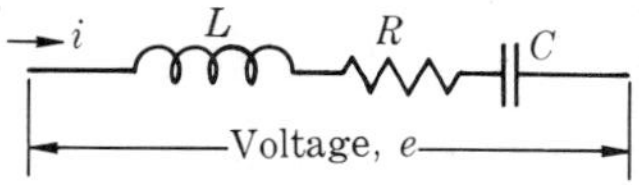	

* "Source" also referred to as "forcing-" or "driving-function."
† Thermal resistance may be for conduction, convection, or radiation.
Reference:
Alexander, J. E., J. M. Bailey: "Systems Engineering Mathematics," pp. 72–88, Prentice-Hall, Inc., Englewood Cliffs, N.J., 1962.
Also See Appendix D, of this book.

Fig. 10.13 Analog computer. Mechanical analog showing the input-output devices at the sides of the chassis and the shafts for ratio and coupling transformations down the center. During problem simulation the machine draws output curves on the plotting boards at the side. (UCLA photo)

Fig. 10.14 Analog computer. Electrical analog widely adaptable to mechanical, thermal, chemical, aeronautic, and other varieties of problems. Adjustable system elements on the vertical panels and the output meters and tapes at the left. (UCLA photo)

The Symbolic Model

Symbolic models are abbreviated abstractions of relevant and quantifiable parts of the problem situation. The symbols x, dx/dt, $=$, A, θ, etc., are convenient shorthand notations to retain generality and to economize time during the analysis. Very little synthesis is done with symbolic models. We all know that mathematical expressions like $f = Ma$, $x^2 + y^2 = r^2$, are symbolic models; the former representing Newton's first law of motion and the latter representing a circle of radius r symmetric to the origin of the X-Y axes. But what do we mean by "abstraction," "relevant," "quantifiable"?

We use the word *abstraction* in the same sense as in semantics, where it means successive steps of generalizing an object or situation into broader and more inclusive classes and categories. In this direction these steps are called *higher levels* of abstraction. For example, my personal typewriter, a Royal portable, model "Futura," serial number A3994841, can be abstracted into many levels, at least as many as the following:

1. *The* typewriter (the actual object)	no abstraction, no generality
2. All "Futura" model Royal portables	first abstraction, slight generality
3. All Royal typewriters	higher abstraction, greater generality
4. Any typewriter	still higher abstraction
5. A ribbon-impression machine	still higher abstraction
6. Permanent record communication means	still higher abstraction
7. Communication means	still higher abstraction (no meaningful upper limit)

As the reader will observe, the succeeding steps include more and more varieties of communication means, and greater numbers of them.

We would abstract the classes differently, depending upon our purposes. This same typewriter could have been abstracted in the categories of metallic mechanical manufactures or paper conversion devices or recording machines, etc. Normally, of course, the abstraction is along the line of the major purpose.

Observe, incidentally, that our types of models are themselves lower or higher orders of abstraction:

Iconic	Lower abstraction
Analog	Intermediate abstraction
Symbolic	Higher abstraction

The engineer abstracts from his problem situation the important, meaningful (relevant) aspects that contribute to the answers. Symbols are then attached to those attributes, so that numbers can eventually be deduced. The less relevant or the irrelevant attributes are neglected. In this way, qualities (such as size, cost, strength) are converted into quantities, or "quantified," and, after analysis, eventuate in concrete numbers (no abstraction at all).

Table 10.2 Summary of the Attributes of Models Used in Problem-solving.

PURPOSES AND POWER

Iconic

A. *Visualize:* Enlarge, shrink, emphasize sense of texture and shadow, show aesthetics.
B. *Establish relationships:* Rank, order, proportion, arrangement. Observe the interactions.
C. *Synthesize:* Make a whole out of parts.
D. *Communicate and record:* With self, clients, subordinates.

Analog

A. *Simulate performance:* Operate in the desired mode and in real or artificial time.
B. *Determine numerical results:* Check quantitative interactions.
C. *Employ various phenomena:* Suggest new areas of investigation.
D. *Interchange variables and parameters, at will.*
E. *Use one piece of equipment (analog or digital computer) to solve many varieties of problems.*

Symbolic

A. *Use maximum generality in attack on problem.*
B. *Economize effort:* Use symbolic shorthand for attributes and operations.
C. *Lead to numeric outcomes,* to explicit functional relations.
D. *Use methods of mathematics, geometry, and other logics.*
E. *Solve many problems economically:* Use only pencil and paper.

LAWS INVOLVED

Iconic

A. *Scaling ratios:* Lines, areas, volumes.
B. *Laws of projection, geometry, etc.*
C. *Rules of seeing:* Limitations both by physiology and by dramatic effect.

Analog

A. *Equivalence of the differential equations,* or other equations of behavior.
B. *Dimensional homogeneity.*
C. *Rules of similitude.*

Symbolic

A. *Self-consistent axioms and laws of the logic used:* Numbers, functions, etc.

LIMITATIONS

Iconic

A. *Limited number of variables:* 1, 2 or 3 (dimensions).
B. *Optimization generally done visually.*
C. *Limited by the engineer's conceptualizations.*

Analog

A. *Fidelity of assumptions:* Simulation increases probability of correct answers.
B. *The range of the specific equipments available.*

Symbolic

A. *The fidelity of assumptions:* Answer must be tested.
B. *Solving-power and skills of the engineer or mathematician.*

The classes of abstractions are useful because within themselves, various descriptive and predictive relations can be discovered. In the instance of the printing machine above, one applicable rule might have been found: "The degree of imprint varies directly with the force applied and with the sensitivity of the imprinted medium." (i.e., Imprint = Force × Sensitivity, or $I = F \times S$.) This rule applies to single-sheet or multiple-sheet printers, to magnetic tapes, clay tablets, etc. Thus, working with very generalized symbolic models, the mind of the engineer can look over many design alternatives, evaluating each one suggested for its advantages.

SUMMARY

We will examine modeling in greater detail in Chapters 12 and 13, but let us conclude this section with the following thoughts. The engineering problem always requires a specific answer, corresponding to some real object, measured by given criteria and meeting a series of limiting requirements. The various models, and particularly the symbolic model, allow the engineer to examine many more classes of possible solutions before settling on the one "best" solution. He is thus guaranteed the widest review before the inevitable returning, back down the "ladder of abstraction" to a final specific object. The natural penchant for falling in love early with a particular embodiment or operating principle is conveniently offset by the discipline of this wide search, this test by many questions. The adequate formulation and use of a model is the engineer's best assurance of project success. Even with the risk of a double meaning, "an engineer's best friend is a good model."

BOOK PROJECT EXAMPLES

1. *Iconic.* In the earlier chapters, we introduced the several examples related to the engineer in the transit-mix concrete plant.

The first example, a foundation base for a two-car garage, led to an iconic model; simply a drawing of the garage floor and footing. This is direct and obvious, so no extended discussion will be needed. The engineer simply made a sketch which included all the dimensions, notes on materials, special instructions, standard parts, etc., which anyone familiar with the notation could read. The drawing was an iconic model.

2. *Analog.* The example in Chapter 9 concerning sand going off the end of a conveyor belt could have been simulated by the analog of water spouting out of an inclined pipe as in Figure 10.15. Since air

resistance is neglected, a text on mechanics or physics will derive the expression showing that the distance d is,

$$d = \frac{V_0^2}{g} \sin 2\alpha$$

or, the distance varies only with initial velocity V_0, the angle α, and the acceleration of gravity g.

Thus, if we needed a demonstration or wanted to make some measurements, we could have aimed a water pipe at angle α, turned the water on to velocity V_0, and the water stream would have gone over the

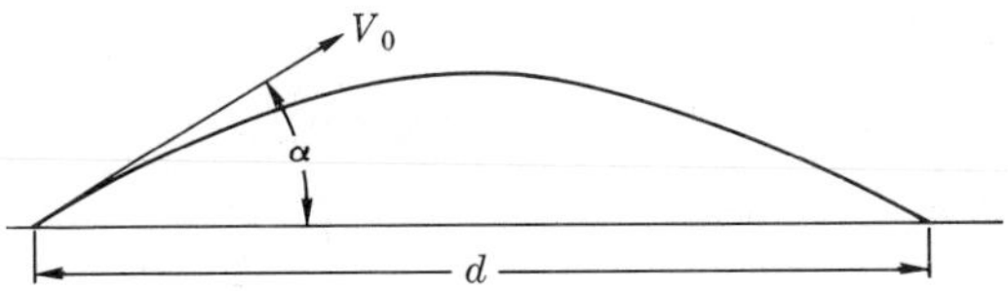

Fig. 10.15 Path of a projectile as a liquid analog.

same path as a stream of sand. This would have given us an analog model.

3. *Symbolic.* The symbolic model was illustrated in the previous chapter in both examples. But, choosing the latter, the conveyor problem, the equation

$$P = Q\left(\frac{V^2}{g} + h\right)$$

was the direct "power model" and was the basis of the solution.

EXERCISES

10.1 Are any of the instruments in an automobile analog models? Speedometer, gasoline gauge, ammeter, oil gauge, water temperature gauge, odometer, magnetic compass? Discuss.

10.2 An auto was purchased six years ago for \$2500. Assuming uniform depreciation to its present resale value of \$700, and annual license and insurance costs at \$25 and \$125 respectively, derive a mathematical expression (symbolic model) for the annual cost of operation, assuming all costs are averaged over the period of use. Other expenses include: Annual repairs at \$75 per year; 13,000 miles of travel per year at 17 miles per gallon (fuel cost—\$0.31/gal.); parking fees and tolls at \$5 per month. What is the average annual cost in dollars?

10.3 Derive a mathematical expression (symbolic model) of the cost to a student for flunking a math course that must be repeated before he can graduate.

10.4 Warehouse and factory walls are frequently made by pouring large reinforced concrete slabs on a flat floor and later erecting them to the vertical by the use of cranes. Write an expression for the cost of this type of construction per square foot of wall, compared to a conventional brick and mortar construction.

10.5 In the water analog of Figure 10.10, what is the relation of the cone angle and the initial height of the water to the constants K and b, for the example of $y = Kx^2 + b$?

10.6 The deflection in the center of a simple rectangular beam is given by the following expression. (See Figure 10.16.)

$$d = \frac{PL^3}{4Ebh^3}$$

where

d = maximum deflection, ft

P = center load, lb

L = length, ft

E = modulus of elasticity, lb/sq ft

b = beam section width, ft

h = beam section height, ft

Suppose you made a beam of a given material to $\frac{1}{10}$ scale, and tested it with a load of $P/10$. What would be the relation between the deflection d of the model to that of the full-scale beam of the same material under load P? What would be a purpose in making a reduced scale model when you already have the above equation?

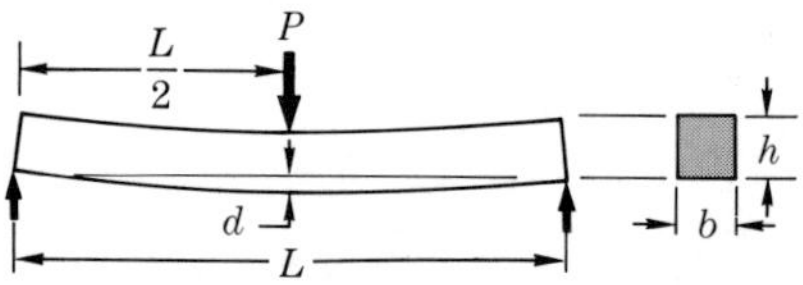

Fig. 10.16 Dimensions of a rectangular beam of modeling.

REFERENCES

1. Asimow, M.: "Introduction to Design," pp. 15–74, Prentice-Hall, Inc., Englewood Cliffs, N.J., 1962.

2. Churchman, C. West, R. L. Ackoff, E. L. Arnoff: "Introduction to Operations Research," John Wiley & Sons, Inc., New York, 1957.

3. Gosling, W.: "Design of Engineering Systems," chap. 2, John Wiley & Sons, Inc., New York, 1962.

4. Johnson, L. H.: "Engineering Principles and Problems," chap. 4, McGraw-Hill Book Company, New York, 1960.

5. Mantell, M. I., M. E. Reeder, J. A. Stevens: "Orientation in Engineering," chap. 8, Prentice-Hall, Inc., Englewood Cliffs, N.J., 1963.

General Aspects of Graphics: The Iconic Model

Eleven

"The eyes believe themselves;
the ears believe other people."
German proverb

Ubiquitous Graphics

Graphics, like problems, constantly rise up and confront the engineer; they are everywhere. All the systems basic to engineering, all the branches and fields of engineering, even the performance of engineering projects depend heavily on graphics for their functioning. In fact, graphics shares importance even with the scientific method and with logic and mathematics in its fundamental usefulness to the engineer. Thus, instead of being a road-block, or a deadweight, graphics turns out to be a welcome, helpful traveling companion.

We mentioned the basic systems of engineering. These have been identified as the following; and as one reviews the list, a little reflection will reveal how graphics is important in each case:

Information systems
Structures and circuits
Machines
Processes and transformations
Living systems

As we have mentioned earlier, all engineering systems integrate work from many branches or fields, and one project may comprise several or all of these systems. Consider a municipal sanitary waste disposal system which depends upon aerobic bacteria to break down the impurities in the water; such a composite system can be shown to intermix all the five basic systems. As different as all these systems are, they all involve *form*, *geometry*, *relationship*, *proportion*, and *embodiment* and each system has its own special language of *symbols*, *conventions*, and *performance curves* for communication purposes. In a word, these systems all involve *graphics*.

The design engineer is less interested in the mechanics of drawing, the use of instruments, and the making of traditional shop drawings, than he is in graphics as a *coding technique* to transmit information. To be sure, he must know and appreciate good drafting. He must be able to read drawings easily and give instructions to the draftsman, but his primary purpose is to communicate; and the lines and symbols of graphics are simply efficient *codes* for communication. He will himself be personally more involved in *sketching* than in *mechanical drawing*.

The sketches are representations of the object, in code form, sometimes only remotely looking like the object. But, the better the coding technique, the better the reader's recognition, and the better the information transmittal. Again, we stress that the drawing is only a coded representation, its success depending upon the reader's knowledge of the code.

We have been creeping up on a useful definition of graphics, one that assumes the dictionary definition, but goes beyond it, to support the work of this chapter. From the dictionary* we have the following:

> *Graphics:* "Art or science of drawing, especially according to mathematical rules, as in perspective, projection, etc; . . ."

But we are using graphics in a broader sense:

> *Graphics:* A coding technique, using the art and science of drawing together with the conventions of line and symbol to transmit information visually from sender to receiver.

Communication and Its Purposes

With our stress upon communication, the importance of the receiver comes to the surface. All engineering communication is for *action*, some kind of action, and the sender's purpose is accomplished only when that action takes place. But, the receiver is the link on which success depends, and when the receiver is a person (not a machine), all the fetishes, foibles, and habits

* "Webster's New Collegiate Dictionary," G. & C. Merriam Company, Springfield, Mass., 1959.

of the race intrude to confuse the message and thwart the sender. Thus, as engineers, we should press for the most efficient and successful coding and the widest and most effective education for its recognition. A few of the types of codings are shown in Figure 11.1, as a sort of smorgasbord to remind us of the divergent graphic conventions we encounter.

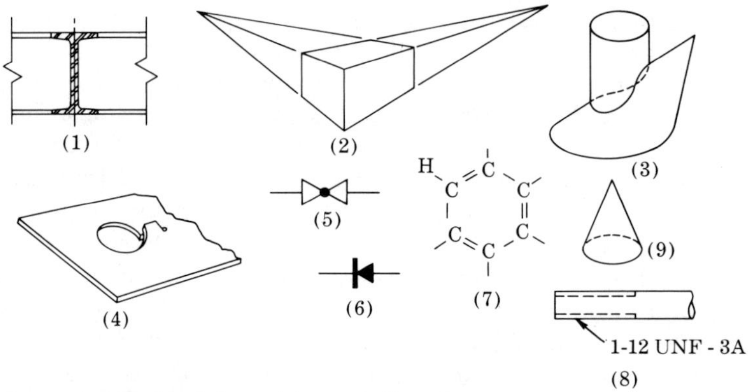

Fig. 11.1 Variety of graphic symbols and conventions. (See footnote for complete legend.)

The reason that the sender (engineer) initiates communication in the first place is one of the following:

To issue instructions or data (or to request them).
To make proposals or to persuade.
To communicate with himself: Synthesize, assimilate, organize, refine, decide.

At first thought, the last item, self-communication, may seem trivial; but as the qualifying words are taken into account, one can see that this item is probably equal in importance to any of the others. In self-communication, the "sender" and "receiver" are the same, but the rules of communication still apply (see Chapter 17). The reader will actually have no trouble appraising the importance of self-communication; for all of us, new ideas and recalled memories are fuzzy, indistinct, and uncertain at first, and require clarifying and sharpening by repeated cycling through the mind. Whenever we develop new ideas, even the "flash" variety of ideas, revision and refinement play a vital role in readying the ideas for "public" view or for use.

* For Figure 11.1: (1) Longitudinal section of I beam; (2) Perspective of box; (3) Intersection of cylinder and cone; (4) Sheet-metal screw thread; (5) Plumbing globe valve; (6) Electrical diode rectifier; (7) Kekule ring for hydrocarbons; (8) Thread: 1-in. diameter, 12 threads per inch, Unified Fine Thread Standard, class 3 fit external; (9) Cone.

Capability of Eye-Mind Team

The recycling of ideas depends almost entirely on inputs from our senses, and all 14 (±) human senses are called upon. However, in sheer communication capability, the eye* far outranks even the ear, and the ear outranks the sense of touch and the rest of them all together. It is the eye's speed and versatility that make it the dominant communication tool. There are other "coding techniques" for transmitting information, such as spoken and written language, gestures, punched cards, tape, and so on. But even with these, the eye is still an agent in most cases. The world functions primarily by visual communication.

The eye carries this importance in both communication and learning because it can almost instantaneously comprehend distance, size, scale, identity, texture, congruence, etc.

Not only can the eye and mind comprehend, but they can routinely perform useful sleight-of-hand at will. Given a simple sketch of an object, the educated mind (with a little practice) can enlarge it, shrink it, distort it, invert it, get inside, view it from a distance, changing from one mode to another in fractions of a second. As an example, consider the right circular cone (9) in Figure 11.1 When first looking at it, we see it as a plane figure, simply an ellipse inscribed in an angle. Then we can see it as a cone. Then when we imagine two triangular flaps on the front, it becomes a wigwam. And, simply by thinking about it, we can picture ourselves walking up to this open entrance, peeking in and looking around at the furnishings. Turn the cone sideways and put a stick on it and we have a biologist's butterfly net.

The point of all these imaginings is that the engineer can look at a graphical form and call upon his memory to appraise a large variety of modifications very quickly. The eye-mind combination thus works with ideas easily and fluently. Using form and symbol, one can imagine new shapes and modify them to meet the constraints of the problem. A further value is in the recovery of ideas, since the past sketch, or note, quickly starts the train of thought over again after an interruption. Also the clean piece of paper on the pad encourages a fresh start, with radical improvements this time. (Incidentally, as a practical hint, leave the sketch or note only partly finished when you put down the pencil; you will quickly shift into gear when you pick up the work again.)

*The eye can receive a television picture at 6,000,000 bits per second (off-and-on units). The ear can receive music on a radio channel at a maximum of perhaps 50,000 bits per second; but these are not coded for significance. In receiving spoken words, the ear can comprehend only about 50 bits per second. Reference: D. G. Fink and D. M. Luytens, "The Physics of Television," p. 22, Doubleday & Company, Inc. (Anchor), Garden City, N.Y., 1960.

It is these capabilities of the mind plus the inevitable occurrence of line and form in engineering works, that prompt us to spend these pages on the generalities and applications of graphics.

As a final reminder, the practicing engineer must satisfy the clamorings of a diverse group of persons who each are trying to *give* or *get* information:

Company or organization management.
Associates and subordinates.
Shop and laboratory personnel.
Industrial designers, patent attorneys, and package designers.
Vendor's salesmen.
Manufacturing and inspection personnel.
Government, armed service, and regulating agencies.
Visitors and general public; students.
Colleagues in professional societies.

Graphics works with each of them.

What Can Be Done by Graphics?

We shall describe and illustrate the following engineering uses of graphical methods:

Specification
Visualization-Synthesis
Determination of Dimensions
Computation
Showing Relation and Emphasis
Persuasion and Selling
Recording

Although some of these items are obvious and familiar, we may be able to convey a new sense of their over-all importance by bringing them together. This is a significant list of capabilities, and adds evidence of the strategic role played by graphics in engineering.

Specification

The most vital use of graphics is *specification*, that is, describing and requesting the "specific." The engineer must be able to describe what he has in mind exactly. Other means are available to specify (words, coordinates, operations, and special machine-symbol systems) and each means has its place. However, for wide usage, and especially when persons are the receivers, graphics dominates. Of course, when machines are the receivers, we communicate specifications to them electrically or mechanically.

What if we couldn't or didn't use graphics? We shall illustrate here three methods of specifying (two without overt graphics), in each case describing a simple small steel bracket.

1. *Specification by Words**

   Title: *Angle Bracket Fabrication*
   Material: Hot Rolled Steel, AISI 1010–1020

   Given a rectangular parallelopiped of the subject strip steel, having a width of 1 in. ($\pm\frac{1}{16}$ in.), a length of 4 in. ($\pm\frac{1}{16}$ in.) and a thickness of $\frac{1}{8}$ in. ($\pm\frac{1}{64}$ in.), perform the following operations:

   a. Drill in the flat plane two holes $\frac{1}{4}$-in. diameter symmetrically located on the longitudinal center-line, 3 in. apart between centers.
   b. Enlarge each of these holes by an 82° included-angle countersink to the full depth of the hole.
   c. Bend the bar at its midpoint to a 90° angle (with a $\frac{1}{8}$-in. vertex fillet) around an axis parallel to the flat plane, said axis also perpendicular to the longitudinal center-line. Other dimensions are $\pm\frac{1}{32}$ in.

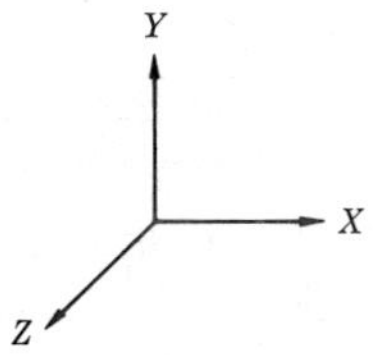

2. *Specification by Coordinates*†

   Same title and material as Part 1 above. Assume conventional (X, Y, Z) axes, in accordance with the adjoining figure.
   The boundaries of the material are as follows:

   Surface #1: Closed boundary: (2, 0, 0)(0, 0, 0)(0, 2, 0)($\frac{1}{8}$, 2, 0)($\frac{1}{8}$, $\frac{1}{8}$, 0)(2, $\frac{1}{8}$, 0).
   Surface #2: Closed boundary: (2, 0, 1)(0, 0, 1)(0, 2, 1)($\frac{1}{8}$, 2, 1)($\frac{1}{8}$, $\frac{1}{8}$, 1)(2, $\frac{1}{8}$, 1).
   Surfaces #3, #4, #5, #6 are defined by lines along the Z-direction connecting the above corresponding pairs of points: (2, 0, 0) to (2, 0, 1), etc.
   Hole #1: Axis (X, $1\frac{1}{2}$, $1\frac{1}{2}$); cone 82° included angle opening in the $+X$ direction, vertex ($-\frac{1}{8}$, $1\frac{1}{2}$, $1\frac{1}{2}$)
   Hole #2: Axis ($1\frac{1}{2}$, Y, $\frac{1}{2}$); cone 82° included angle opening in the $+Y$ direction, vertex ($1\frac{1}{2}$, $-\frac{1}{8}$, $\frac{1}{2}$)

3. *Specification by Sketch*

   Figure 11.2 carries the same information as Parts 1 and 2 above.

As a commentary on these three methods, note the frustration in the first two cases caused by not knowing anything about the general shape except that from the title: "Bracket." The descriptions were accurate for duplication but not for *understanding*. Also consider the relative readability, the time to prepare, the danger of error in the three methods. Project, if you will, the tangle that would be created by such descriptions of a really complex part, such as an automobile engine block.

* Even though more difficult to visualize, this style is used and required in writing patent specifications.

† This method is used, when appropriate, for numerically controlled automatic machine tools. Notice the tolerances were omitted. They would be necessary on each coordinate.

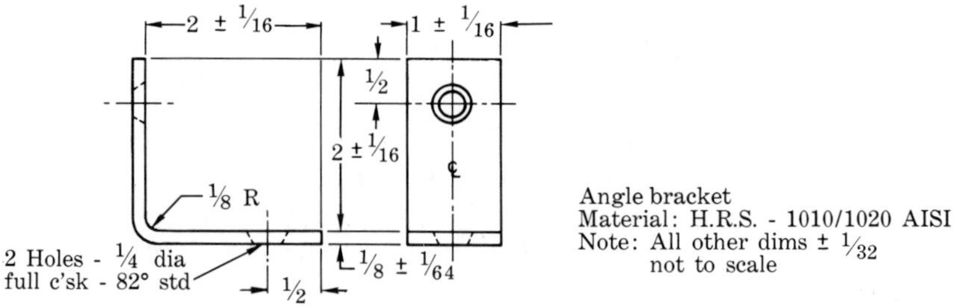

Fig. 11.2 Sketch of a bracket.

Yet, to read even the elementary sketch of Figure 11.2, one uses the prior knowledge of a coding method called "orthographic projection," a method closely related to our "rules of seeing," which we shall discuss next.

Visualization and Synthesis

The next most important engineering use of graphics is *visualization.* Everyone learned the rules of distinguishing objects and of sensing distances during infancy and childhood. For all of us, the learning period has been long forgotten. No one is normally conscious of these rules. But, experiments[1] with adults who, by a surgical operation, first received their eyesight at maturity, demonstrate clearly that one cannot "see" without painstakingly learning the "rules." These adults reported a difficult, prolonged learning period before they could distinguish visually between even such simple shapes as a circle, a square, and a triangle. Learning shapes in one color or over one background did not guarantee they could identify the shapes in another. Long accustomed to feel, they insisted on fingering the shapes to be able to "see" them.

Extended research has shown that children develop the perception of distance, size, and form very slowly. Even in middle childhood, they have trouble with rotations of figures and with perspectives. The concept of space is built up by means of long and tedious experience, repeated over and over again, in many contexts.

Persons with normal vision always "see" from the viewpoint of this past experience. Our eyes scan the surface and run along the boundaries of strange new objects, and memory simultaneously searches for familiar aspects. When enough similarities are found, some expected missing pieces are mentally filled in and we "recognize" the object by category. The eye motions and the impressions are affected by convergence, obliquity, shadow, comparison, texture, color, and of course, our store of previous "seeings."

As an added complication, people see objects in the near distance by the stereoptic effect; that is, the slightly different images and angles received by the left and right eyes, and by the changes in visibilities of the component parts as we move our heads. The simplest graphic equivalent for simulating such distance is *perspective*, or the convergence of parallel lines toward the horizon. A one-eyed person sees the convergence of these lines, but only on a plane and not "in space." We make these references to the "rules of seeing" because of the close relation to the engineer's two-dimensional representations.

If engineers are to communicate their ideas accurately, they must represent their ideas and devices according to exactly the same rules that are used by the eyes and mind.

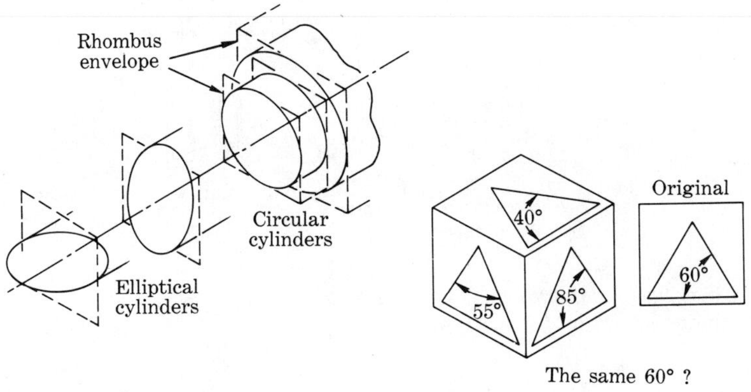

Fig. 11.3 Effect of obliquity on "seeing."

In Figures 11.3–11.5 we illustrate some of these ideas, related to seeing by obliquity (seeing at an angle), convergence, shade and comparison-to-knowns.

If the drawing is an *isometric* as in Figure 11.3, only those measurements made on or parallel to the three axes are true. All others are long or short. In this figure we see how obliquity distorts angles, cylinders and dimensions. To be "seen" as a circle correctly, the circular end of the cylinder must be an ellipse diagonally inscribed in the rhombus. However, with exactly the same ellipse at other orientations as shown at the lower left, the cylinder is "seen" as elliptical, not circular.

The cube illustrates how the eye is misled by angles. The equilateral triangle, inscribed on three sides isometrically, actually is a 40°-55°-85° triangle. Rotate the page sideways and prove it, even though the eye still says its 60°-60°-60°.

Seeing at an angle leads us to the seeming convergence of parallel lines. In Figure 11.4 we see to the left a normal perspective view of a three-pane window. Because of careless or inaccurate drawing, the eye is forced to "see" different windows, presumably not what was intended. Some of the supposedly undesired variants are shown together, and a quick glance at each shows how differently the eye "sees" each one, even though there may be relatively minor differences between them.

As a part of convergence, the stairway drawn with parallel steps in Figure 11.5a can be inverted at will and held in the mind indefinitely right side up or upside down. But the one drawn in perspective is so distorted when mentally inverted that the image is difficult to hold. The mind rejects it as unreal.

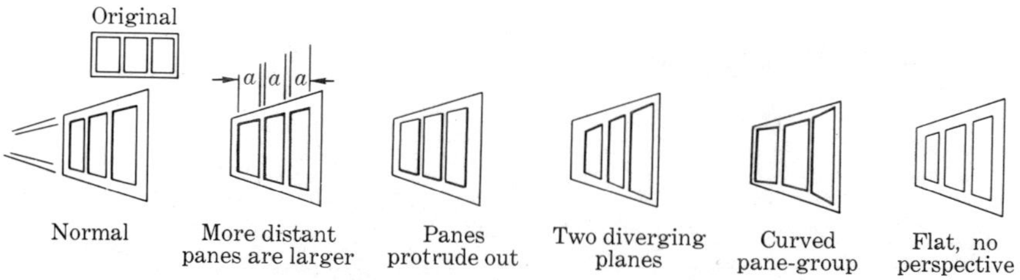

Fig. 11.4 Effect of convergence on "seeing." Impact of "minor" errors in delineation.

Shadowing, as in Figure 11.5b, is a natural aid to seeing, and the eye always assumes that light comes from the top of the picture. Thus the large circles on the block at first appear as recesses and the small ones as protrusions. Turn the page over and they reverse. They retain their depths if you look at them while slowly turning the page, but they then snap into reverse topography if you blink your eyes.

Comparison to known sizes is illustrated in Figure 11.5c where the viewer thinks he knows the size of the distant post *A*. However, the post suddenly becomes unbelievably small when seen transposed to the reference figure in the foreground.

Thus, these various characteristics of the eye-mind combination, with its strengths and weaknesses, combine to form the circumstances under which visualization occurs.

Many engineers find visualization easy; some do not, and must work at it. They should watch for all opportunities to sketch, draw, shade, and section in a self-improvement campaign. Visualization is a skill that can be improved with study and practice.

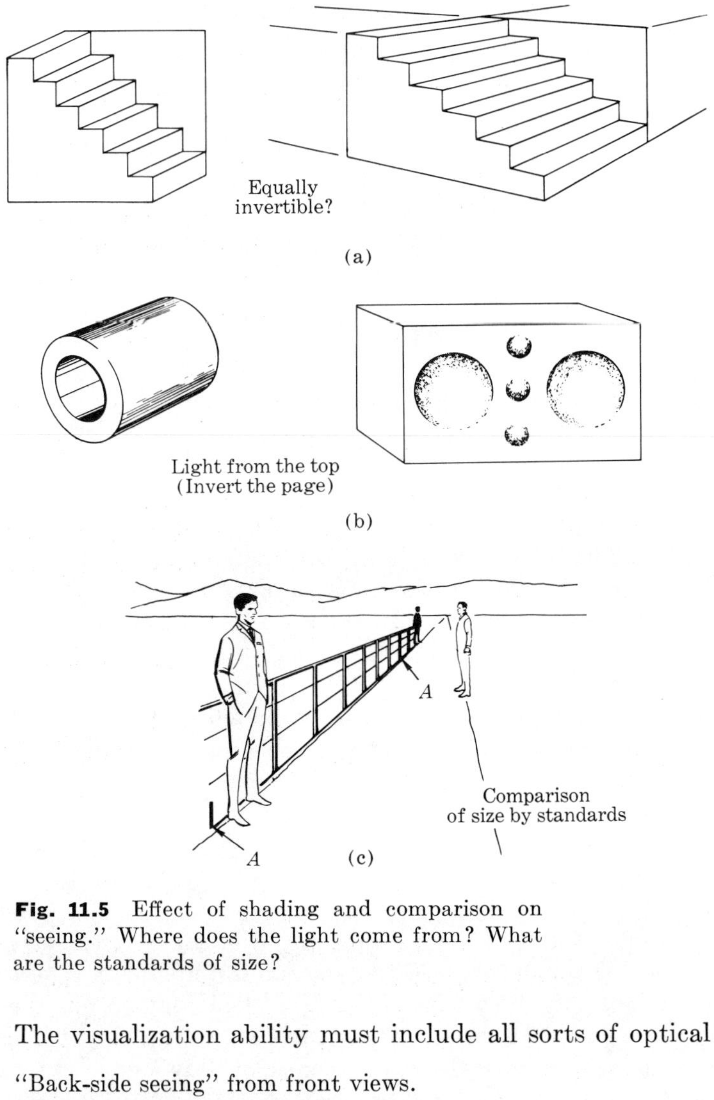

Fig. 11.5 Effect of shading and comparison on "seeing." Where does the light come from? What are the standards of size?

The visualization ability must include all sorts of optical gymnastics:

"Back-side seeing" from front views.
Seeing a solid as a transparent whole.
Extending lines and edges to their piercing points.
Shape and continuity of hidden curved surfaces.
Close-up look at local details.
Tracing a closed surface, or edge, from start around to start again.

In effect, the engineer must "comprehend" and "understand" the object from the drawing, or the drawing from the object. This ability, incidentally, seems rather mysterious, because we are told that when we see a square, nowhere in the brain do the neuron reactions form a geometrical square; rather, a code has been established which signifies "straight line," "squareness," etc. The mental process is much like that involved in learning a new language. For instance, the usual American will respond with a totally blank look to the sight or sound of "jin" or its symbol "人"; yet every literate Chinese person unfalteringly recognizes the sound and symbol as representing "man." All people can learn the language of graphics, just as all people can learn Chinese (if they want to). As with any new language, learning occurs easiest and fastest when the beginner is immersed in it every day and converses in the language with experts. Graphics is such a language, and visual fluency in it is critical to the professional development of the engineer.

Synthesis of geometric forms develops out of visualization. Here we refer to the engineer's assembling in his mind a lever-and-fulcrum to push something, a train of gears or pulleys to reduce speed, a wiring or piping circuit to feed several loads, or a frame to support the device. Here, graphics is the means by which the idea first comes to form and is mentally tested for function. First, what shapes will do the job? Then how do they meet the various requirements? At this point, the visualization "gymnastics" mentioned above are of the greatest use. It is safe to say most of the engineer's new ideas come out as lines and symbols, not as words, and are subjected to test in their geometric forms.

Once the ideas are synthesized, they must be reduced to proper dimensions in order to fit where desired and perform their jobs adequately, or preferably, optimally.

Determination of the Dimensions

The next usage of graphics is establishing the *actual dimensions* by a layout. Pieces can be related together in the mind, but unless they can fit in the space available, the idea falls on its face. Thus, laying out the total assembly to scale on paper is a mandatory early step. The "layout" can be sketched by an engineer with scale and triangles, or drawn on a 15 by 20 in. desk-size drafting board with a drafting machine. Or it can be laid out by a professional drafting designer to the usual sharp-pencil accuracy of $\pm\frac{1}{64}$ in. A good "layout" illustration is that of the Gemini spacecraft in Appendix Figure B-5 where the astronaut, the cabin, and the controls are fitted together. Large objects are naturally laid out in special views to reduced scales, 2 to 1, 10 to 1, etc. The spacecraft layout is shown at about 20 to 1 scale. It cannot be emphasized too much, however, that much time can be wasted in conjecture about the fit of mating pieces, when a quick layout will settle the argument conclusively. This scaling or layout activity also brings to attention the forgotten corners, those incidental but necessary details that the designer had pushed temporarily to one side or overlooked. The total embodiment of the idea gets a good review at this time.

Computation by Graphical Methods

One of the valuable engineering uses of graphics is *computation.* This usage is suggested because frequently it is sufficiently accurate and *much faster* than other computational methods, and the engineer must be aware of diverse methods open to him. We list a variety of types below:

1. *Values of coordinates from empirical curves* (interpolation, extrapolation).
2. *Arithmetic proportions* (ratios, multiplication, and division).
3. *Trigonometric calculation* (sides and angles in triangles).
4. *Vector calculations* (addition, subtraction).
5. *Nomographic constructions* (alignment charts).
6. *Integration* (finding areas or summations).
7. *Differentiation* (finding slopes or rates of change).
8. *Curve fitting* (algebraic approximations).
9. *Solution of algebraic equations* (single or simultaneous equations).
10. *Optimization in geometric situations* (subjective optimization).

We illustrate these in several following figures. Coordinates and arithmetic ratios appear in Figure 11.6, trigonometry and vectors in Figure 11.7, and nomographs in Figure 11.8. The rest are in Figure 11.9, 10 and 11. At the end of this chapter we shall list various applicable reference books.

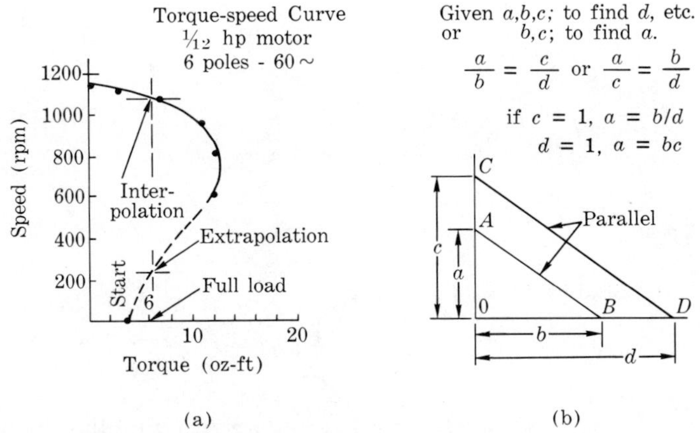

Fig. 11.6 (a) Interpolation and extrapolation on curves determined experimentally; (b) Using triangles for multiplying by fixed arithmetic ratios.

1. *Interpolation and extrapolation* are familiar to the readers. Figure 11.6a shows these applied to the curve drawn for measured data from a $\frac{1}{12}$ hp a-c motor. Note that the full load torque for rated horsepower output (approximately 6 oz-ft for $\frac{1}{12}$ hp) corresponds to a speed of 1080 rpm, as interpolated from the curve drawn in to match the data points. Similarly, the curve is extrapolated by the dashed lines so that we find that the speed is 250 rpm where the torque is again 6 oz-ft. Notice that graphic inter- and extra-polations are most appropriate on empirical curves, where calculation is not practical and the visual "fit" must suffice.

2. *Arithmetic ratios* $a/b = c/d$ and other equivalent operations can be done by similar triangles. If three values are given: a, b, c, they are simply laid out to scale on coordinate paper, and the line CD is drawn from C parallel to AB extending to the intersection at D. Multiplication and division can be done if line d is made equal to one unit ($a = bc$) or c is made one unit ($a = b/d$), respectively. These processes are occasionally useful in place of the slide rule, or on scales which take care of the decimal point.

3. *Trigonometric computations* can be done graphically, whether right-triangle or general-triangle. The two-sides-and-an-angle or two-angles-and-a-side are laid out to the given values, and the unknown sides or angles are scaled from the drawing. The given values are shown in heavy lines in Figure 11.7a.

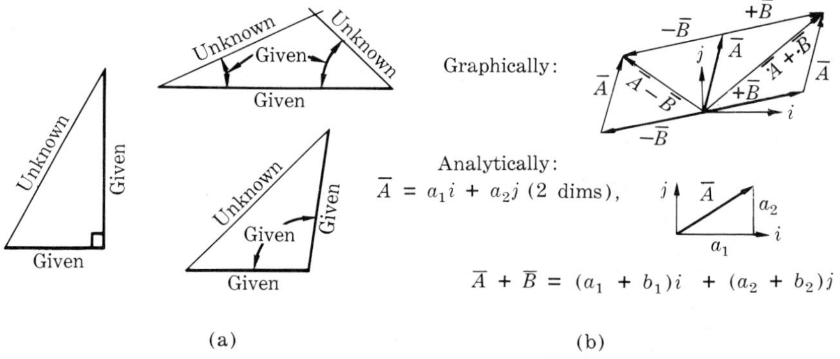

Fig. 11.7 (a) Graphical solution to trigonometric problems; (b) Vector addition and subtraction. (See Appendix E for products of vectors.)

4. Graphical *addition and subtraction of vectors* are familiar. These are shown in Figure 11.7b in two dimensions for simplicity. Also, see Appendix E-19 for complete reference details on addition, subtraction, and multiplication of vectors.

5. *Nomograms*, or alignment charts, are linear scales which have the property of solving simple equations (or certain types of complex ones). Nomograms are especially useful for repeated problems within a given range of values of the variables. These examples are shown in Figure 11.8:

$$C = A + 3B \quad \text{(simple sum-product)}$$

$$V = \pi r^2 h \quad \text{(volume of cylinder)}$$

$$\frac{1}{c} = \frac{1}{a} + \frac{1}{b} \quad \text{(sum of reciprocals)}$$

In each case, the solution is found on one scale (C, V, or c scale) when a straightedge is laid upon the respective scale points representing the

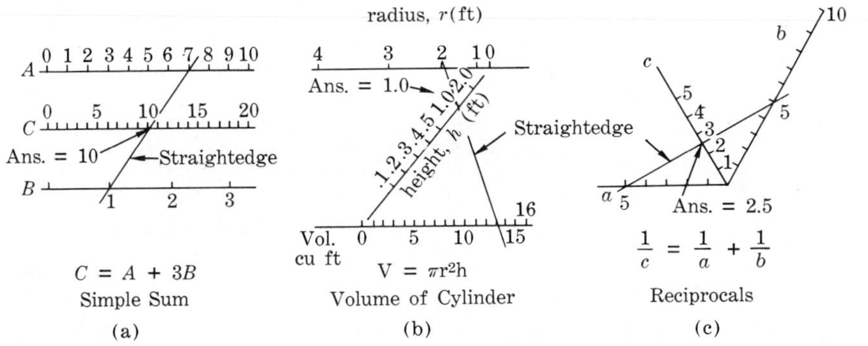

Fig. 11.8 Solution of equations by nonograms or alignment charts.

other two numbers. Of course, given any two of the variables, the value of the third can be ascertained.

6. *Graphical integration* is shown in several examples in Figure 11.9: three using the parabola $y = x^2$, and one using the curve of torque vs. angle for an automatic clotheswashing machine.

In the first case the integration simply produces the area under the curve; in the second case, it gives the amount of energy required per half cycle of the agitator shaft, and eventually the motor power required.

Integration is the summation of unit-width blocks whose variant edges approximate the given curve (see Figure 11.9b,c,d). These blocks may be trapezoids or rectangles, and may connect computed or estimated points. Trapezoids or rectangles yield greatest accuracy when the edge is drawn so that the excess area beyond the curve visually equals the incomplete coverage inside the curve (see Figure 11.9d,e where the "+triangles" equal the "−triangles"). Even if trapezoids are made by simply connecting adjacent

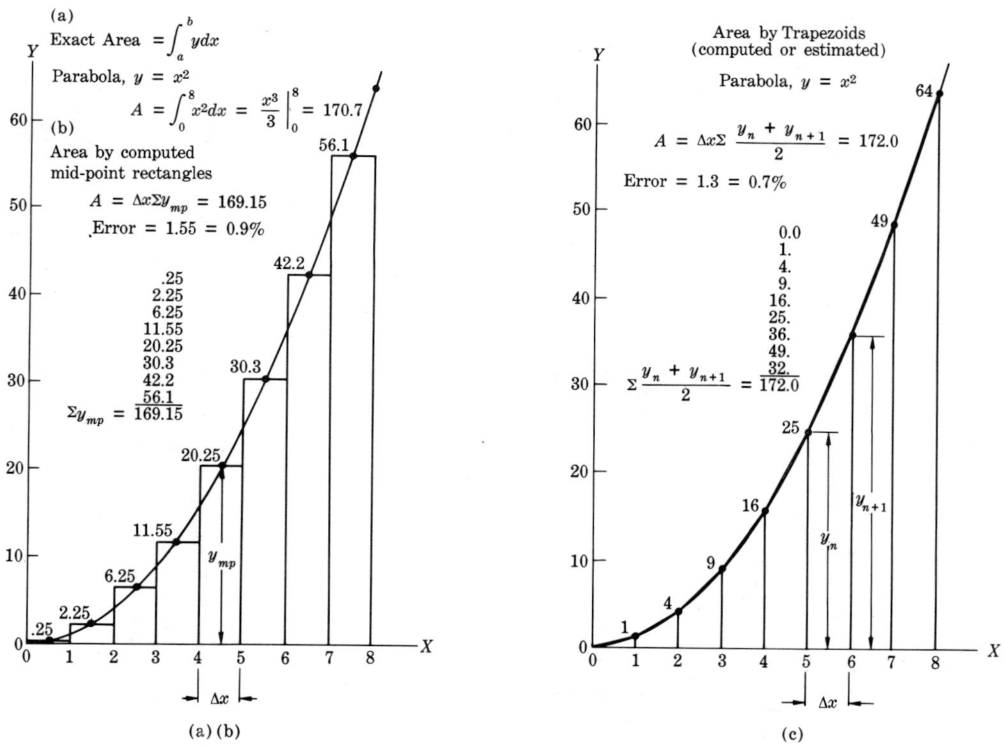

Fig. 11.9 Graphical integration. (a) Exact area under parabola using calculus; (b) Area under parabola $y = x^2$, by rectangles at computed midpoint value of x_i,

$$\int_0^8 y\,dx = \Delta x(y_{01} + y_{12} + y_{23} + y_{34} + y_{45} + y_{56} + y_{67} + y_{78}$$

(c) Area under parabola $y = x^2$, by trapezoids, computed at integral values of x_i,

$$\int_0^8 y\,dx = \Delta x(y_0/2 + y_1 + y_2 + y_3 + y_4 + y_5 + y_6 + y_7 + y_8/2)$$

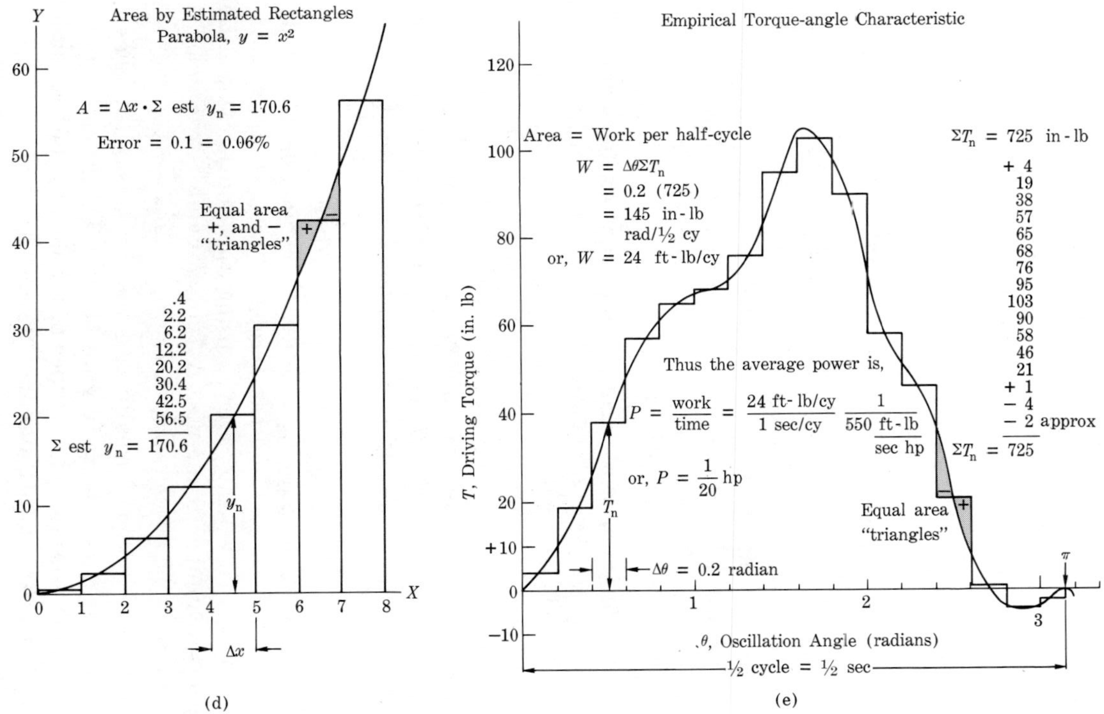

Fig. 11.9 Graphical integration (continued). (d) Area under parabola $y = x^2$, by rectangles whose heights have been chosen such that the + and − "triangles" have equal areas. (e) Computation of work and power by the area under a curve, with the example of a measured torque-angle characteristic of a clothes washing machine agitator-shaft. Measurements were taken from high-speed motion pictures. Net area under the curve is work per half cycle.

ordinates, the error is still surprisingly small. In the curve shown, $y = x^2$, the trapezoidal sum is 172.0 units, only 0.7 per cent higher than the theoretical value of 170.7 units. If, as in Figure 11.9d, rectangles are drawn through the respective ordinates so as to visually equate the outside and inside triangles, the accuracy is further increased to 0.06 per cent (sum = 170.6 units). This method is surprisingly accurate due to the cancellation of plus and minus errors. And it is the simplest because each rectangle, having unit width, is represented by one number, its height. Of course Simpson's rule* for irregular areas can be applied. However, it requires more arithmetic than our previous simple summations and benefits only by slightly greater accuracy.

From the second illustration, Figure 11.9e, energy per half cycle for a pulsating load and the consequent motor driving power can both be calculated. In this case, we find the load on the machine is $\frac{1}{20}$ hp, a light load for the usually provided $\frac{1}{4}$-hp motor. Notice, incidentally, that negative work is taken into account conveniently by the algebraic sign of the few ordinates.

7. *Graphical differentiation* is a slightly more complex technique. By definition, the derivative dy/dx of a function $f(x)$ is the slope of the curve at the given point. This slope can be approximated by drawing tangential lines at the desired points and measuring the ordinate for a unit abscissa; i.e., along the tangent line, measure Δy, for $\Delta x = 1$.

In Figure 11.10a this has been done for $y = x^2$, at the respective points $x = 0, 1, 2, 3, 4, 5$. The values of the slopes are shown plotted on the accompanying graph (b) and the resulting straight line checks the theoretical slope-curve, $dy/dx = 2x$.

In Figure 11.10c the velocity curve of a 6000-hp diesel-electric locomotive and train is shown under starting conditions. The acceleration is the derivative of velocity ($a = dv/dt$) and the acceleration curve (d) is plotted correspondingly below the velocity curve, using the above tangent-line method to measure a number of points. A typical value is expanded at $t = 4$ min where we find the acceleration is 7.0 mph/min.

8. *Curve-fitting* can be done visually (graphically), using families of curves such as those shown for $y = x^n$ (Figure 11.11a). This group depicts a wide range of values for n. Many other types of families (including log-log plots) can be found in mathematical handbooks, but the above exponential $y = x^n$ is generally useful. The given points can be plotted to the same 1:1 scale on a transparent sheet and overlaid to discover any whole or partial sections which match.

* See any elementary calculus text, or "Standard Mathematical Tables," p. 373, Chemical Rubber Publishing Co., Cleveland, Ohio, 1957.

9. *Solution of algebraic equations* is done graphically by finding intersections. In the case of single equations, the intersections occur at $f(x) = 0$, so that x_1, x_2 . . . are found by plotting $f(x)$ in the region of the X-axis crossing points. See Figure 11.11b.

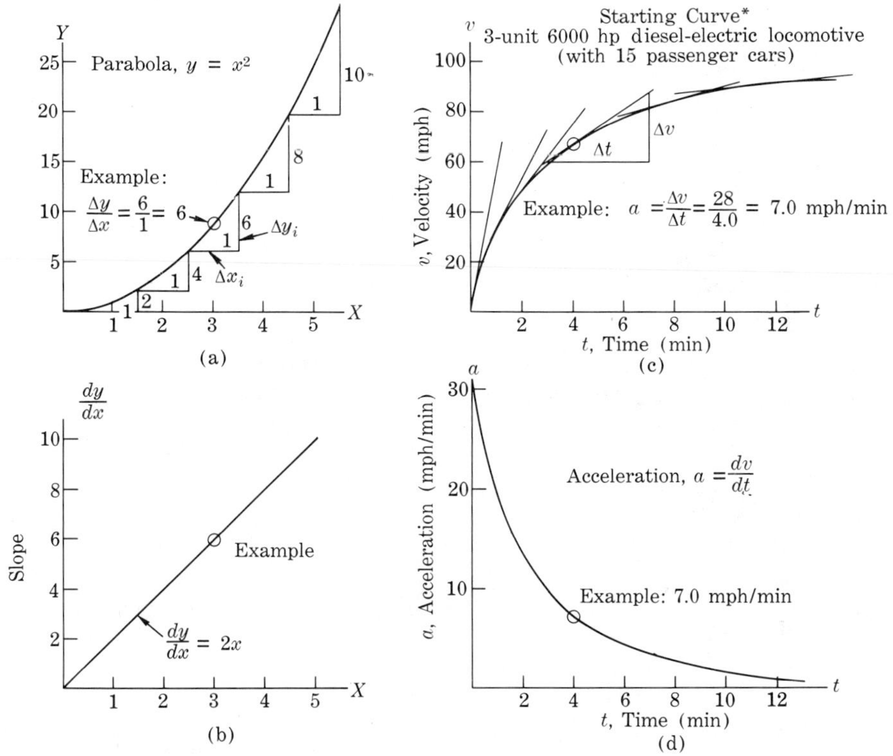

Fig. 11.10 (a, b) Slope of parabola by triangles whose hypotenuses are tangent at the given points; (c, d) Determining acceleration from a velocity curve. Diesel-electric locomotive starting characteristic. (*From "Marks Mechanical Engineer's Handbook," 6th ed., McGraw-Hill Book Company, New York, 1964.)

Simultaneous equations are solved by finding the intersections of two curves. See Figure 11.11c. For straight lines or curves, we have solutions wherever they cross; and the crossing region can be plotted to enlarged scales for accuracy.

The solution, of course, is at the point (x_1, y_1) which satisfies both of the equations.

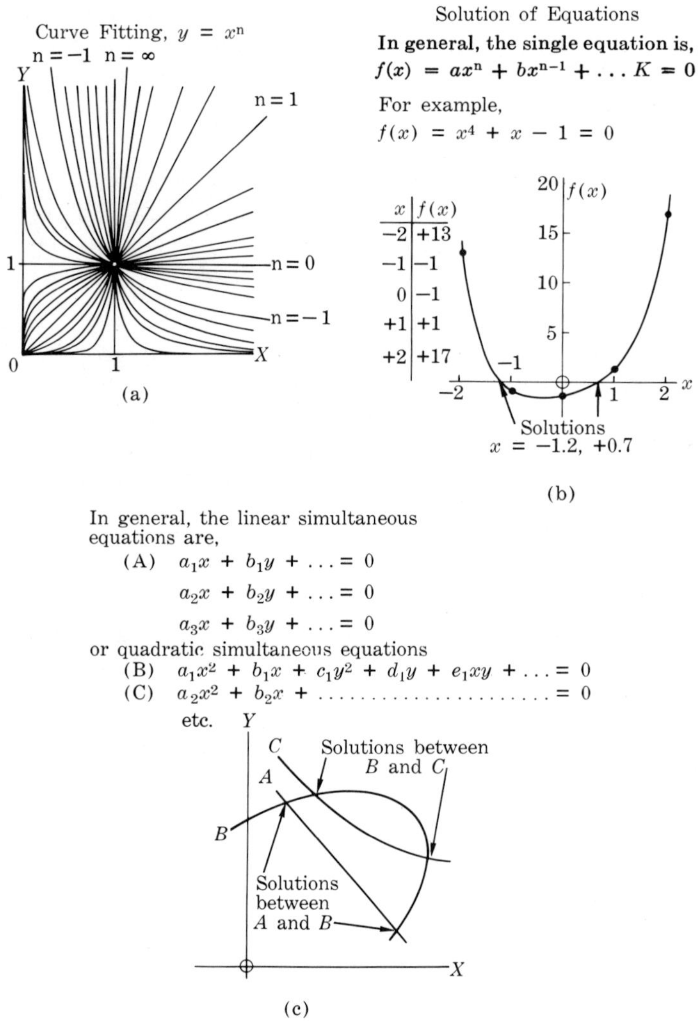

Fig. 11.11 (a) Curve fitting by the characteristic form, $y = x^n$; (b, c) Solution to equations, single or simultaneous, by finding intersections graphically.

10. *Optimization* will be treated more extensively in Chapter 15, but much optimization is subjective and done visually. The parking lot illustration of Chapter 7 was a simple example. Remember that all constructive sketching includes some degree of graphic optimization.

We shall leave graphic computation now, with the recommendation to use it when it best serves the purpose; and we shall now proceed with the remaining engineering uses of graphics.

Showing Relation and Emphasis

Graphics can help the engineer to *relate* components and *emphasize* the relative importances or sequences for given purposes. The engineering organization diagrams shown in Figures 1.2, 1.3 illustrate relationships of individuals, groups, and sections. Solid lines show authority going down and responsibility coming back. Dashed lines between individuals should show advisory status. Relationships are also shown in flow and circuit diagrams* and in exploded drawings,† which show the sequence of assembly. Phantom drawings show a given part in heavy lines with all the others generally faded down to a light gray to de-emphasize them.

Changes in viewing angle also allow the engineer to relate desired parts to each other or to emphasize others. When we swing around or move away from an object, the view changes, that is, our visual image changes. Notice, however, that only certain aspects of this image change, others do not. As an aid to this "change of view" sketching, we list below the "invariants" and "variants" which pertain to changing views, such as occur in going from orthogonal to isometric or to perspective views.

Invariants	*Variants*
(These do not change with viewpoint.)	(These do change.)
Straightness of lines. (St)	Circles (Circ) and ellipses (Ell) (vary and interchange).
Tangency points. (T)	Angles ($<\alpha$) (increase, decrease).
Center line and concentricity. (℄)	Lengths (L, a, b) (increase, decrease).
Symmetry. (Sym)	Ratios of lengths (a/b) (vary).
Piercing points. (Pp)	Tangent perpendicularity to radius ($R\perp$) (not true).
Flatness of planes. (Pl)	Areas (decrease in perspective).
	Shape of envelopes (Env) (squares become rhombi or trapezoids).

Figure 11.12 has been drawn to illustrate some of these elements. The individual items have been given symbols to allow the reader to compare them. This list serves primarily as a checklist of concepts and principles to remember while reviewing a sketch.

* For example, Appendix Figure B.11, Waste water.
† See Figure 10.1b, Exploded drawing.

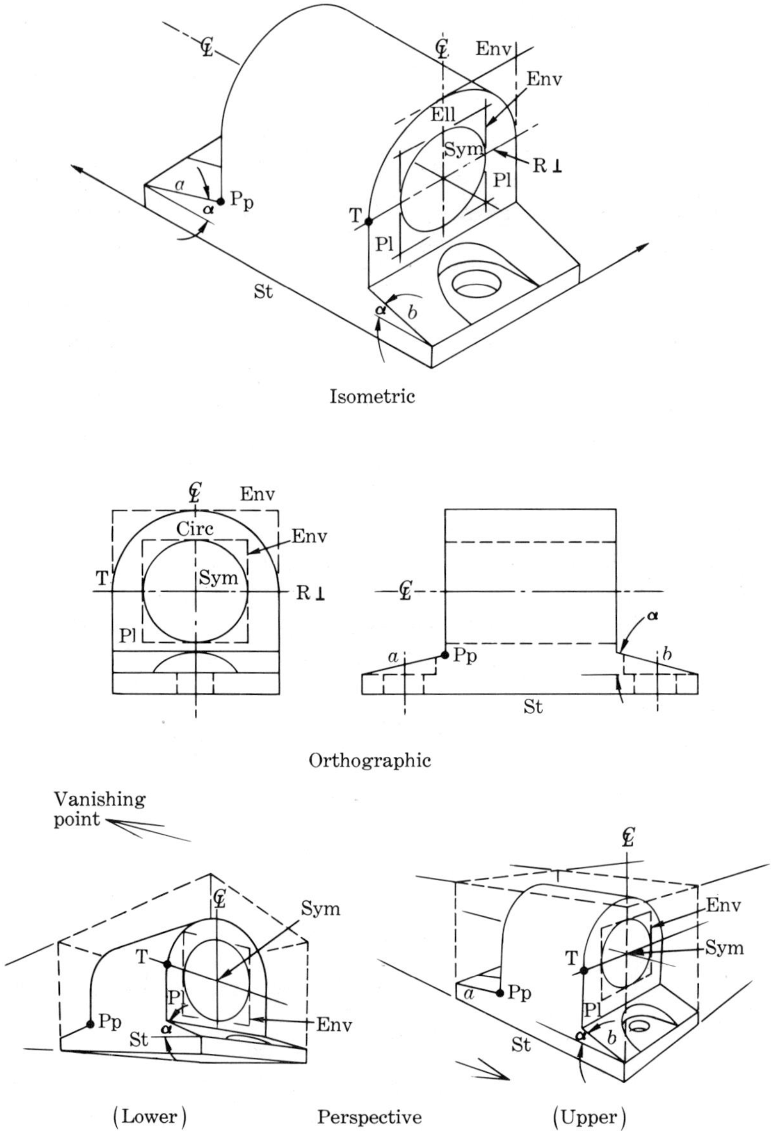

Fig. 11.12 Demonstration of variants and invariants in effect as the viewing point changes. (Orthographic, isometric, and perspective projections. Although not shown, oblique projections follow the same principles.)

Persuasion and Selling

Graphics is important in selling ideas and products. A more accurate than usual translation of the favorite Chinese proverb is, "One showing is worth a hundred sayings."[2] The simple half-hitch knot illustrates this dramatically. Which alternative would you choose: Demonstrating with a piece of string, or describing the hand motions solely by words?

Visual aids in the form of graphs, diagrams, working models, and drawings all are much more persuasive than the words alone. Part of this rests on basic psychology and reasoning: The listener is more likely to be convinced when the speaker shows accomplishment of *detailed progress*, beyond what might otherwise be interpreted as glib sales-talk. The graphics of prepared charts and other visual aids also makes any presentation more interesting and more easily remembered, in addition to providing the reassuring ready-made prompter. We shall discuss oral communication more thoroughly in Chapter 18.

For any presentation, remember that the pictures, the setting, and the people in the chairs are all much more indelible in the memory than are the words spoken.

Recording

As our final use of graphics, we come fittingly to the longest-lasting use. Our ideas, our designs and specifications must be carried from office to factory or construction site; and they must be maintained from year to year. No small effort is put into making drawings *permanent* and *reproducible*. The paper, the reproduction methods, the fireproof files, the elaborate index and numbering systems all attest to the importance of good records. Repeated use of drawings in the factory and the changes that inevitably accompany any continued manufacture all leave their mark by a stream of "alteration notices" flowing through the engineering department. It is not unusual to see lists of "change" entries filling up the side of a drawing, in columns 5, 10, or 20 items long. Burdensome as such changes are, they are the necessary concomitants of the practical world of errors, substitutions, economies, new vendors, improved processes, added tooling operations, and changed tolerances.

Records for patent purposes (establishing dates, inventorship, and coverage) were briefly discussed in Chapter 6; these records depend heavily upon graphics. The importance of graphics is recognized by the development of a special branch called Patent Drafting, with its own standards, conventions, and symbols.

Before leaving the "uses," we shall mention Appendix E, where a helpful digest of instructions is reprinted from a recently published problem-set on "Engineering Graphics and Design."[3] These worksheets provide an excellent reference to use while graphics problems are under study or when a design project needs graphics support. Also, an auto-instructional text[4] has been prepared to allow the individual to prepare or refresh himself in a number of graphics subjects: Sketching, Orthographics, Isometrics, and Descriptive Geometry with applications of Analytic Geometry and Vectors.

Iconic Models: Scaling

Uniform proportionality: The first type of model referred to in Chapter 10 was the Iconic Model, which "looked like" the original or the reality. The characteristic that makes a representation "look like" the original is *scaling*, or the making proportionate of all the features of the outline and detail parts. If the representation was 1/4 scale, *every dimension* was reduced to 1/4 of its original value. This applies to all the called-out dimensions as well as all other possible dimension measurements. This is simple as long as the only concern is length; every possible dimension has its proportional new value: The old one multiplied by the scale ratio.

However, whenever some linear dimensions are multiplied by others, the resulting model equivalent is no longer simply proportional. As a first instance, notice that the areas are not proportional to r, the scale ratio, but to r^2. If the scale ratio of lengths if 1/5, then the areas are 1/25; or if 10/1 in length, then 100/1 in area. All the "area-sensitive" variables change the same way, also:

Force	(= pressure × area)
Pressure	(= force × 1/area)
Strength	(= stress × area)
Heat transfer	(= rate × area)

Similarly, the volume changes with r^3. If lengths are 1:5, areas are 1:25, and volume 1:125. The "volume-sensitive" variables are:

Weight	(= density × volume)
Mass	(= weight/g)
Material cost	(= cost/lb × weight)
Buoyancy force	(= weight of liquid displaced less weight of object)

Thus a 1/100 scale model of the 1000-ft dam referred to earlier would be 10 ft long (r), have 1/10,000 the area for pressure of water on its upstream face (r^2), and weigh 1/1,000,000 as much (r^3).

Nonuniform proportion. To exaggerate an otherwise minor effect, the scale in one direction (vertical, for example), can be greater than in other directions (horizontal). This is usually done for profiles of long lengths of roads, canals, railroads, etc. Many Rocky Mountain area railroad timetables have a vertically exaggerated scale on the picture profile of the track elevations reached. These are used for pictorial purposes only, because computations with different length and width scale ratios are unnecessarily complex.

One final comment on this type of model: A change of scale "downward" allows large objects or large areas to be brought within manageable and

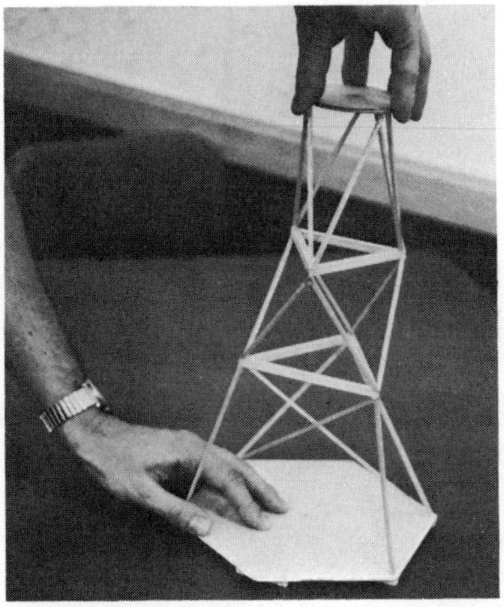

Fig. 11.13 Three-dimensional iconic model. Balsa wood scale model of Mariner I superstructure. (Courtesy Jet Propulsion Laboratories, California Institute of Technology)

comprehendible sizes. Fortunately, the eyes can see very fine detail, so that much is still saved even in gross reductions.

Contrarily, the finer aspects can be amplified as much as desired to establish clearly the desired shapes. The micro-circuits of new miniature electronics are drawn much expanded (10 × size, for instance), and photographically reduced for manufacture.

Three-dimensional iconic models. Another aid both to visualization and to strength- and stiffness-analysis is the miniature, the balsa-wood or cardboard representation. The design of the superstructure of the Mariner spacecraft was first tried out as a balsa model by Jet Propulsion Laboratory engineers (see Figure 11.13). This was more than a "look like" model; its stiffness made it an "acts like" or analog model.

However, this example reminds us of the whole field of pocketknife carving, using not only balsa but clay or soap. It also brings to mind models using wires, string, paper, cardboard, sheet plastic, and cement. These are definite practical aids, as we have seen in the case of Mariner, not disdained by practicing engineers, and more than a little recommended to the student engineer. It may take a little courage to put together and experiment with a cardboard and toothpick mock-up; but it practically always is worth the effort.

Abstract graphics problems, *per se*, are much more easily solved with carved clay miniatures, which can be picked up and rotated in front of the eyes. Use straightened paper clips for lines and make grease-pencil projections on clear plastic viewing planes.

Only be sure that the scaling is reasonably faithful.

Skills Necessary

Part of the skills of graphics are mental, in the form of concepts and problem-solving abilities. The rest of the skills are primarily muscular, representing coordination between mind and finger, forcing the pencil along the intended path.

Coordination, as any artist, musician or athlete will testify, comes only with long and intensive practice. The hand and fingers just do not obey the mind and eye until they have tried, erred, re-tried, and corrected endless times. In our case of the young engineer, he too must learn new languages like the telegrapher and the violinist, and, like them, must develop his dexterity and accuracy. Part of this required dexterity is essential to adequate "sketching communication," and part is simply demanded for the pursuit of excellence. Even a poor sketch, with apologies, can be used to convey an idea, when accompanied by words and pieces of hardware and the waving of arms. But after the sketcher has gone, uncertainties and incompleteness may remain to leave the listener with guesses and questions. As an engineer's career progresses, he really has no excuse for poor sketching.

The strongest motivation for improvement in either the mental or muscular graphics skills usually occurs, quite naturally, when the project problem is beyond the depth of the young engineer, or when he looks at the quality of his sketching and drafting and realizes it is "pretty sad." Also, naturally, the instructor or supervisor will not be reticent in pointing out the deficiency and suggesting remedies.

The tools of engineering sketching are deceptively simple. In their basic form, they need only be a medium pencil (HB to 2H), a soft eraser, and paper. The softer pencil (HB) is used for shading. A small drawing board such as referred to above, plus scales, compass, triangles, and handbooks are the next most useful tools.

But whether a sketch is strictly free-hand or made with a straightedge, the final effectiveness results from the combined mental and muscular skills of the engineer.

A variety of graphics textbooks are listed at the end of this chapter. Whether used in mechanical drawing courses, or as extracurricular references, such books provide ample opportunity for exercise. Although each book has some standard reference data tables and symbols, there has been published a special book[5] bringing together in one place the pertinent work of the American Standards Association (ASA) and other groups.

The alphabet of line and usage has been summarized on the first page of our Appendix E, chart E.1.

The communication aspects of graphics are well explained and related in the latter pages of a fresh new book, "Engineering Communications."[6] This work includes good expositions on engineering uses of pictorial drawing and perspective.

SUMMARY

In this chapter we have established the wide occurrence and the essential nature of all forms of graphic communication in engineering. Note that graphics constitutes a language, a shorthand code of lines and symbols to communicate efficiently with the engineer's associates. As with any language, the symbols and usage must be learned, and they must be used habitually if they are to be kept available and fluent. There are many uses of these tools of graphics and it is simply up to the engineer to pick the one most suitable for his specific need. For a given problem, the engineer must choose between the computer, the slide rule, the table of integrals, a pencil and straightedge, a balsa model, or a laboratory test. Each tool has its place and time.

EXERCISES

Sketching

11.1 Make four free-hand sketches of a 2-in. cube with views progressively ranging from nearly front face to full diagonal view. Use oblique projection.

11.2 Repeat Exercise 1 above using "two-vanishing-point" perspective. See sketch (2) in Figure 11.1.

11.3 Make an isometric sketch of a square nut for a 1-8UNC screw; also of a hex nut for the same screw.

11.4 Make an isometric sketch of a sphere resting on a cylinder of the same diameter and of a height equal to that diameter. Show how you obtained the height of the sphere's visual (apparent) center.

11.5 Draw a free-hand perspective sketch to $\frac{1}{10}$ scale of an elliptically-cross-sectioned automobile gasoline tank 30 in. long. The major axis of the ellipse is 17.5 in. and the minor axis is 6 in.

Computation

11.6 By graphical measurement find what height on a building wall can be reached by the end of a 90-ft power ladder mounted on a fire engine. The center-line of the truck is in the street 20 ft away from the building and the base of the ladder is 7 ft above the pavement.

11.7 In Problem 6, how high up the building will the ladder reach if the ladder, which can pivot about a vertical axis through its base, is turned through a horizontal angle of 45°? If the fire engine tips slightly on its springs due to the overhanging weight, will this produce much or little effect on the contact height up the side of the building?

11.8 Consider the wind pressure of 1000 lb concentrated three quarters of the way up a 350-ft vertical radio mast and in line with one of the three guy wires, all of which are attached to the mast two thirds of the way up. The guy wires are anchored in the ground at 120° horizontal spacings around the mast 125 ft radially out from the base. What is the increment of force on each wire due to the wind? What would those forces be if the wind direction were exactly opposite?

11.9 Use vectors to determine the ground speed of an airplane flying from Chicago to New Orleans on a course of S40° W (i.e., 220°, with north as 0°) at 610 mph air speed in a jet-stream wind of 100 mph, directed E20° N.

11.10 Construct, by adjacent numbers along a line, a conversion scale for degrees Centigrade (−40° to +120°C) to degrees Fahrenheit.

11.11 Construct a Z-scale nomogram (Figure 11.8b) to read miles per gallon from one scale of miles (0–1000) and another scale of gallons (0–100).

11.12 Integrate the area under a sine curve, from 0 to π, using the summation-of-rectangles approximation. Check by computing the theoretical integral.

11.13 Find graphically the real (not imaginary) solutions to the equation $x^4 + 2x^2 - 5x + 1 = 0$.

11.14 Two ships are sailing courses described by $6x - y = 5$ and $\frac{1}{4}x - y = -2$. Find the coordinates of the intersection of their paths graphically.

Models

11.15 Make $\frac{1}{15}$ scale cardboard models of a 10-in. I beam representing several lengths: One each 5 ft, $7\frac{1}{2}$ ft, 10 ft, respectively. Use the same weight cardboard for all parts and join the flange and web with "airplane cement." If the parts have been made uniformly and carefully, find, by experiment, the maximum concentrated center load each length will carry before buckling, when simply supported at the ends. Use the adjacent shapes, cementing the center web pieces together in the flat before adding the outer flange plates.

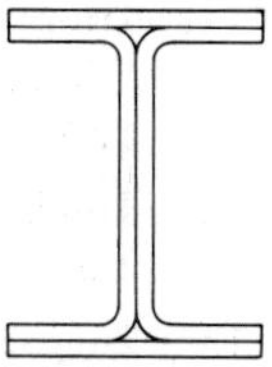

11.16 Carve a clay or soap replica of the cylinder-cone intersecting solids shown as sketch (3), Figure 11.1.

11.17 Using thumb-tacks, paper clip wire and cardboard, make an operating model of a single-claw film advancing device used in movie projectors.

Drawings, Diagrams

11.18 Make an organization diagram of a specific engineering department as directed by the instructor. (Industry, university, consulting firm, etc.)

11.19 Make an exploded view of a ballpoint pen, showing assembly relations.

11.20 Draw schematically the electric circuit of the ignition system (only) of a stock automobile.

REFERENCES

1. Young, J. Z.: "Doubt and Certainty in Science," p. 67, Oxford University Press, Inc., New York, 1951.
2. Watts, Alan: "The Way of Zen," Pantheon Books, Inc., New York, 1959.
3. Hill, P. H.: "Engineering Graphics and Design," Worksheets, The Macmillan Company, New York, 1964.
4. McKee, R. B.: "Graphics Auto-instructional Text," John Wiley and Sons, Inc., New York, 1966.
5. Arnell, Alvin: "Standard Graphic Symbols," McGraw-Hill Book Company, New York, 1963.
6. Rosenstein, Allen B., R. R. Rathbone, W. F. Schneerer: "Engineering Communications," Prentice-Hall, Inc., Englewood Cliffs, N.J., 1964.
7. Levens, A. S.: "Graphics with Introduction to Conceptual Design," John Wiley & Sons, Inc., New York, 1962.
8. Rule, J. F. and S. A. Coons: "Graphics," McGraw-Hill Book Company, New York, 1962.
9. French, T. E. and C. J. Vierck: "Fundamentals of Engineering Drawing," 9th ed., McGraw-Hill Book Company, New York, 1960.
10. Paré, E., L. Francis, J. Kimbrell: "Introduction to Engineering Design," Holt, Rinehart & Winston, Inc., New York, 1963.

11. Hammond, R. H., C. P. Buck, W. B. Rogers, G. W. Walsh, H. P. Ackert: "Engineering Graphics for Design and Analysis," The Ronald Press Company, 1964.

12. Hoelscher, R. P. and C. H. Springer: "Engineering Drawing and Geometry," John Wiley & Sons, Inc., New York, 1962.

13. Zozzora, F.: "Engineering Drawing," 2nd ed., McGraw-Hill Book Company, New York, 1958.

14. Springer, R. D., L. G. Palmer, W. A. Kleinhenz, P. W. Bullen: "Basic Graphics," Allyn & Bacon, Inc., Boston, 1963.

15. Giesecke, F. E., A. Mitchell, H. C. Spencer: "Technical Drawing," 4th ed., The Macmillan Company, New York, 1958.

16. Mochel, M. G.: "Fundamentals of Engineering Graphics," Prentice-Hall, Inc., Englewood Cliffs, N.J., 1960.

17. Giachino, J. W. and H. J. Beukema: "Engineering-Technical Drafting and Graphics," American Technical Society, Chicago, 1961.

18. Luzadder, W. J.: "Basic Graphics," Prentice-Hall, Inc., Englewood Cliffs, N.J., 1962.

19. Smith, R. J.: "Engineering as a Career," 2nd ed., chap. 14, McGraw-Hill Book Company, New York, 1962.

Twelve

Conceptual Representation: Symbolic Models and Modeling

"Logical consequences are the scarecrows of fools and the beacons of wise men."
Thomas Hardy

Introspection

We all agree that symbolic modeling, or mathematical representation, is important to the engineer. Our education, our experience and intuition all reinforce this feeling; even though we are not sure exactly what the symbolic model will be or how it can specifically next help us as individual engineers.

We have alluded to models and modeling in earlier chapters, and there we developed a few simple examples. Here we will concentrate on symbolic modeling, emphasizing the aspects of synthesis and decision rather than the more frequently presented techniques for analysis. Of course, some interplay of synthesis and analysis is going on at all stages of design, and we urge the engineer to arm himself to do battle with either. Actually, synthesis depends more upon experience and judgment than upon formal education, because synthesis is less regulated by rules, less understood, and less predictable in outcome. But let's set about getting some of that experience right now.

System Description

When an engineer tries to describe a system, before, during, or after construction, he uses both quantitative and qualitative relationships. He writes down as many of the former as he can, then he finishes up as far as he can go with the latter. Numerical statements have the double advantage of 1) manipulation by mathematical rules and 2) finer precision in conclusion and outcomes. Qualitative statements can certainly be made in degree and by comparison (i.e., "very safe" or "a belt is safer than a chain") but they suffer in precision, measurability, and universality, and they are semantic booby-traps; the reader *thinks* he knows what the qualitative statement means.

One way of visualizing the quantitative versus qualitative descriptions is by the symbolic statement:

$$\text{System behavior} = (Q_1 + Q_2 + \cdots) + (\Delta_1 + \Delta_2 + \cdots)$$

where Q_n is a quantitative relationship and Δ_n is the residual, qualitative relationship.

For instance, we describe the behavior of a materials handling system in terms of tons per hour, kilowatts of power, feet of length, years of life, dollars of cost—all numerical relations. But we also must relate safety, maintainability, appearance, legality of code compliance, public nuisance of dust or noise—primarily qualitative relations. We add more and more statements of both kinds and we approach the actual system as a limit. Thus, our description becomes complete in the limit.

$$\sum_{n \to \infty} (Q_n + \Delta_n) = \text{system behavior}$$

where $\Sigma\, Q_n$ constitutes the quantifiable or scientific content and $\Sigma\, \Delta_n$ constitutes the not-yet- or not-at-all quantifiable, human content.

Both sets of these statements require engineering "art" or professional skill, beyond the bounds of science, to appreciate and to compose them. The very words used above suggest the breadth and sensitivity called for in the engineer: "safety-appearance-legality-nuisance."

Since our "*n*" above never becomes very large, much less approaching infinity, our thought models are always very limited idealizations of a few of the major manifestations of a system. Our science advances, however, and with that progress, a few of the Δ descriptions move over into the Q category.* But the intricacy of our value systems and the difficulty of such research make these transfers sporadic and infrequent.

* For instance, scheduling development and construction now with PERT and CPM. (*P*rogram *E*valuation and *R*eview *T*echnique and *C*ritical *P*ath *M*ethod.)

Further examples of the Q and Δ categories range from the space frontiers to the daily commonplace. Consider space rockets first.

Q-items	*Δ-items*
Propellant thrust.	Political advantage of being on the moon first.
Combustion temperature.	At what system reliability do we trust a man in it?
Structural forces.	Effect of prolonged weightlessness upon man.
Orbital velocities, etc.	Psychology of interplanetary trip idleness, etc.

Some topics, like the effect of micrometeorites on components, may not be known and hence presently thought to be qualitative, but they are susceptible to quantification (probabilistically) and are really Q-items.

In the realm of the commonplace, consider urban planning.

Q-items	*Δ-items*
Population densities.	Should the state support low-income school districts?
Racial distribution.	Should gasoline taxes be used to subsidize public transportation?
Transportation capabilities.	How many acres of park per square mile of city?
Tax revenues and sources.	In which direction will the city grow?

Whereas many of these are not direct engineering questions, they all certainly bear on engineering systems and typify the ramifications the engineer must be prepared to deal with and accept.

Thus, we illustrate verbally a little of the range and variety of aspects that make up whole systems. Naturally, only very large elaborate models can hope to represent such large systems as the rockets and cities mentioned above, but upon demand such models have been synthesized and used in analysis. Whenever a model is synthesized, the engineer puts all he can in quantifiable form, and he reserves the qualitative information as critic and modifier of the model's output. The numerical results of the model are practically never accepted verbatim; they are shifted in some degree or direction, as dictated by qualitative criteria, since the project objectives, not the model, are ruling. Appendix H discusses the model of a complex system, the supersonic transport aircraft.

We shall proceed with an example in a moment, after refreshing our memories on the reasons for a symbolic model and the "laws" which the engineer uses to assemble such a model. A symbolic model is used because,

it provides the maximum generality in attacking problems,
it simplifies complex problems into smaller ones,
it uses existing analytical, mathematical, or logical methods,
it produces numerical results.

The last two reasons remind us of the ready-made answers awaiting us from the relevant mathematical or "logical laws" of algebra, trigonometry, and calculus, as compared with the empirical answers from "physical laws."

Physical vs. Logical Laws

The engineer most frequently works with "physical laws." Such laws are statements of relations between variables and parameters extracted from a phenomenon, these relations being more or less accurate for a range of values. Newton's first law is an example of this concept: *Acceleration of a body is directly proportional to the applied force* ($a = F/m$). This relation has been tested by astronomers, scientists, and engineers over 200 years, by experiments of great precision, and both this law and its consequences have been found to be "true" over wide ranges of values. How precisely it holds at the subatomic levels of the nucleus or at the astronomic scale of galaxies has yet to be resolved.

There are other "laws" men have identified which apply over different ranges of values. The law of static friction, for instance, $F = f \times N$ (tangential force F equals the product of the friction coefficient f by the normal force N), holds only for a narrow range of applications, and not for waxes or very flat surfaces. The Stefan-Boltzmann law of energy radiation (heat radiated varies as absolute temperature to the fourth power) holds only for black bodies, certain surface conditions, and certain temperature ranges.

In all these cases the "laws" are verbal statements of how several variables relate to each other. They may be deduced from fundamental principles (or hypotheses) or they may be simply fitted curves chosen and limited so as to describe the observed phenomena reasonably well.

However, there is one other kind of "law" the engineer meets, and this is the *constructed law* of various kinds of logic,* mathematical and other. It is possible to imagine a perfect circle without having seen one, or a perfectly straight line, or a polygon with a hundred equal sides. Although one cannot truly imagine a space of more than three dimensions, he still can use symbols for such a space analogous to those he uses in the familiar space of three dimensions.

This leads us to the "perfect" laws of geometry, of numbers, of imaginary numbers, of binary logic, or statistics, relativity, etc. These laws are perfect because we make them perfect. We define the sine of an angle, the square root of a number, the shape of a parabola, and so on. We set up a

* Logic: (from the Greek, *logos*, word, speech, reason). According to Webster, "The science that deals with the . . . criteria of validity in thought and demonstration; the science of the formal principles of reasoning." ("Webster's New Collegiate Dictionary," 2nd. ed., G. & C. Merriam Company, Springfield, Mass.)

logic based on a consistent set of assumptions* and then develop the "logical" conclusions. Many such types of man-made logic exist not only in mathematics but in other fields as well. The usefulness of these idealized laws lies in their approximation to observed reality. Lines, spheres, hexagons, ellipses, number series, and statistics all approximate what we see in nature. We use the idealizations when they are helpful, or when combinations of them describe with fair accuracy a phenomenon in which we are interested. But we must always remember to keep the phenomena separated from our attempts to describe them. "The map is not the territory." Our only connection with real phenomena is through observation and experiment—never by direct "personal" experience. We can experience only our own feelings and senses.

Notice that all the preceding laws provide operating relationships by which we can deduce any desired specific case. This is true whether the "laws" were derived from physical origins or conceived as pure abstractions. For example, from the physical relationships of Hooke's law on elasticity,

$$F = k \times d$$

or, the force on a spring = a constant multiplied by the extension of the spring, with k as the constant and d the extension. We can thus (presumably) compute algebraically the force for *any* desired displacement, large or small, positive or negative.

Or, from the geometric law of rectangular area, Area = base × altitude, ($A = b \times a$), we are able to compute the area for *any* base-altitude combination. The answers to *all* cases lie waiting in the statement of that law.

The words and ideas that precede the symbolic expression constitute the essential contribution, even though we as engineers normally go immediately to the symbols and consider them alone.

Laws and Implicit Answers

The important fact about all these logical laws, however, is that all possible answers to any question of analysis about the given system are implicit in the statement of the law. Referring again to the rectangle law, $A = b \times a$, we see nothing spectacular about this statement because we completely comprehend it. We have a "feel" for all the variations: the effect of equal sides, of zero or very large sides, of all the aspects and limitations of the various alternatives in between. We can compute a specific case in

* For example, the "logic" of geometry is based upon the axioms of Euclid; the "logic" of differential calculus is based upon the theory of limits of a function.

our minds quickly and fairly accurately. So far, so good; however, if we consider a "rectangular" field on a sphere (area bounded by a fence of four perpendicular planes), we slow down abruptly. The one extra variable, the curvature of the sphere, makes it so complicated we can no longer compute the area in our minds. Expressed as an integral, this curved area is

$$A = 4R \int_0^{a/2} \sin^{-1}\left(\frac{b}{2\sqrt{R^2 - x^2}}\right) dx$$

where R is the radius of the sphere, and a and b are altitude and base of the rectangle, and x is the distance from the origin to the element of area $b\,dx$, respectively. Yet, even with this very involved expression, the areas of all possible such spherical rectangles can be calculated, because they are implicit in the complete and self-consistent laws of geometry.

Whereas this type of computation is frequently termed "problem-solving" for the engineer (finding hidden but discoverable answers), let us recognize that this is one of his smaller tasks. His larger one, in uniqueness, difficulty and challenge, is that of synthesizing new design solutions. While undertaking this synthesis, he has no guarantee than his sought-for answers are obtainable, no assurance that suitable analytic models have been even constructed before. The symbolic models we are setting about to discuss here are intended to help him in both capacities: Creator and analyzer.

As with any professional, the engineer uses the various tools of the available logics when and where they are appropriate and useful. There is no certainity that the next problem he faces will have a ready logic or model to fit it. Continuing work is in process on the broad front of developing models for the new problems generated by advancing technologies.

Symbols, Expressions, and Equations

In creating any conceptual model, the engineer assigns symbols; he then groups the symbols into expressions and finally assembles them into equations. Before we go on to the equations, let us take a brief look at the symbols themselves.

A symbol is simply shorthand;* it is a quickly recognizable label which stands for a specific object, a class of objects or their attributes, for an operation, a state of nature, or simply a number. A dollar bill is a symbol of value, and although traded, it is not the value itself. We assign and use

* In one form or another, all these varieties of symbols appear in design expressions: *Specific object or class of objects:* b metal bar, Δ triangle, R chemical radical, M motor. *Attributes:* d dimension, ρ density, c cost, U utility, L life. *Processes:* $\sqrt{\ }$ square root, Σ summation, ! factorial, d/dt, $\rightarrow$ flow. *States of nature:* V voltage, F force, H magnetic field, L liquid, °C temperature. *Numbers:* n, π, 10^6, 0.11111, ...

symbols for the sake of simplicity, convenience, and generality: but in the context of modeling, the commanding reason is generality.

When we write $e_g = n(d\phi/dt) \times 10^{-8}$ for the electric potential induced in n turns of a generator winding by changing magnetic flux ϕ, we succinctly compress a very general relation into a few hieroglyphics. This expression holds for *any* arrangement of the turns and *any* value of time rate of flux change, and incidentally it employs all the above varieties of symbols (object, attribute, process, state, number).

In design our ultimate purpose is the selection of "best" values of system parameters to satisfy the project's various needs: The greatest effectiveness, the maximum strength-to-weight ratio, the largest capacity per unit cost, the fewest parts, the best looking, or a combination of all of these. Generalized or literal symbols carry their singular importance because, as noted earlier, they allow the engineer to examine most simply all possible variations. He can do this without wasting materials more valuable than ordinary paper. In seeking these "best" answers he must inevitably do much searching and proposing, much questioning and deciding. This process is demanding enough with literal symbols; if specific numbers were used instead, the optimizing process would be impossibly tedious, error-ridden, and uncertain.

How to Get Started

In specifically describing a proposed system or its components, the engineer *collects* variables; *assigns* symbols, *eliminates* the less important variables (through assumptions), *identifies* the variables, parameters, constants (or constraints), and *assembles* pertinent expressions and equations. In any sophisticated problem the engineer frequently goes through the sequence of an iconic or pictorial model first, possibly an analog model, finally arriving at the symbolic model. Incidentally, the amount of work going into the construction of the models *increases* as the sequence progresses; and curiously, the difficulty of manipulation *decreases* with the amount of work put into the model construction.

In summary, symbolic models need this preparation:

1. Collect all variables and assign symbols.
2. Eliminate the less important through simplifying assumptions.
3. Identify the variables, parameters, constants, and constraints or boundary conditions.
4. Assemble the initial expressions and equations of structure, behavior, and effectiveness. Equations frequently represent "state" ($PV = mRT$), "flow" (in = out plus storage) or "summation of components" ($\Sigma F_x = 0, \Sigma F_y = 0$).
5. Reduce and simplify the expression to that form describing the most important aspects of the system.

As we said early in this chapter, the description of a system combines numerical and non-numerical expressions. Naturally, we want to convert as many non-numerical expressions as possible into numerical ones. In any case, we must retain the non-numerical statements for use as supplemental criteria at the end of the numerical work.

Characteristics of a Good Model

A given model being put together may prove to be reasonably faithful in representing the problem situation, but it may or may not be easily understood or manipulated. The characteristics of the good model are:

1. Realism in making predictions.
2. Minimum complexity (fewest number of terms, simplest mathematics).
3. Separate terms for separate actions or phenomena (to identify the different effects).
4. Direct manageability of expression (susceptible to familiar operations).
5. Ease in substitution of known checking situations.

BOOK PROJECT EXAMPLE: Mathematical Model for a Water Tank

When the engineer starts a math model for a design situation, he doesn't know how it will come out. Will it be successful? Will the effort have been worthwhile? Much of synthesis is subjective. Whether or not to attempt a math model is also subjective. Yet it is always best to build a *large complex* structure or idea out of *little* and *simple* pieces. And it's always best to follow rational and repeatable methods rather than solely inspirational and random ones. So much for the philosophy.

We have been told to design a separate water-carrying tank to be added onto a transmit-mix concrete delivery truck. The water is to be used for thinning the mix to specification when necessary in the field and for cleaning the chutes and accessories. We are undertaking the design as a check on the job-shop to whom this job is being sub-contracted. We have looked at the truck and decided the only place for the tank is on brackets over and to the rear of the truck cab and those space limitations are 5 ft maximum length by 2 ft maximum height and width. The tank should hold 50 gallons, be fast filling and emptying, and have adequate corrosion protection for economic life. Also we are told to consider weight, support, appearance, and access. Since the production is to be very low, simple fabrication (no production tools) will be relied upon. Without special tools we are restricted to geometrical shapes which can be cut, rolled, bent, and welded out of flat sheet stock. Other data are listed below.*

We consider three shapes initially: Sphere, cylinder, and square-section box. See Figure 12.1. The sphere is promptly found to be too large (diam. = 2.3 ft for vol = 6.7 cu ft). We know that a cylinder has less wall area than a rectangular box of the same volume, so we shall choose the cylinder.

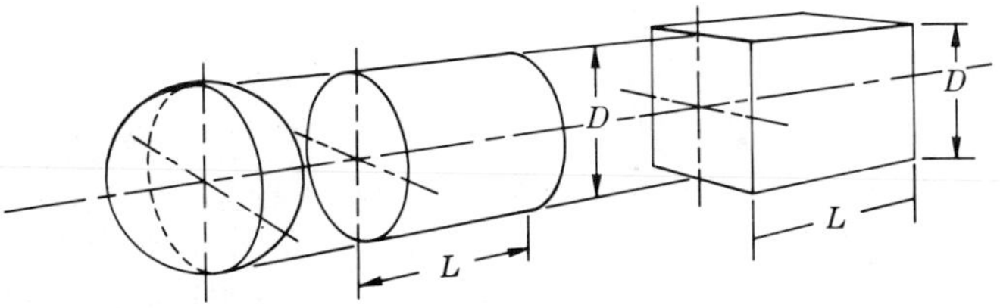

Fig. 12.1 Water tank example. Sphere, cylinder, and box alternatives.

Cost seems the most likely place to start, so we choose variables and put together small equations from which we can assemble the equation for total cost.

Following the sequence of steps named earlier, we name the following variables, parameters, and constants (or constraints) as we now see them.

1. *Identify the variables (etc.) and assign symbols:*

Variables:

- C = Total cost
- C_s = Total steel cost
- C_w = Total welding cost
- C_g = Total galvanizing cost
- W = Weight of steel
- w = Length of weld
- A = Area of surface

Parameters:

- L = Length
- D = Diameter
- T = Steel thickness
- E = Area of end cap if not flat

Constants and Constraints:

- V = Volume = 6.7 cu ft min.
- c_s = Unit steel cost
- c_w = Unit welding cost
- c_g = Unit galvanizing cost per side
- $L_{\max}$ = 5 ft
- $D_{\max}$ = 2 ft

* Pertinent information:

$\frac{1}{16}$ in.-thick H.R. steel costs \$0.25/sq ft (\$0.10/lb).
Galvanizing is \$0.05/sq ft, one side.
Cutting and welding are \$0.20/linear ft each.
50 gal = 6.7 cu ft.

2. *Eliminate or combine in order to simplify:*

   Start with a flat end cap. Disregard E.
   Assume a normal wall thickness, $T = \frac{3}{32}$ in.
   Combine costs of material and galvanizing (both sides), $c_s + 2c_g = c_c$.
   Temporarily disregard tank support, appearance, fast fill/empty, etc.
   Assume one side-weld and two end-weld seams.

3. *Assemble into pertinent equations*

   When we use area ($A = 2(\pi D^2/4) + \pi DL$) and volume ($V = \pi D^2 L/4$) relations, and eliminate length L, the cost equation comes out in terms of parameter D

$$C = c_c(\pi D^2/2 + 4V/D) + c_w(2\pi D + 4V/\pi D^2)$$
$$C = 0.74D^2 + 1.26D + 12.6/D + 1.7/D^2 \quad \text{dollars (shop cost)} \qquad (1)$$

Now we have an equation which we think gives us reasonable decision-making information. We think that it is realistic and that it should point to the diameter D which yields the lowest cost.

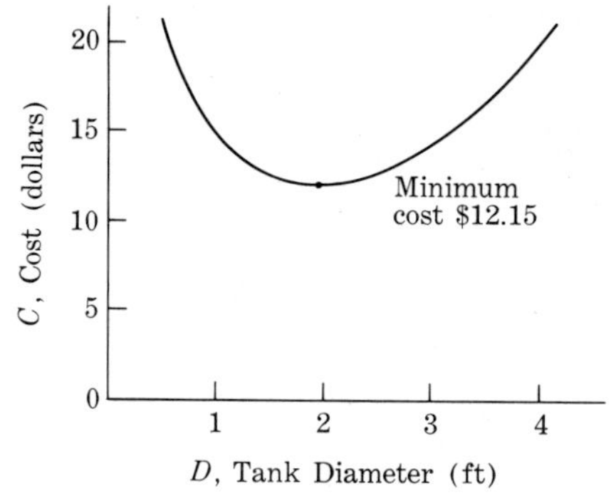

Fig. 12.2 Water tank example. Cost vs. tank diameter. Plotted from Equation (1) above.

Equation (1) has only one variable, the parameter D; and even though a fourth-power equation, it is solvable for the minimum value of cost C. The solution can be attempted analytically, but it is much easier graphically, plotting C vs. D, as in Figure 12.2, where $D = 2$ ft at the value $C_{\min} = \$12.15$. This diameter is within the limit although it happens to be the maximum value allowable. The volume leads to an acceptable length ($L = 4V/\pi D^2 = 2.15$ ft), so it seems that we have the "best" solution. The cost seems low ($C = \$12.15$), when we test it against expectations. Rough quotations from possible suppliers are in the range of \$35–50, which would imply a shop cost of \$25–35, so we may be off in our unit costs.

But, more important questions than cost questions arise when we look back at the original criteria. We plunged off into cost, looking for minimum material and labor; but what about weight, support, appearance, fast-fill/empty? How important are they? How do we measure them? Who can compare and evaluate the intangibles, assuming we could quantify the tangibles like weight, fast-fill, etc.?

If we could talk to the owner of the transit-mix company, he would tell us the relative importances; but in his absence, we will have to try it

Table 12.1 Successive Attempts to Evaluate Relative Importance of Criteria. Truck Water Tank Example.

Criterion	*No. 1*	*Argument*	*No. 2*	*Argument*	*No. 3*	*Tentative Conclusion*
Weight	10 %	Low proportion of truck weight.	0 %	500 lb/30,000 lb = neglig.	0 %	Negligible effect.
Support	15	Need special brackets made.	10	Simply a bracket design problem.	5	Vibration is the only real problem.
Appearance	5	Tank is not a dominant feature.	10	Tank can be seen; should not detract.	10	Should follow reasonable practice to preserve image
Access to fill	20	Requires operator labor if not accessible.	20	Same as before.	5	Doesn't take any truck time because of sand loading time.
Quick fill/empty	10	1½ in. diam. pipe empties tank in 2 min. Enough?	40	Realize that the customer and truck can be held up waiting for water.	65	Truck costs $7.50/hr; 1 min. wasted time/load equals $150/yr.
Cost (up-keep and replacement)	40	$ first cost is most obvious.	20	Cost includes repair and replacement due to corrosion.	15	Primary cost is loss of service of truck.
	100 %		100 %		100 %	

ourselves. One way of doing it is to assign percentages to the various criteria reflecting our judgments of one factor versus others and versus the total. After one trial, we think about it some more and try again, and then a third time.

The tabulation in Table 12.1 reports the sequence of the engineer's thinking as he evaluates the relative importance of the several criteria, together with the reasons he states to himself for changing the particular percentages as he progresses. The columns of numbers represent the percentages of proportional importances.

The three successive attempts shown indicate how the engineer's thinking matures as he further penetrates into the sub-problems from the more or less superficial level he usually starts with.

Attempt No. 1 in the table shows the engineer's initial emphasis on cost; but No. 2 shows how the realization of truck waiting time is sinking in concerning "quick-fill/empty." The truck costs \$7.50/hr to operate, so that a delay of just 1 min/load, waiting for the water to empty, accumulated over a year, adds up to 20 hr or \$150.* As the engineer progressed from No. 1 to No. 3 his thinking changed about the other factors as well, so that, at the third iteration his "conclusions" were:

1. The tank and water weight were only 2 per cent of the total and will not affect wheel loading or highway codes. Item is negligible.
2. Support bracket is a straight-forward design problem, with attention needed only on bracing to minimize tank vibration and weld failure. Item relatively unimportant.
3. The appearance of the tank and brackets must not detract from the unified appearance of the truck as a whole.
4. The water fill takes no extra time because it always occurs during the sand and gravel loading, and the water is piped in under pressure.
5. The water-emptying period is most important because operator, truck, and customer can do little or nothing during that time.
6. Tank first cost is overshadowed by cost of replacement and loss of truck service. Replacement should be rapid because the total cost for the loss of service for one day is probably \$50/truck.

Column No. 3 in Table 12.1 translates the above conclusions into what are thought to be appropriate percentages. Now we need to produce and review alternate designs which take into account this new emphasis on "quick-empty." In order to provide this now required feature we may find it necessary to make the tank considerably different. Incidentally, the engineer may or may not use a table and a sequence of rankings, as above, but he certainly should proceed through the thinking indicated. The table simply records the thinking-step refinements taken, whether three or thirty.

This review of criteria re-orients the whole situation and points out that we should have done some reviewing at the beginning of the tank analysis. Since our needs analysis had this loophole in it, let us introduce a "basic design equation" which should prevent such near-sightedness in the future. Then we will return for a revision of the tank example.

A Basic Design Equation

Our early misplaced emphasis on material cost reminds us that *effectiveness* is the final reason for and measure of a design . . . *total effectiveness*

* Cost = 5 loads/day × 250 days/yr × $\frac{1}{60}$ hr/load × \$7.50/hr.

in meeting the needs of the client, customer, or boss. This effectiveness is made up of the sum of the individual effects and the product of their interactions.* It is beyond our present scope to include the interactions in an equation, but their presence should be known and acknowledged as we attempt to maximize the total effectiveness of our candidate systems. As mentioned later, in the chapter on optimization, *all* the constituent criteria can never be maximized simultaneously. That is, the particular tank designs which empty the fastest will not also be the lowest costs and the best looking, except by an impossibly rare coincidence. The various desirable (and undesirable) features and qualities must be blended together by "trade-offs" in order to reach the maximum over-all effectiveness. We will come back to those decisions on our tank example in a moment.

We will measure the over-all effectiveness of a given design concept by an expression called the "criterion function."† In our context, the essential work an engineer does in design is the choosing of alternative configurations in such a way as to maximize‡ the value of this criterion function:

$$CF = \sum_{i=1}^{n} a_i x_i \qquad \text{(dimensionless, or in homogeneous units)} \qquad (2)$$

where

a_i = weighting coefficients, measuring respective importances
x_i = criterion variable, as a system function or a parameter

This CF is the sum of n individual constituent effects, x_i, each weighted or proportioned by the coefficients, a_i, so as to balance the respective x_i importances in the final result. In other words, as the engineer chooses candidate designs and their components in turn, he must remember to combine them in such a way as to obtain the highest effectiveness out of the whole system. Even the subsystem alternatives should each be sub-optimized before being compared with competing concepts, in order to make the comparisons and decisions most fairly. Whether or not the engineer does it consciously, as we urge, he always *does* make choices; He eliminates and adds, he modifies and assumes. We here recommend that he decide issues (1)

* *Synergism,* or the benefit of the interaction of two factors beyond that of their simple addition, is illustrated by success and wealth, each additively desirable, and each increasing the other. Physiologically, there are drugs whose effects together are greater than the sum of their individual effects. Examples in engineering include the grid and plate of a vacuum tube and the cement and steel of reinforced concrete beams.

† Criterion function is also given other names in the literature: "Objective function," "measure of effectiveness," "utility," etc.

‡ Of course, the criterion function may measure "cost" rather than "benefit" or "value," in which case the object is to minimize the CF. If both cost and benefit terms are present, their contrast must be taken care of by the algebraic sign, by a place in the denominator, or otherwise.

consciously, (2) in small identifiable parcels, and (3) in writing, for review, revision, and later presentation.

Decisions, the Value System, and the Criterion Function

The *decisions* with respect to the criterion function are the critical steps. It is easy to assemble an expression of variables and coefficients; but it is very demanding to make assumptions, to evaluate, and *decide*. The engineer uses all the analysis and quantification he can command; but in the end, the decisions are made *subjectively;* and there is no avoiding it.

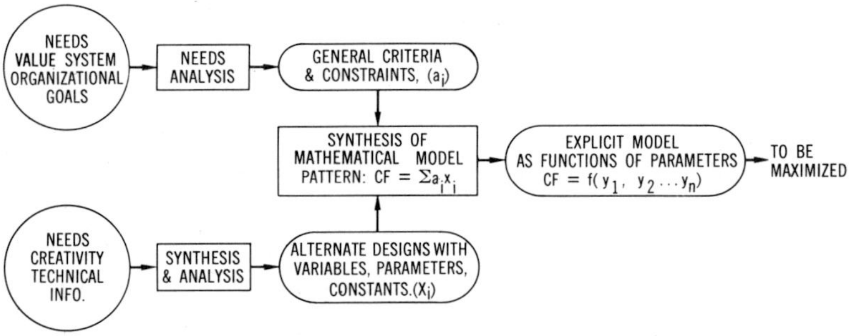

Fig. 12.3 Sources of inputs leading to the formulation of a criterion function.

Either the engineer takes responsibility and makes decisions, or he abrogates the right, and someone else (usually less technically qualified) makes the decisions for him. It is this thrust-upon responsibility which gave the engineer the right to make the series of subjective evaluations in our current water-tank example. Naturally, along with the decision-making right, goes the responsibility to be, or to become, qualified to make those decisions.

As we look behind the scenes, we find major influences on decision-making coming from the individual's own value system, from that of his organization, and from the culture, as well as from the technology. In fact, these two streams of influence converge at the point of making a mathematical model of the system. These are diagrammed in Figure 12.3.

The top stream in the figure comes from those needs which originated the project, and from all the persons related to the project. This influence establishes the relative importance of the various sub-needs and the corresponding system functions. The top stream also supplies the engineer's knowledge of the applicable value system. In a word, it is the source of the value ideas in a_i, the weighting coefficients, and the assumptions concerning system constraints.

The lower stream is the creative response of the engineer to the stated primitive need, by way of his general technical background and specific new information. Along this path the engineer creates alternate system concepts and configurations, which lead to the suggested operating functions, to the criterion variables x_i, and finally to the parameters.

The upper half of the schematic Figure 12.3 comprises the needs analysis, pointing out to the designer the objectives and the relative values of the sub-objectives. The value system of the designer, the customer, and the culture are each involved, but together they lead to some specific numbers for a_i.

The lower half of the figure comprises the synthesis of alternate solutions. The system is conceived in order to perform certain functions which are labelled and measured by criteria x_i. The x_i in turn are composed of parameters (z_k later) whose values the engineer seeks to fix at an optimum. For instance, in a traffic signal, the criterion *visibility* is composed of parameters lamp size, brightness, height over the street, direction, etc.

This formulation may, at first blush, seem over-elaborate and over-formal; but it represents a logic and a rationale which must pervade all system development if the basic objectives are to be met economically and promptly. Do practicing engineers and operating departments use such methods? A few do; most don't, but should. Large companies and the armed services* are now preparing instructions which follow substantially this reasoning.

If this expression, $\Sigma\, a_i x_i$ is a reasonable symbolic model of the system, at what point do all the other mathematical models of system behavior fit? How about circuit analysis, thermodynamics, feedback control and stability, structural integrity, materials and process analysis? Are all these rigorous mathematical tools to be subjugated to subjective decisions from the top? The answer, is of course, that they *are* subjugated; but that as tools they do fit in at lower levels of the design process, aiding in producing adequate decision-making information. National and international politics decide the annual United States federal budget of over 100 billion dollars. Law decides court cases of millions of dollars; banks decide loans of hundreds of thousands; corporate middle management decides the projects of tens of thousands. Science and logic by themselves decide nothing. They only recommend. They are simply two extremely useful tools in a kit which also comprises past experience, law, necessity, personal preference, empathy, and self-interest.

* For instance, see AFSCM 375-5, Air Force Systems Command Manual.

Working with the Criterion Function

To return to our modeling, certain rules of convenience apply to the terms a_i and x_i. Normally, a_i represents the proportional importance of the various criterion variables, x_i; so each a_i is less than one and the sum of all a_i *is* one. Mathematically, this is shown as

$$\Sigma\, a_i = 1 \quad \text{and} \quad 0 < a_i < 1 \tag{3}$$

For instance, the criterion variables for a new "heavy-duty" type of mechanical pencil might be suggested as follows, together with one proposal on their respective weightings:

Criterion Variable	*Proposed Weight*
x_1, adaptability to the hand	$a_1 = .2$
x_2, length of line before refill	$a_2 = .3$
x_3, amount of erasing before refill	$a_3 = .1$
x_4, manufacturing cost	$a_4 = .1$
x_5, reliability	$a_5 = .2$
x_6, attractiveness, flair, style	$a_6 = .1$
	$a_i = 1.00$

Thus,

$$CF = 0.2x_1 + 0.3x_2 + 0.1x_3 - 0.1x_4 + 0.2x_5 + 0.1x_6 \quad \text{(to be max)} \tag{4}$$

If the pencil were aimed at glamour appeal instead of usefulness, a_2 and a_6 probably would be numerically interchanged. (i.e., then $a_2 = .1$ and $a_6 = .3$).

The above x_i would eventually be evaluated numerically, to maximize the criterion expression; so the units of x_i must be homogeneous or non-dimensional to be meaningful. If the x_i are non-dimensional, they are ratios of the candidate x_i to some standard (i.e., x_i/x_s). If they are homogeneous, the x_i must be in dollars, or some defined unit of value or utility (utiles). If the criterion is effort, foot-pounds will serve; if it is certainty, we could use probability units. In the pencil case above, the mixture of line length, reliability, and appearance requires the units to be non-dimensional, that is, to be "normalized" by dividing by a standard for each variable.

The trick now, of course, is to measure these criteria meaningfully in terms of other variables and parameters; so as to derive some useful symbolic expression to optimize. Length-of-writing-line-before-refill can be more or less readily converted into diameter, length, hardness, and finger pressure on the writing graphite (or other expendable). So also with erasing, manufacturing cost, and others. But how about adaptability-to-hand, attractiveness, etc.? Saying that we can't represent them readily doesn't solve the problem; it simply forces us to use ingenuity in providing some method of measurement, particularly when that particular criterion is weighted "important."

The criterion variables x_i are functions of dependent variables y_j which in turn depend upon the parameters z_k. Mathematically we say the set of x_i is a function of the sets of y_j, z_k, and constants c_m,

$$\{X_i\} = f\{Y_j, Z_k, C_m\} \tag{5}$$

Let us explain these words and their relationships with the further illustration of a simple lawn rake.

1. X_i is the set of criterion variables, x_1, x_2, x_3, in the given case, including all the measures of effectiveness for the final system we are considering:

   x_1 = raking effectiveness

   x_2 = convenience to person's hand and back

   x_3 = cost

2. Each of the above variables is a function of further variables, whose set Y_j consists of

   y_a = tooth length-to-separation ratio

   y_b = tooth stiffness

   y_c = cost of teeth and support bar

   y_d = cost of handle, etc.

3. These dependent variables in turn depend upon the parameters and constants, whose set Z_k and C_m include:

   z_r = length of handle

   z_s = diameter of handle

   z_t = number of teeth

   z_u = width of tooth support bar

   z_v = length of teeth, etc.

   c_m = strength of steel

   c_n = cost of plastic

   c_p = cost of labor, etc.

Of course, the object of the design, as we have stated before, is to fix the value of the parameters z_k so as to maximize the criterion function. The variables and parameters are assigned no more extensively than necessary to reach quantifications. This is treated in the next chapter.

In our present situation the engineer proposed different alternative rake designs A_1, A_2, A_3 and uses the appropriate y_j and z_k to fill in one *CF* equation for each alternative A_i. That is, there is an equation

$$CF_1 = (a_1x_1 + a_2x_2 + a_3x_3) \qquad \text{for alternative } A_1 \tag{6}$$

where

$$x_1 = f_1(y_j, z_k, c_m), \text{ etc.}$$

Quantitative solutions are reached by substituting numbers and parameter symbols so that a *CF* value can be computed for each alternative. Of course, the alternative A_i with the highest *CF* is the winning or optimum design. Presumably, this leads to fixing the values for z_r, z_s, z_t, etc. The *CF* equation (6) above is made up in any fashion that meets the need. For instance, raking effectiveness will be related not only to tooth length-separation-ratio but to bar width and tooth length as well.

A part of the problem of optimizing lies in natural or imposed limits on the various variables or parameters. These must be met in the maximizing process. For instance, the handle length limits probably are: 4 ft $< z_r <$ 10 ft; the handle diameter limits probably are: 0.5 in. $< z_s <$ 2 in., etc. The finally accepted design must of course be within all these given ranges.

There is no easy, paved road to the quantification of these difficult, debatable, little understood, frequently subjective criteria. Much of the time, they cannot be adequately expressed meaningfully. At least when they are listed, they are on the table for all to see. Eventually, some relevant variables can be adduced to help. Certainly time is on the engineer's side, since more knowledge means more quantification or at least more common factors, and hence more quantification.

We will return to our illustrations after one more observation. The engineer is boss of his criterion function. It is *his* tool and he can do anything to it that is logical. If he wants to follow the suggestions of Equations (3), fine; if he finds it more convenient, he can make the a_i sum greater than "one" or even make the a_i "digits" instead of decimals. He would usually do this because he needs to add a_i terms later. The choices are listed below, together with the advantages.

$0 < a_i < 1$;	$\Sigma a_i = 1$	(proportions most easily recognized)	(3)
$0 < a_i < 1$;	$\Sigma a_i > 1$	(convenient to modify)	(7)
a_i digits;	$\Sigma a_i \gg 1$	(approximate relations are indicated)	

Of course, if an a_i is added to an original list, the relations are shifted (slightly or much). In the pencil case above, for example, suppose the criterion "impact resistance" (a_7 = .2) was included by later decision. The original a_1 through a_6 are effectively reduced 20 per cent by the simple addition of the new term. In certain stages of the work such an approach may be preferable to the continued re-adjustment of the original a_i.

Note that, after all the a_i are assigned, the only way they can exert their respective influences occurs when the numerical values of the various x_i are nearly equal. If, in one design alternative, x_2 was found to be 1000 and x_5 found to be 10, either the respective units were unreasonable or the influence of x_5 was trivial (i.e., $a_2x_2 = 0.3 \times 1000$ vs. $a_5x_5 = 0.2 \times 10$) Thus the engineer must adjust the maximum or mean values of his criterion variables so the various a_i have due significance in evaluating the numerical value of the final *CF*. This is called a sensitivity check and will be further detailed in the next chapter.

One way of controlling this sensitivity is for all the maximum values of x_i to be adjusted to unity. Another way is for the mean values of x_i to be made equal. These two alternatives are stated

$$x_{i\max} = 1 \tag{8}$$
$$\bar{x}_i = \bar{x}_2 = \bar{x}_n \tag{9}$$

The engineer has only to be sure that his successive iterations toward a reasonably faithful model make good sense in the light of his knowledge of the total case.

Now by means of the design equation (2), we attempt to bring the two parts (the quantitative and the qualitative) of our earlier-mentioned system description together, blending them into a coherent set of requirements and measures. We can never make a full blend; so, certain of the qualitative measures will remain to one side, immiscible.

An example of Equation (8) will now be shown, $x_{i\max} = 1$. In Chapter 13 we will present one of Equation (9), ($\bar{x}_n = \bar{x}_1$, etc).

Revision of the Tank Example

More design alternatives must now be conceived and submitted to the new water-tank project criteria, of the form of Equation (2), to determine which values of which design alternative maximize the criterion function.

Let us review the requirements again, noting that the a_i have already been tentatively chosen; and they are shown as Column 3 in Table 12.1.

Quick-empty could be accomplished by using a 2 in.-diameter pipe, which, by reference to handbook formulae, we find is able to drain the tank in about 1 minute (a 1 in.-diameter pipe would take 4 minutes). If we put in a 2 in.-diameter fill inlet and a 2 in. quick-release emptying valve with hose on the accessible end of the tank, we will meet both quick-empty and accessibility requirements at the same time. But, the heavier pipe connections will require thicker end walls ($\frac{3}{16}$ rather than $\frac{3}{32}$ in.) both to weld and to carry the pipes in service.

To improve the appearance, we must consider the total tank surroundings, including cab, bracket, and mixing drum. It is beyond our present scope to discuss more than the tank improvement. We find we could secure enough appearance change of the tank by using domed ends in place of the uninspired flat ones. We would attempt to locate a sheet-metal vendor who could supply standard-sized domed ends of suitable appearance and fit. Such a vendor would save us the cost of production tools, yet allow us to upgrade the tank enough to satisfy an industrial styling designer. We have no internal tank pressure; the doming is strictly for compatible appearance.

As a first step toward a redesign of the tank, we will make a comparison between the earlier "cost-oriented" approach and the latter "effectiveness-oriented" approach. The comparison serves our purpose here because of the change in design viewpoint; although such an exercise may or may not serve the engineer normally. Our object is to project details of the new implementation mentally to estimate how well it meets the requirements we originally started out with. We go down the list of criteria one by one and estimate in decimals how well the new design measures up. If the condition is fully met, we assign $x_i = 1.0$, if half met, $x_i = 0.5$, etc. This simply collects our thinking in one place, takes a reading on progress and exercises the $\Sigma a_i x_i$ for practice. Table 12.2 below lists our estimated x_i values for the original and the new designs respectively. The a_i are copied from Table 12.1, Column 3, and are multiplied by the x_i to yield the $a_i x_i$ columns.

Table 12.2 Estimates of Improvement in Criterion Function with Revised Tank Design.

			Original "Cost Oriented" (1 in. pipe, flat ends)		*New* "Effectiveness Oriented" (2 in. pipe, domed ends)	
		a_i	Est. x_i value	$(a_i x_i)$ old	Est. x_i value	$(a_i x_i)$ new
x_1	weight	.00		.00		.00
x_2	support	.05	.95	.048	.95	.048
x_3	appearance	.10	.5	.05	.8	.08
x_4	access to fill	.05	.9	.045	.9	.045
x_5	quick empty	.65	.1	.065	.9	.585
x_6	running cost	.15	.95	.142	.7	.105
		$\Sigma a_i = 1.00$		$CF_o = 0.35$		$CF_n = 0.86$

With the figures we have estimated, we see the criterion function has increased to double the original (0.35 to 0.86).

This simply tells us that the ideas of our new design are more relevant than the earlier ones were. We must now go ahead and produce the results indicated. As listed above, the x_i values are non-dimensional now, simply normalized to maximum values of unity.

The reader's natural questions are, "Does this exercise really accomplish anything?" "Should I do this for my project?" "How can this help to quantify the parameters?"

The answers come in two parts. First, do this if you want a preview and an estimate of likely outcome. It is always better to estimate the values of the smaller pieces than to try an over-all estimate of a complex system as a whole. Errors tend to cancel out when component pieces are looked at. Also, estimate the results this way, in the unlikely event that there is no other way of doing it. Similarly, remember that some sort of evaluations are going on continuously, whether conscious or not, if any project progress is being made at all.

Second, the parameters should be quantified by whatever appropriate means are at hand. Even the projected figures of x_i in Table 12.2 came from quantitative estimates that were available, such as material costs, time consumed, etc. The next step is the symbolic representation of the parameters to see if optimum values do exist. As a project develops, appropriate sections of the work can be segregated and subjected to suitable mechanical, structural, or other analyses.

A Quantitative Solution

To return to our water-tank problem, its quantitative solution has been worked out on the basis of the expression

$$CF = a_1x_1 + a_2x_2 + a_3x_3 + a_4x_4 + a_5x_5 + a_6x_6 \quad \text{(to be maximized)} \tag{10}$$

Since $a_1 = 0$, and both a_2 and $a_4 = .05$ (very small) and x_3 (appearance) was a constant with respect to diameter, Equation (10) reduced to two terms

$$CF = \underset{\text{(quick empty)}}{a_5x_5} + \underset{\text{(costs)}}{a_6x_6} \quad \text{(to be maximized)} \tag{11}$$

All additive constants can be eliminated from the maximizing expression.

The considerable work necessary to produce a quantitative expression for the criterion function must be justified for any project. Such work for our water tank would only be warranted if large production were expected, providing the broad base across which to spread the engineering cost of hundreds or thousands of dollars.

We assume the work had been approved, so we modify the original Equation (1) to take into account the head of water affecting quick-empty, the extra wall thickness, and extra area of the tank end.

Without going through the steps here, we develop an equation in T, wall thickness, and D, tank diameter, based upon the relations:

1. Volume is 6.7 cu ft.
2. Ratios of quick-empty to initial and replacement cost is .65/.15 (a_5/a_6).
3. Drain pipe diameter is proportional to wall thickness.
4. Drain time is proportional to square root of tank diameter.

The equation resulting is

$$CF = 120T^4\sqrt{D} - (0.12D^2 + 1/D) \qquad \text{(to be maximized)} \qquad (12)$$

The family of curves graphically demonstrating Equation (12) with T shown at different constant values ($T = .1, .2, .25$ in.) is shown in Figure 12.4.

As we choose a solution point within the permissible range, we consider the qualitative conditions of marginal improvement, practicality, and original objectives. Our first inclination is to go directly to the CF_{max} of 15 at $T = 0.25$, $D = 2.0$; but a more suitable choice is $D = 2$, $T = 0.2$, giving a drain pipe diameter of about 2 in.

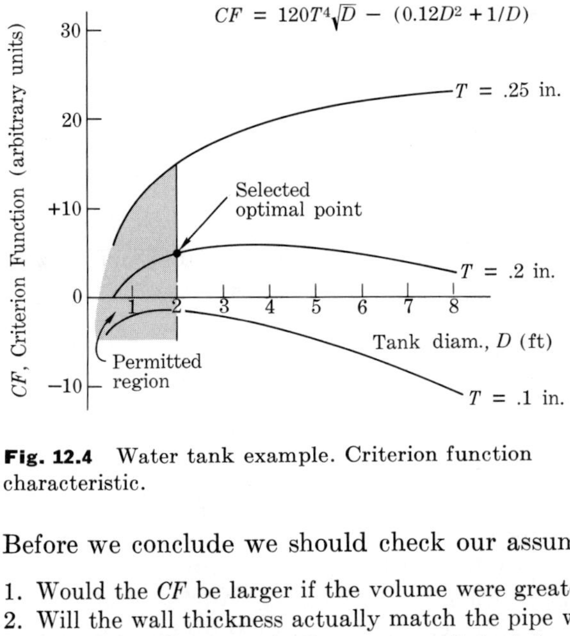

Fig. 12.4 Water tank example. Criterion function characteristic.

Before we conclude we should check our assumptions:

1. Would the CF be larger if the volume were greater than 6.7 cu ft?
2. Will the wall thickness actually match the pipe wall for welding?
3. Are the coefficient ratios in agreement? (a_5/a_6), etc.

The next chapter will deal more fully with the explicit evaluation of the criterion function, after the initial form has been assembled in the same fashion as above.

SUMMARY: Check List for Assembling Criterion Function

1. Sketch "black boxes" and connecting lines showing the elements and interrelations for your proposed system. If more than one alternate is worth considering, do the same for each one. (Iconic model)

2. List all the variables (dependent and independent), parameters, constraints, and constants that seem to be applicable.
3. List the functions to be performed by the system as criterion variables x_i, and the dependent variables y_j which determine x_i.
4. Write separate equations including the variables of item 2 and the phenomena of item 3 above, incorporating simplifying assumptions as necessary.
5. Decide whether the variables should be continuous or discretely valued.
6. Decide what are the limits or ranges tolerable for each significant variable or parameter.
7. Write a criterion-function equation measuring the respective effects of the various controllable aspects.

EXERCISES

12.1 Write a criterion function expression, $CF = \Sigma a_i x_i$ for the packaging container for shipping apples across the country.

12.2 Write a criterion function for an electric fan.

12.3 Write a criterion function for an automotive bumper jack.

12.4 What are the criterion variables to consider in starting the design of downtown parking structure?

REFERENCES

Asimow, Morris: "Introduction to Design," chaps. 5, 10, 11, Prentice-Hall, Inc., Englewood Cliffs, N.J., 1962.

VerPlanck, D. W. and B. R. Teare: "Engineering Analysis," chaps. 3, 7, John Wiley & Sons, Inc., New York, 1954.

Chestnut, Harold: "Systems Engineering Tools," John Wiley & Sons, Inc., New York, 1965.

Alexander, J. E. and J. M. Bailey: "Systems Engineering Mathematics," Prentice-Hall, Inc., Englewood Cliffs, N.J., 1962.

Wilson, Ira G. and M. E. Wilson: "Information, Computers and System Design," John Wiley & Sons, Inc., New York, 1965.

Thirteen

Expansion of the Criterion Function

"Experience is the best of Schoolmasters, but the school fees are heavy."

Carlyle

Resume

We left the criterion function expression in the last chapter, in a partially assembled state. We discussed the total equation, the criterion variables x_i and the weighting coefficients a_i; but we didn't show how to establish relative units, how to preserve sensitivity and compatibility between terms or how to evaluate and maximize the total expression. These several waiting subjects will receive our attention in this chapter.

To reiterate, we construct this design-measuring expression, the criterion function, for two purposes: 1) to record all the important criteria which will measure the acceptability or degree of excellence of the proposed solutions, and 2) to provide a mathematic model whereby the optimum value of the parameters can be chosen. In a major system development, this expression becomes both elaborate and complex, frequently being split into affinitive parts for convenience. In the usual highly technical system, the original interdisciplinary character practically always disappears by separation into the respective electrical, mechanical, thermal, or chemical sub-problems. Each of these fields uses other math models for its own purposes. At the end, of course, the system must be put together again and the original questions of over-all function and performance asked (and answered) all over again in review.

Our present purpose is to take the relatively simple form,

$$CF = a_1x_1 + a_2x_2 + \ldots a_nx_n \qquad \text{(to be maximum, or minimum)} \qquad (1)$$

and to show how the x_i are expanded into useful constituent dependent variables y_j and parameters z_k, and how the original relative proportions between the a_i are preserved during this expansion. We shall illustrate how this may be applied to a number of competing alternatives, so as to choose the best.

One other purpose is to illustrate the evaluation of the criterion function by permutation of alternatives using discrete values of the variables as against the continuously varying values which were considered in the recent water tank example.

General Form of the Function

In constructing our model we have been advised in the last chapter to be realistic and simple. We should start only with necessary and valid measures, embroiling ourselves in complexity only when and as the logic impels us. The general form of the model is demonstrated by the simplest form; so will start our discussion there.

The simplest form of the "design equation" is the first term of Equation (1), wherein x_1 = cost, (i.e., $a_1 = 1$, nothing else is important). So $CF = x_1$ and it is to be minimized. Here we assume that all candidate alternatives equally fulfill the purposes of the design. The cost is naturally a function of other variables (y_j, dependent) and of the parameters (z_k). In this case, the dependent variables would be such quantities as first cost, maintenance cost, and possibly salvage value. The parameters are unit costs, dimensions, numbers of units, process costs, assembly or installation costs, overhead, etc. The parameters are also called independent variables; because we can change them at will and observe the effects upon the other variables and upon CF itself. Thus we have the progression,

x_i = criterion variable, which depends upon

y_j = dependent variable, which depends upon

z_k = parameter (independent variable) (2)

Of course, x_i may occasionally depend directly upon z_k (e.g., when z_k is a quoted total installation cost). These successive dependencies lead to the expansion,

$$CF = a_1 f_1(y_a, y_b, \ldots) + a_2 f_2(y_a, y_b, \ldots) + a_n f_n(y_a, y_b, \ldots) \qquad (3)$$

where $f_1(y_a, y_b, \ldots) =$ functions of the dependent variables y_j and

$$y_a = f_a(z_r, z_s, \ldots)$$

functions of parameters z_k.

This general equation (3) is of little practical help; but it does indicate that each term probably is a different function of all the dependent variables and all the parameters.

The typical analytical problem in engineering economics is symbolized by the cost criterion function,

$$CF = x_1 = y_a + y_b - y_c \qquad \text{(dollars, to be minimum)} \qquad (4)$$

where

$y_a =$ present purchase cost of the system

$y_b =$ present value* of operating and maintenance costs, over the life of the system

$y_c =$ present value* of the salvage value, when the system is scrapped.

Here the total cost x_i is simply the algebraic sum of the constituent costs or credits. When operating decisions are at stake, the operating cost variable y_b may be the only one considered. For instance, when additional processing facility or additional power is needed, the engineer simply looks at the incremental cost of equipment A versus equipment B, or power plant A versus power plant B, and divides the load appropriately.

Then the cost criterion function reduces to the increment of operating cost, y_b, or

$$CF = \Delta y_b \qquad \text{(dollars, to be minimum among A, B, C, available alternative systems)} \qquad (5)$$

This may seem very elementary, but the division of usage or demand among the various capabilities in order to implement Equation (5) led to the development of the techniques of linear programming (see Chapter 15, Optimization).

* Present value is the technical expression for the number of dollars, which compounded at going rates of interest from the present until a given future date will equal the principal amount referred to at that future date. If there is a sequence of future dates (i.e., one per year), the several sums are considered individually and added together as of the present. See references at the end of Chapter 8: Barish, Morris, Grant and Ireson.

Cost Plus One Physical Variable

We shall move on to an example of a criterion function with two criterion variables. Our illustration is the wire connector, a simple tubular metal sleeve used by the millions to join abutting wires for assembling electrical equipment. Figure 13.1 shows how dimples are crimped in the side for mechanical joint strength and electrical contact. In cramped applications,

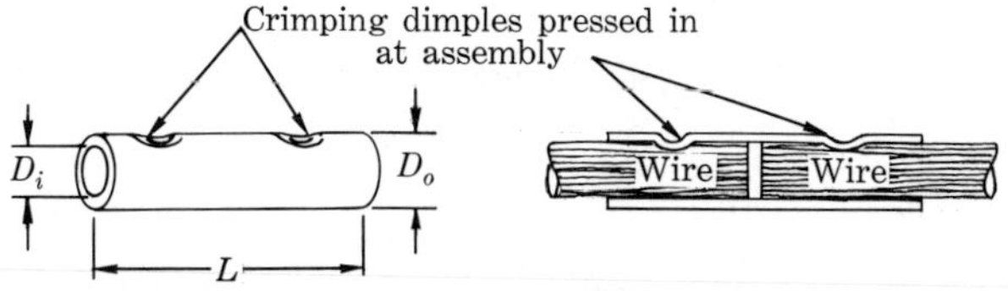

Fig. 13.1 Crimped sleeve wire connector.

size as well as cost is a criterion, assuming that the joint function is performed satisfactorily. Thus, we have

$$CF = a_1x_1 + a_2x_2 \qquad \text{(units, to be minimum)} \tag{6}$$

where

x_1 = cost
x_2 = size; (i.e., volume, since both diameter and length are pertinent)

Both cost and size are dependent upon the same parameters, diameter and length; so, as shown below, the equation immediately reduces to one of two variables, D and L. The two criterion variables, x_1 and x_2, are:

$$x_1 = (c\$/\text{lb})(d\ \text{lb/cu in.})(\pi/4)(D_o^2 - D_i^2)L\ \text{cu in.} \qquad \text{(dollars)} \tag{7}$$

$$x_2 = K_2(\pi/4)(D_o^2 - D_i^2)L\ \text{cu in.} \tag{8}$$

Thus, using Equations (7) and (8) in (6), we have

$$CF = [a_1cd(\pi/4) + a_2K_2(\pi/4)](D_o^2 - D_i^2)L \qquad \text{(dollars, to be minimum)} \tag{9}$$

or

$$CF \sim (D_o^2 - D_i^2)L \qquad \text{(to be minimum)} \tag{10}$$

where

c = unit cost of material, \$/lb
d = density of material, lb/cu in.
D_o, D_i = outside and inside tube diameters, respectively, in.
L = tube length, in.
K_2 = proportionality constant to compensate for different units

The bracketed first term of Equation (9) contains only constants, so, for maximizing or minimizing purposes, that term can be disregarded and only Equation (10) be considered. Notice that neither constant multipliers nor constant additions have any effect on the criterion function; hence such terms can be omitted when the final explicit form of *CF* is written. Thus, in this particular case, the relative importance of cost vs. size (i.e., the ratio of a_1/a_2) does not enter because the only variables appear as a common factor, $(D_o^2 - D_i^2)L$ in both criterion expressions.

In this case so far there has been no scale factor or units problem between criterion variables. If, however, the size criterion had been diameter only,

$$x_2 = K_2 D_o \qquad \text{(units)}$$

the constants a_1, a_2, K_2 would require evaluation for the criterion function to reduce to the explicit form shown in Equation (10).

In the latter suggested case, $x_2 = K_2 D_o$, the criterion function becomes

$$CF = a_1 cd(\pi/4)(D_o^2 - D_i^2)L + a_2 K_2 D_o \qquad \text{(to be minimum)} \qquad (12)$$

We must now first relate a_1 and a_2. If cost and size are equally important, $a_1 = a_2$, and they disappear because they are constant common factors. If related as a ratio, $a_2/a_1 = m$, the *a*'s disappear, but m remains, multiplying the latter terms.

Let us evaluate the K_2 factor, assuming equal importance between cost and size ($m = 1$),

$$CF = cd(\pi/4)(D_o^2 - D_i^2)L + K_2 D_o \qquad (13)$$

Eventually, we must substitute numbers for all the symbols in Equation (13), in order to compare competing alternative designs. This expression of *CF* will be meaningful only if the effects of x_1 and x_2 are numerically equal terms in the range of solution values. The nature of the functions, $(D_o^2 - D_i^2)L$ and D_o, precludes equal numerical effect at *all* values of the variables, D_o, L, so we must fix K_2 in the median range of the D_o and L variation. Unless the equations fortuitously permit, we can only secure an approximate value of K_2; but even with these inevitable approximations, the *CF* can yield useful conclusions.

The ranges of the variables and constants, and assumed medians are:

$0.05 \text{ in.} < D_o < 0.15 \text{ in.}$

$0.25 \text{ in.} < L < 0.75 \text{ in.}$

$0.10 \text{ \$/lb} < c < 0.30 \text{ \$/lb}$

$0.1 \text{ lb/cu in.} < d < 0.4 \text{ lb/cu in.}$

Let

$$\bar{D}_o = 0.1 \qquad \bar{L} = 0.5 \qquad \bar{c} = 0.2 \qquad \bar{d} = 0.25$$

and

$$D_i = 0.04$$

For the numerical values to be approximately equal in the operating range, both terms must be equal, or

$$a_1\bar{x}_1 = a_2\bar{x}_2 \tag{14}$$

Thus

$$\bar{c}\bar{d}(\pi/4)(\bar{D}_o^2 - \bar{D}_i^2)\bar{L} = K_2\bar{D}_o$$

We find,

$$K_2 = 0.0016 \text{ \$/in.} \tag{15}$$

This allows us to write the expression (13) over again,

$$CF = cd(\pi/4)(D_o^2 - D_i^2)L + 0.0016D_o$$

$$\text{(dollars, to be minimum)} \tag{16}$$

Here, K_2 is used as a scale factor between the x_1 and x_2 terms and as a units conversion factor between them (i.e., from \$/cu in. to \$/in.). We can now proceed to use Equation (16) to measure various submitted designs to find the optimum values of D_o and L. As the best values of D_o and L are nearing selection, small changes in the CF will be significant; but both terms will have nearly equal importance, as our original condition required.

Multiple Criteria

We will now present a case with several criteria to demonstrate how one might start from the beginning with an input-output analysis, select the primary and secondary objectives of the system and then identify the criteria for assembly into the criterion function.

Presently there are a number of manufacturers of home intracommunication systems; but with the current million new "housing starts" a year in the United States, the potential market is large. Hence the optimization study of such an "intercom" is worthwhile.

Using the pattern set in Chapter 4, an input-output analysis is tabulated in Table 13.1. This analysis presents known limitations as inputs, and features to optimize as outputs, while the system which transforms the

Table 13.1 Input-Output Analysis for a Home Intercom System.

INPUTS (*Limitations*)		*System Alternates:*	(*Features to Optimize*) OUTPUTS	
Intended:	*Environmental:*		*Desired:*	*Undesirable:*
1. Voice, phono, radio. 2. Children's low level noises (monitoring). 3. Voices of all ages. 4. Operators: Youths and adults. 5. Large and small homes. 6. New and old homes. 7. Mount in or on walls (stud, brick, concrete, stone). 8. Power on continuously or intermittently. 9. Switchable to different stations in the house. 10. Pass Underwriters Lab for electrical safety. 11. Reasonable cost. 12. Use conventional materials and processes. 13. Operate during power outage (?).	1. Typical home wall basic materials, coverings, finishes. 2. Typical temperatures, dust, humidity, wind, abuse. 3. Salt spray if near ocean. 4. Insects, rodents, fungi. 5. Cleaning utensils, chemicals. 6. Nearby noises, electrical static. 7. Line voltage variation. 8. Inside vs. outside wall mounting (?).	1. Tubes of air. 2. Microphone-speaker, no amplifier. 3. Vacuum tube amplifier, etc. 4. Transistor amplifier, etc. 5. Wireless; powerline carrier; separate low-voltage lines.	1. Accurate general voice communication. 2. Microphones and controls accessible. 3. Monitoring children at play or asleep. 4. Good coverage of rooms in house, amplified if necessary. 5. Instant start (?). 6. Hi fidelity. 7. Entertainment. 8. Simple operation. 9. Good volume and control. 10. Low first cost; low maintenance; long life; easy service. 11. Good appearance.	1. No privacy. 2. Clicks, hum, buzz, vibration. 3. Not used much. 4. Must get too close to use. 5. Must tear up house to install. 6. Bulky, protrudes. 7. Few trained service men. 8. Can't see in addition. 9. Sudden voice may scare a visitor.

Constraints:

115 or 230 v a.c. if on 60 cy. power line.
$1\frac{1}{2}$ to 6 v. d.c. if on battery power.
Conventional plug jacks from phono or radio input.

Voltage: +10%–25% variation.
Walls: 2 × 4 studs, lath and plaster, 5–6 in. brick or concrete.
Height: 54–66 in. from floor for microphone.
Temperature: −40°F to 130°F
Humidity: 0–100% R.H.
Wind: to 75 mph.

inputs to outputs (the intercom) assumes *any one* of the plausibly conceivable forms or implementations. The constraints at the bottom are presently known definite numerical limitations and boundary conditions. Such an input-output analysis is a demanding but necessary and useful exercise of the engineer's knowledge and imagination in preparation for a project.

Our concern here is to identify the criteria which are important enough to symbolize in the criterion function. One other aid in selecting the criteria is a listing of the primary and secondary objectives, even though they presumably are merged together under "Desirable Outputs" in Table 13.1. In this way we can decide whether a given criterion *is* the feature or function performed or the *measure* of its degree of accomplishment.

The criterion variables shown in Table 13.2 are a selected list of the five considered most important from the group of eleven in the "Desired Outputs" of the previous table. We can subsume sensitivity and volume under communicability, and reduce the number of variables from five to four. Cost is shown negative in this case for illustration. It may be given a plus sign if the cost variable is put in the denominator, (i.e., $+a_4/x_4$).

Thus a criterion function built of these four would be

$$CF = \underset{\text{communi-cability}}{a_1x_1} + \underset{\text{accessi-bility}}{a_2x_2} + \underset{\text{appear-ance}}{a_3x_3} - \underset{\text{costs}}{a_4x_4} \quad \text{(to be maximized)} \qquad (17)$$

Under "dependent variables" in Table 13.2, we have shown a series of quantities which fully or partly determine the criterion variables. These latter, in several cases, might be varied continuously to search for an optimum; but another method is to arrange discrete sets of alternatives which are permutated in viable combinations to locate an optimum.

Looking forward to the problem of evaluating Equation (17), we anticipate that the appearance, and possibly the accessibility will be intangibles very debatable and difficult to quantify. Whether or not we assign numerical values to them, we cannot escape judging the alternate solutions: Either "better" or "worse," or "much better" or "much worse." Such judgments in this category do not usually call upon engineering technology; they are more likely to call upon competence or at least familiarity in the arts, humanities, and social sciences.

One solution for the intangibles, of course, is to put them to one side until the end, and at that time make peripheral improvements according to the respective dictates. Common sense tells us, however, that appearance, accessibility, safety, and convenience cannot realistically be pasted-on or squeezed-into a design after the basic configuration has been settled. All the desirable attributes must be built in from the very beginning. A face-lifted design shows its patches and baling wire for all the world to see; but a thoughtfully-integrated design looks and behaves like a symphony. Thus

Table 13.2 Primary and Secondary Functions and Criterion Variables for a Home Intercom System. (Numbers correlate with those of "Desired Outputs," Table 13.1.)

Primary Functions or Objectives	*Secondary Functions or Objectives*	*Criterion Variables x_i*	*Dependent Variables, y_j or Parameters, z_k*
Talk (Broadcast)	2. Be quickly available.	1. Communicability.	1a. No. of stations/no. of rooms.
1. General communication (inside house or outside).	4. Amplify voice if necessary.	2. Accessibility.	1b. Volume.
4. Call loudly, if necessary.	Save walking.	3. Sensitivity.	1c. Fidelity.
Listen (Receive)	Sound emergency alarm.	4. Volume and control.	2a. No. of calling stations.
1. General communication, entertainment.	Use as public address system for crowd.		2b. Location of calling stations.
3. Monitor children, answer callers.	Connect with neighbor (?).		3a. Gain/stage.
	Listen for intruders.		3b. No. of stages.
	Amplify TV (?).		3c. Sensitivity of microphone.
	Not detract from room's appearance.	10. Cost.	10a. Central station cost.
			10b. Unit cost × no. of modules.
			10c. Installation cost.
			10d. Power cost.

Table 13.3 Dependent Variables y_j and Ranges of Value for Home Intercom System.

Criterion Variable x_i	*Dependent Variable* y_j	*Range Values* y_j	*Average Value* $\bar{y}_j$	*Proposed Design Alternative* A_1 y_{j1}	*Decision Sources of* y_{j1} *for* A_1
x_1 = Communicability	y_a = No. stations/no. rooms.	0.1 to 1.0	0.4	0.75	Upper middle class home ($\frac{6}{8}$).
	y_b = Audio power, watts.	0.1 to 10.0	1.0	2.0	Equipment chosen.
	y_c = Lowest voice heard, db (above threshold).	20 to 60	35	30.0	Equipment chosen.
	y_d = hi-freq/low-freq response.	$\frac{600}{200}$ to $\frac{5000}{50}$	40	40.0	($\frac{2000}{50}$)
x_2 = Accessibility	y_e = No. calling stations/total.	$\frac{1}{5}$ to $\frac{15}{15}$	0.5	0.67	Assumed $\frac{4}{6}$.
	y_f = Location priority.	0.05 to 1.0	0.4	0.7	Subjective evaluation of the calling station.
x_3 = Appearance	y_g = Decor compatibility.	0.05 to 1.0	0.5	0.7	Optional coverings, flush boxes.
x_4 = Cost	y_h = Components and installation.	\$50 to \$500	200	300	Manufacturing quotations.
	y_k = Operating and service (present value; for 30 yr).	\$300 to \$1000	500	300	Projections.

we must judge very early the capacity for and promise of good appearance, convenience, etc.

Returning to the example of the home-intercom, we review and evaluate the respective importances so that with normal cost, the value of CF is about 1.0. Thus, one way of writing the expression is

$$CF = \underset{\text{communi-cability}}{0.55x_1} + \underset{\text{accessi-bility}}{0.35x_2} + \underset{\text{appear-ance}}{0.4x_3} - \underset{\text{costs}}{0.3x_4} \qquad \text{(to be maximum)} \quad (18)$$

The preceding decimal values for a_i represent the cognizant individual's careful judgment in weighing the respective criteria. In other words, the decimals indicate how important a customer would feel these various aspects are. Naturally these a_i may be modified if new information comes to light as the solution proceeds.

Developing these four variables into their y_j components and referring back to the several parameters of Table 13.2, we think out the following ranges of values which are plausible for the proposed system's attributes and characteristics. Where we don't know enough (and there are usually many such occasions), we must stir ourselves to obtain the necessary help or information. Table 13.3 shows the expansion of the criterion variables into parameters y_j.

Having listed these several related variables, we search for mathematical combinations that will fairly represent the actual operating relationships. The following are questions the engineer asks himself in establishing the relationships* between x_1 and y_a, y_b, y_c, y_d, to start with. The answers are given in parenthesis.

1. Is communication x_1 zero when the ratio of stations to rooms is zero? That is, is $x_1 = 0$ when $y_a = 0$? (The answer, of course, is yes.)
   Is x_1 proportional to y_a? (Yes.)
   Is x_1 proportional to y_b? (More or less.) Would $x_1 = 0$ if $y_b = 0$? (Yes.)
   How does x_1 vary with respect to y_c? (The lower the picked-up voice level, the better the communication is: $x_1 \sim 1/y_c$.)
   How does communication vary with respect to the ratio of high-frequency to low-frequency response limits? (Very important in music; not so important in voice communication. $x_1 \sim y_d$ approx.)
2. How does accessibility vary with the ratio of "calling" stations to total stations? ($x_2 \sim y_e$.)
   How does "calling" station location enter in? (Let us assume a priority of: 1) kitchen, 2) family room 3) center hall near the bath, 4) near front door, 5) etc; $x_2 \sim y_f$.)

* The engineer's natural reaction is to plot sample curves of relationships between x_1 and y_b, for instance. Such graphic displays help in deciding which mathematical function to use: Sum, product, inverse, exponential, etc.

3. What constitutes good appearance? (Assume appearance is highest when flush mounted *and* when having provision for textile or other decor-matching surface covering to be easily assembled on site. Appearance is lowest when surface mounted in bulky, non-matching pre-painted boxes. $0 < x_3 < 1.0$. Each case proposed will be decided on its merits.)
4. How is first cost balanced against operating and maintenance costs? (Let us add component and installation costs to the "present value" of 30 years of operating and maintenance costs; $x_4 = y_h + y_k$.)

The culmination of this questioning is the assembly of descriptive sets of y_j variables, to replace the x_i. Following our above series of decisions on proportionality, we substitute the equivalents we mentioned in the answers above:

$$x_1 = K_1 y_a y_b y_d / y_c \quad (K_1 \text{ here assumed unity})$$

$$x_2 = K_2 y_e y_f$$

$$x_3 = K_3 y_g$$

$$x_4 = K_4(y_h + y_k) \quad (\text{no separate } K\text{'s for } y_h,\ y_k\text{; because both are in \$})$$

where K_2, K_3, K_4 are constants which offset the different units, making the terms compatible. Now the criterion function is ready for the evaluation of the K's and reduction to the final form.

$$CF = 0.55 y_a y_b y_d / y_c + 0.35 K_2 y_e y_f + 0.4 K_3 y_g - 0.3 K_4 (y_h + y_k)$$

(to be maximum) (19)

We are able to evaluate the K's by separate equations taken at the average values of the y_j, with the first criterion, x_1, being the basis. We have prepared the figures in the "average value" column of Figure 13.3 for this purpose. Therefore, finding the average x_1, equating it to the average $x_2 (= K_2 y_e y_f)$ and the others in turn. we have,

$$\bar{x}_1 = \frac{\bar{y}_a \bar{y}_b \bar{y}_d}{\bar{y}_c} = \frac{(0.4)(1.0)(40)}{35} = 0.46 \qquad K_1 = 1$$

and

$$K_2 \bar{y}_e \bar{y}_f = 0.46 = K_2(0.5)(0.4) \qquad K_2 = 2.3$$

$$K_3 \bar{y}_g = 0.46 = K_3(0.5) \qquad K_3 = 0.86$$

$$K_4(\bar{y}_h + \bar{y}_k) = 0.46 = K_4(200 + 500) \qquad K_4 = 0.00066$$

Therefore,

$$CF = \frac{0.55 y_a y_b y_d}{y_c} + 0.80 y_e y_f + 0.34 y_g - 0.0002(y_h + y_k)$$

(to be maximum) (20)

The constants in Equation (19) have now been assigned values such that the chosen relative a_i weightings will hold true when realistic values for the several parameters are tested in Equation (20) for its maximum. The latter equation is now ready for use on the proposed solution-designs.

Note that the maximum value of CF would very simply result from choosing the largest values of y_a, y_b, y_d, y_e, y_f, y_g and the smallest values of y_c, y_h and y_k, if in fact we could so choose them. However, the best performance and the lowest price never go together; so such an ideal set of choices as we imagined above cannot occur. The values substituted for the various y_j must be those corresponding to a real candidate system.

From among the many possible intercom systems, let us choose the following one (alternative A_1, in Table 13.3) and put the CF to work in this one case. We choose an eight-room upper-middle-class home and assign the instruments as follows (S-speaker, M-microphone, C-central control board)

Living room (near entrance)—S, M	Bedroom 1—S
Dining room—none	Bedroom 2—S, M
Kitchen—none	Bedroom 3—S, M
Family room—S, M, C	Bath (hall)—S

Also we choose a transistorized amplifier at each microphone, a.c. or standby d.c. battery operated, low voltage wiring to each station, and flush mounted in each stud-wall. These conditions now explain how the column y_{j1} was filled in from its "source column" for A_1, in Table 13.3.

Putting the numbers into the CF of Equation (20), we find

$$CF = \frac{0.55(0.75)(2)(40)}{30} + 0.8(0.67)(0.7) + 0.34(.7) - .0002(300 + 300)$$

$$CF = 1.10 + .38 + .24 - .12 = 1.60 \qquad (21)$$

This figure above, $CF = 1.6$, becomes meaningful only as it is compared to those of other designs. In this case, we work with "catalog" equipment until we find reasons to refine components for special needs. Such refinement becomes necessary only as we discover significant improvements possible but not available. There is no reason to design and produce special devices when "optimum" system performance can be obtained otherwise. In the above case, probably speakers, microphones, switches, transistors, resistors, batteries, and circuits would be standard items. The etched-circuit-boards, the boxes, and mounting would need special design. We have listed all the y_j as separate variables, but obviously the parameters z_k tie several of the y_j together. For instance, the amplification factor strongly affects lowest voice level y_c, the first cost y_h and operating cost y_k. Also the size of the speakers affects the power y_b, the frequency response y_d and the first cost y_h. As sophistication can be increased, these interrelations can be quantified, but they are beyond our present scope.

There is still one important missing question. How important to the final value of the *CF* are each of the several y_j terms that make up the x_i? Could we leave one or more out of consideration without noticeable effect? This question is called a "sensitivity test" and is asked by making say, 10 per cent changes in each y_j, while noting the change in *CF*. We will not pursue this topic here, deferring that illustration until our last example.

To summarize the results of this example, useful criterion functions for a given design project can be thought out, assembled and used to judge the degree to which proffered solutions meet the original objectives. Such mathematical expressions are created *by* and *for* the engineer and have no dignity or glory except as tools to aid him in optimization. If they so help him, they are useful. If they don't, he should seek other methods and aids. How does he know the solution is optimum? Only by changing (perturbing) one or more of the parameters and looking for poorer results thereby. He can thus only be sure of "local" optima. As described so far, the engineer cannot discover "better" values in unknown or distant varieties of solutions. He would have to explore a wide variety of considerably different systems to assure himself that there were not other, greener fields. This is life, however. There may be a hidden gold mine nearby!

Permutation of Alternatives: Educational Toy

This past example gave the appearance of permutation of alternatives, but it was not, because only certain combinations of variables could be made to work together. Only certain microphones and amplifiers, for instance, can combine with certain speakers and lines. In a more illustrative example of permutation almost any combination should be employable usefully; the appraisal by the criteria simply measuring the felicity of the choice.

Criterion variables. Our next illustration concerns the design of an educational toy in which the parameters are relatively independent, each permutation presenting a more or less acceptable combination whose criteria can be computed or estimated.

Investigations of the various needs for educational toys will reveal opportunities in language, geography, mathematics, history, art, mechanics, etc. But toys in any of these fields may be measured by similar criteria, the most important of which are listed on the next page, together with our estimate of their weighting coefficients for our particular application.

$\{X_i\}$ *Criterion Variables*		$\{A_i\}$ *Weighting Coefficients*
x_1	play value	$a_1 = 1.0$
x_2	educational value	$a_2 = 1.0$
x_3	manufacturing cost	$a_3 = 1.0$
x_4	durability	$a_4 = 0.8$
x_5	simplicity	$a_5 = 0.5$
x_6	attractiveness, novelty	$a_6 = 0.3$
x_7	ease of shipping	$a_7 = 0.05$ (to be neglected)
x_8	safety* = constant	$a_8 = \ldots$
		$\Sigma a_1 = 4.6$

* Safety minimums: Nontoxic materials and paints; rounded corners; smallest pieces can't be swallowed; parts do not have sharp cutting edges when broken; no dangerously heavy pieces, etc.

The coefficients depend upon the type of toy, the anticipated age group, the objectives of the educator and manufacturer. The values listed above were chosen for a snap-joint "Erector Set" construction toy directed to the 6–10 year-old child. The set consists of structural bar members of several lengths, provided at each end with spring rivets to snap into cooperating holes of other parts. The rivet connectors can be used with brackets and plates or in the body holes of other members. Sketches illustrating the idea and some of the pieces are shown in Figure 13.2. The bars are wide enough so that any twisting motion will pull the rivet points out of the mating holes without deformation or breakage. The object of the design is to encourage the child to learn some of the fundamentals about pin-jointed structures through the easy construction of many "business-like" varieties and shapes. Trusses, bridges, towers, platforms, arches, and derricks are types of structures possible. The toy should not deteriorate appreciably in several years of use.

Notice that we listed a_8 as a blank; this corresponds to safety = a constant. The criterion of safety could be considered a variable, but we elect to meet all safety minimum limits and can omit a_8. Also a_7 will be considered zero because of its relatively low importance here. The cost is shown as a positive criterion along with the rest. This doesn't mean that we want the cost to be high, simply that the cost criterion will be written here so that a_1x_1 increases as the manufacturing cost decreases, or x_1 max = 1, at very low cost. The weighting $a_1 = 1.0 = a_3$ means that equal efforts are justified on cost reduction or improvement in play value. If all the variables are ideally implemented (i.e., each $x_i = 1$), the sum would add up to 4.6. We have chosen the a_i in this example so as *not* to add up to 1.0, in order to demonstrate that case in which the designer wants freedom to add other criteria later, without changing the relation between those criteria previously decided upon.

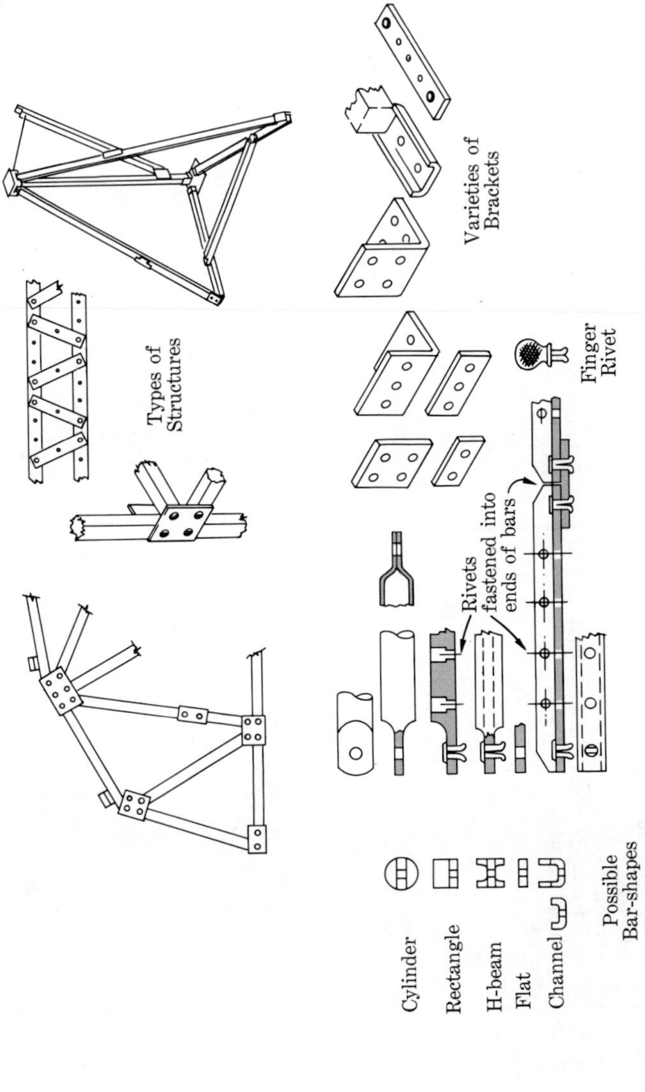

Fig. 13.2 Educational toy. Spring snap-joint construction set.

Table 13.4 Primary and Secondary Functions and Variables for a Construction Toy.

Primary Function	*Secondary Function*	*Criterion Variables* x_i	a_i	*Dependent Variables* y_j	*Parameters* z_k *Constants* c_m	*Range of* z_k	*Mean Value* $\bar{z}_k$
Entertain: Use hands and mind creatively. Present rewards. *Teach:* Develop familiarity with structures through examples. Reinforce learning by the resulting structures' inherent strength in carrying loads. Improve muscular coordination	*Persuade* the buyer. (low cost, endurance, simplicity, safety) *Attract* the child. (satisfy need for pictorial instructions) *Provide* packing and shipping ease. etc.	x_1 Play value	1.0	y_a Number of possible structures	z_a Number of bars	25–100	60
				y_b Number of kinds of parts	z_b Number of brackets	25–100	60
					z_c Number of kinds of bars	4–8	5
					c_c Number of kinds of brackets	6	. . .
		x_2 Educational value	1.0	y_a Number of possible structures			
				y_c Authenticity, realism			
		x_3 Manufacturing cost	1.0	y_d Cost of bars	c_a Avg. cost/bar	\$0.04 each	. . .
				y_e Cost of brackets	c_b Avg. cost/bracket	\$0.01 each	. . .
				c_f →	c_f Cost of rivets and assembly	\$0.20	. . .
				c_g →	c_g Cost of packing, book and overhead	\$0.80	. . .
		x_4 Durability	0.8	y_f Bar strength	z_f Bar thickness	0.2–0.6 in.	0.4 in.
					z_g Bar material tensile strength (plastic, aluminum, steel)	10–75 K psi	40 K psi
		x_5 Simplicity	0.5	y_h 1/(no. kinds of parts) = $1/y_b$	. . .		
		x_6 Attractiveness	0.3	y_k No. of colors = z_c			
				y_c Authenticity	c_h Constant if bar is channel section		
		x_7 Shipping	. . .	. . .			
		x_8 Safety	. . .	. . .			

Now as we expand on the criterion function, we need to identify the other variables y_j and parameters z_k, as we did previously in the home intercom example. Table 13.4 summarizes the objectives of the toy and the criteria, and gives suggested expected ranges of values for the parameters.

The reason for the listing of the primary and secondary functions is to establish the name and priority of the criterion variables. If the former are well thought out, the x_i of the third column in Table 13.4 can simply be copied down in order. The list is continued until the variables become so unimportant, relatively, that the a_i can be neglected, or assigned the value zero.

Parametric expressions. For each of the criterion variables x_i let us now look for and define the cause and effect relationships with respect to the other variables y_j and parameters z_k.

1. The play or *entertainment value* x_1 of the toy depends upon motivating the child, upon releasing his creativity and providing rewards for his effort. These three are interrelated and we believe jointly dependent upon the following sub-conditions (and their measures):

Simplicity of instructions	(inherent)
Speed and ease of assembly	(inherent)
Validity of the resulting structures	(inherent)
Number of possible structures	(y_a = number of bars = z_a)
Number of kinds of parts	(y_b = number of kinds of bars and brackets = $z_c + c_c$)

The preceding conclusions are explained by our assumptions that, as a first approximation, the number of building varieties constructable varies as the number of bars z_a in the kit. (It may be realistically more like the number of bars *factorial*, $z_a!$, but that large number counts predominantly trivial variations.) The number of kinds of parts y_b is simply the total of the kinds of bars and brackets, the latter being relatively minor and assumed constant under our expected conditions.

We believe the two variables y_a, y_b affect x_1 proportionally; so for a first approximation again, we put the relation in mathematical form as a product, using the symbols listed in Table 13.4.

$$x_1 = K_1 y_a y_b = K_1 z_a (z_c + c_c) \tag{22}$$

where

$K_1 = 1$	(K is assumed unity for the first criterion)
z_a = number of bars	(mean value $\bar{z}_a = 60$)
z_c = number of kinds of bars	(mean value $\bar{z}_c = 5$)
c_c = number of kinds of brackets	(assumed constant $c_c = 6$)

The mean play value, $\bar{x}_1$, is obtained by substituting the respective mean values in Equation (22). Thus,

$$\bar{x}_1 = 1 \times 60(5 + 6) = 660 \tag{23}$$

and

$$a_1x_1 = 1 \times z_a(z_c + c_c) = z_a(z_c + 6) \tag{24}$$

We will also add the reasonable proviso that the number of brackets must at least equal the number of bars:

$$z_b \geqq z_a \tag{25}$$

2. The *educational value* x_2 depends on the clarity of the instructions, both the implicit and explicit kinds. It also depends on the number of varieties constructable and their authenticity or realism. Successful structures that are able to bear loads will reinforce the learning and reward the efforts. The former qualities of clarity must be built into the toy; but we will assume the number of varieties and the realism can be varied, the former with y_a ($=z_a$, number of bars), and the latter by the type of bar section chosen. In Figure 13.2, the most realistic bar-sections are the H-beam or the channel. We will fix on the channel because of the easy access to side and center-web holes for versatile use. Thus the y_c, authenticity, becomes a constant (although we could have assigned arbitrary values to it, say 1 to 5, corresponding to the range of realism from "round" to "H-beam").

We conclude by identifying x_2, educational value, as the product of "number of varieties" and a factor containing "realism."

$$x_2 = K_2y_a(1 + y_c) = K_2z_a \times \text{constant} = K_2z_a \tag{26}$$

where

K_2 = proportionality constant
y_a = number of varieties constructable = z_a (mean $\bar{z}_a = 60$)
y_c = measure of realism = constant

The mean educational value $\bar{x}_2$ is similarly obtained by substituting in the component mean value and equating to the mean value, $\bar{x}_1$, from equation (23). Thus,

$$\bar{x}_2 = K_2 \times 60 = \bar{x}_1 = 660 \quad \text{or} \quad K_2 = 11 \tag{27}$$

and,

$$a_2x_2 = 1 \times 11z_a = 11z_a \tag{28}$$

This Equation (28) adjusts the value of the x_2 term so that it will now contribute its proportional share to the *CF*.

3. The *manufacturing cost* criterion x_3 is now to be formulated in such a way that it has a maximum value of 1 (approximately) at the lowest expected cost, and decreases to a very small value as the cost increases. Of course, the cost itself includes bars, brackets, rivets and assembly, instruction book,

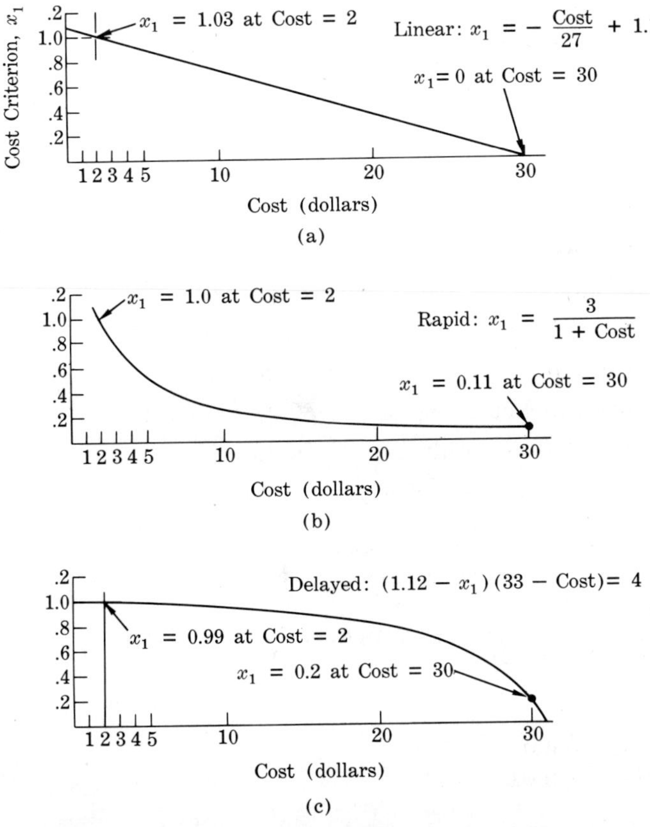

Fig. 13.3 Cost criteria, x, showing three different methods of decrease from 1 to 0 (approx.) as cost goes from \$2 to \$30. (a) Linear; (b) Rapid; (c) Delayed.

packaging, and shop overhead, but not development cost, administration, or profit. We would expect the cost to range from a minimum of \$2 to a maximum of \$30. Now, we can make x_3 drop through such a range linearly, quickly or slowly, as shown in Figures 13.3a, b, and c, respectively.

Of the three shapes, the quick decrease seems most representative of customer reaction to increase of price (i.e., half as many would buy at \$5 cost as at \$2 cost). Hence, we shall take the expression for x_3 from that curve:

$x_3 = 3/(1 + \text{cost})$. Thus,

$$x_3 = \frac{3\ K_3}{1 + y_d + y_e + c_f + c_g} = \frac{3\ K_3}{1 + z_a c_a + z_b c_b + c_f + c_g} \tag{29}$$

or

$$x_3 = \frac{3\ K_3}{1 + 0.04 z_a + 0.01 z_b + 0.20 + 0.80}$$

$$x_3 = 3\ K_3/(0.04 z_a + 0.01 z_b + 2) \tag{30}$$

where

K_3 = proportionality constant
y_d = cost of bars = $z_a c_a$
y_e = cost of brackets = $z_b c_b$
z_a = number of bars (mean, $\bar{z}_a = 60$)
z_b = number of brackets (mean, $\bar{z}_b = 60$)
c_a = average cost/bar (c_a = \$0.04 each)
c_b = average cost/bracket (c_b = \$0.01 each)
c_f = cost of rivets and assembly (c_f = \$0.20)
c_g = cost of packaging, etc. (c_g = \$0.80)

The mean value of the cost criterion, $\bar{x}_3$, and its final form $a_3 x_3$ are obtained by substitution, as before. Thus,

$$\bar{x}_3 = K_3 \times \frac{3}{(2.4 + 0.6 + 2)} = \bar{x}_1 = 660 \quad \text{or} \quad K_3 = 1100 \tag{31}$$

and

$$a_3 x_3 = 1 \times \frac{3300}{(0.04 z_a + 0.01 z_b + 2)} \tag{32}$$

This is the expression for $a_3 x_3$ to be used in the *CF*. Note that in Equation (29) above, no K_i constants were needed in the several terms of the denominator. The reason, of course, was that all terms were in dollars already. If they had been in different units, a procedure would be necessary similar to that used above in evaluating K_2 and K_3 from K_1.

4. *Durability*, x_4, is proportional to the bar's breaking strength, y_f,* because wear will be negligible, and we shall assume high quality rivets and thick bracket material ($\frac{1}{16}$ in.). Thus, considering the bar as a beam in bending, x_4 is proportional to material strength (z_g, tensile strength) and the

* Maximum bending moment supported by a rectangular bar is $M = \frac{1}{6}(Sbh^2)$, where S = breaking strength of the material, b = bar width, h = bar depth or thickness.

square of the thickness (z_f^2) of the bar. We shall proceed to evaluate the K_4 at the mean values of z_g and z_f as before.

$$x_4 = K_4 y_f = K_4 z_f^2 z_g \tag{33}$$

where

K_4 = proportionality constant
y_f = bar strength
z_f = bar thickness (mean, $\bar{z}_f = 0.4$ in.)
z_g = material tensile strength (mean, $\bar{z}_g = 40.\ K$ psi)

The durability mean value $\bar{x}_4$ is found as before, by substitution.

$$\bar{x}_4 = K_4 \times 0.16 \times 40 = \bar{x}_1 = 660 \quad \text{or} \quad K_4 = 103 \tag{34}$$

and

$$a_4 x_4 = 0.8 \times 103 z_f^2 z_g = 82 z_f^2 z_g \tag{35}$$

5. *Simplicity*, x_5, will be assumed to vary inversely as the sum of the different bars z_c and different kinds of brackets c_c. Thus

$$x_5 = K_5/(z_c + c_c) = K_5/(z_c + 6) \tag{36}$$

where

K_5 = proportionality constant
z_c = number of different kinds of bars (mean, $\bar{z}_c = 5$)
c_c = number of different kinds of brackets ($c_c = 6$)

The simplicity mean value, $\bar{x}_5$, is hence found to be,

$$\bar{x}_5 = K_5/(5 + 6) = \bar{x}_1 = 660 \quad \text{or} \quad K_5 = 7260 \tag{37}$$

and

$$a_5 x_5 = 0.5 \times 7260/(z_c + 6) = 3630/(z_c + 6) \tag{38}$$

6. *Attractiveness*, x_6, will be assumed to vary with the quality of packaging, the interest created by the instruction book and by the number of bar colors. The former two will be inherent in the design; and it is felt worthwhile to include number of colors as a variable, choosing one color for each bar length. The plastic colors would be internal, the aluminum colors anodized and the steel would have baked-on colored enamels. Thus,

$$x_6 = K_6 z_c, \quad \bar{x}_6 = K_6 \times 5 = \bar{x}_1 = 660 \quad \text{or} \quad K_6 = 132 \tag{39}$$

where

K_6 = proportionality constant
z_c = number of kinds of bars (mean $\bar{z}_c = 5$)

and

$$a_6 x_6 = 0.3 \times 132 \times z_c = 40\ z_c \tag{40}$$

The full criterion function. We may now collect together all the criterion variables, from Equations (24–40) for the full expression of the *CF*.

$$
\begin{aligned}
CF = \; & z_a(z_c + 6) && x_1 \\
& 11\, z_a && x_2 \\
& \frac{3300}{(0.04z_a + 0.01z_b + 2)} && x_3 \\
& 82\, z_f{}^2 z_g && x_4 \\
& \frac{3630}{z_c + 6} && x_5 \\
& 40\, z_c && x_6
\end{aligned} \tag{41}
$$

Only the first and second terms combine usefully when we simplify Equation (41) above.

$$
CF = 17z_a + 40z_c + z_a z_c + \frac{3300}{0.04z_a + 0.01z_b + 2} + \frac{3630}{z_c + 6} + 82 z_f{}^2 z_g \quad \text{(to be max)} \tag{42}
$$

Our immediate observation is that parameters z_f and z_g might as well be chosen at the top of their ranges because they appear in the numerator only. Also except for Equation (25), $z_b \geqq z_a$, the parameter z_b should be as small as possible.

Sensitivity check. To further find out if all the terms contribute significantly to the *CF*, in fact whether they should even be left in the equation, we will vary each parameter, z_k, by a small amount, say a few per cent, and find its effect on the *CF*. In addition to the importance of the parameter, we can find out which ones we may most freely modify or adjust if we later need to.

We choose a 10 per cent variation of each parameter independently, and compute the resulting effect upon the value of *CF* at its mean value of $\overline{CF} = 3035$.

	Mean value	+10%	*CF*
z_a	60	66	+3.5%
z_b	60	66	−0.26%
z_c	5	5.5	+1.2%
z_f	0.4	0.44	+3.8%
z_g	40	44	+1.75%

Thus, we see that z_b has an insignificant effect, and z_c and z_g each have very little effect on *CF*.

Computation of the maximum. Each remaining active parameter may be given several values (say four each), ranging from the lowest to the highest, in order to find the maximum *CF*. With three parameters (z_a, z_b, z_c) and four values each, the resulting 64 different combinations are best left to the computer. An illustration of a computer solution for this general type of equation is shown as Figure 16.5, where five permutations of three variables are worked out by a Fortran Program.

We can try for a *CF* value at random by choosing some numbers and computing *CF*. Let $z_a = 90$, $z_b = 90$, $z_c = 8$ and the other two at their maximum $z_f = 0.6$, $z_g = 75$. The resulting computation yields $CF = 5550$; but of course we yet have no idea what the maximum really is, or what *values of* z_k *lead to it*. Actual arithmetic must tell us these answers.

SUMMARY

1. The suggested beginning for quantitatively evaluating a criterion function (*CF*) is to repeat the Input-Output analysis so as to identify the primary and secondary functions and the criteria which meaningfully measure them. The criterion variables should copy the name and the priority of the primary and secondary functions (or objectives) of the design.

2. Proceed from the criteria to the supporting variables and parameters assigning appropriate symbols to those not eliminated by simplifying assumptions. Assign further variables only if the given variables cannot be directly quantified, or assigned numbers.

3. Build relations between variables which "explain" or "predict" the relationships provided by natural law (functional constraints) or those necessitated by the problem situation or by man (regional constraints).

4. Set upper and lower limits to the variables and parameters, representing the widest reasonable values.

5. Build the criterion function of all the variables and their constituent parameters.

6. Evaluate the scale-factor proportionality constants at mean values in the range, for compatibility of term-to-term.

7. Collect the terms now to make an explicit expression for *CF*.

8. By whatever reasonably indicated means, maximize the *CF* within the constraints.

9. Test the sensitivity of the constituent parameters, by perturbing each with a small percentage difference, noting the resulting effect upon the *CF*.

10. Select the final design values of the parameters in the light of (but not necessarily bound by) the above-determined maximum *CF*. Make sure all objectives of the original requirements are met, or at least rationalized.

Large system model. An example of a large system is given in Appendix H, demonstrating the use of mathematical modeling in current industrial-military-commercial practice.

EXERCISES

13.1 Choose the following criteria for a prosthetics device by which paraplegics could pick up objects from the floor or down from a shelf: 1) reach distance, 2) ratio of largest to smallest dimension graspible, 3) ratio of hand force to claw force, 4) versatility (string, thumbtacks, fruit, hat), 5) convenience, and 6) cost.
Write and expand a criterion function for the above.

13.2 What are the criteria for the transmitter-receiver assembly to be built and operated by a radio-amateur? (Licensed "ham" operator.)

13.3 What are the criteria for a food-waste and trash disposal system for an ordinary residence for five persons? For a farm residence of the same size?

13.4 What are the criteria and the full criterion function for a typical college student's living quarters? (Dormitory room, fraternity house room, or room in private home.)

13.5 Make an Input-Output analysis for a small workshop in your home, identify the criteria and write the *CF*.

13.6 In spite of precautions, there still are explosions due to gas appliances. Consider the gas water heater, and write a criterion expression for a safety system to prevent explosions from either gas or steam.

REFERENCES

1. Ostrofsky, B.: "Introduction of the Design Process to Freshman Engineering Students." EDP Report No. 5-63, UCLA, Department of Engineering, Los Angeles, October 1963 (also personal communications).
2. Rosenstein, A. B., W. B. Heinz: "The UCLA 1965 Summer Engineering Design Workshop." EDP Report 2-65, UCLA, Department of Engineering, Los Angeles, Fall 1965.

Checking in Engineering Design

Fourteen

"Nothing is so firmly believed as what we least know."
Montaigne

The Role of Checking

The established, practicing engineer is frequently part of a team, but he usually works individually. He has his own responsibilities to discharge, and is held accountable for his own work and decisions. Even in large companies, where much care is taken routinely to avoid error, the engineer is expected to take his own precautions to insure that his work is dependable. Of course, in small companies his dependability is even more critical, because no other qualified person may be available to verify his technical results.

In the world of work, there are no answer books, no associates doing the same problem, and no one is within reach who has ever worked the exact problem before. Furthermore, the results of a task group, even the results of a division or company, hinge on the dependability of the individual's personal answers; so do the engineer's own reputation and chances of advancement.

Such is the case with student and junior engineers as well. Their recognition and acceptance depends on the accuracy and credibility of *their* results. In all cases, there exist the perils of error, of sloppy or incomplete work, or "good" alibis. The problem situation always carries with it too much or too little data; the engineer himself may lack familiarity with some of the fundamental principles, and pressure usually is applied to get the job done quickly. The situation usually is further complicated by the uncertainties of communication with the supervisor or customer. The work is hampered by inevitable clerical errors by subordinates, and by typographical errors in reports or even in published books.

Every transaction in the project is a potential source of error, large or small. The more clear-cut sources of error include:

Original misunderstanding of the problem;
Erroneous data;
Invalid or over-extended assumptions;
Faulty reasoning; illogical conclusions;
Incorrectly stated principles or laws;
Improperly set-up or conducted experiments;
Typographical or copying errors; erroneous table look-up, etc.;
Slips in the routines of arithmetic, algebra, slide rules, or calculator.

With so many chances of making a mistake, it is a wonder that any answer can really be believed and trusted. In fact, it is safe money to wager that any unproven answer is wrong; all the probabilities must be violated for that answer to be right!

Hence we are led inexorably to the vital necessity of checking: Checking at the start, checking part way through, checking after completion. The engineer must check each individual step as well as the broad over-all results. Only through such precautions can he develop the confidence that is necessary for him to stand by his predictions.

Necessary But Not Sufficient

Unfortunately, while checking gives added assurance of eventual success, checking does *not* disclose all that is *true;* it usually only guarantees what is *not true*. It usually tests by *necessary* but not *sufficient* conditions. For example, an order-of-magnitude check might show an error in the following handbook formula for the allowable load P on a structural column:

$$\frac{P}{A} = 65{,}000 - 15\left(\frac{L}{r}\right) \text{ psi}$$

where A = cross section area, L = column length, r = radius of gyration, but that check would *not by itself* show you that the digits had been interchanged and the correct sequence was really $15{,}000 - 65(L/r)$. Rereading the handbook, if it were printed correctly, would supply the needed correction; and that action itself would be carrying out one of the needed checks. The important check, however, would be verifying the handbook. Finding the same formula in another book (not cross-referenced) is *some* added assurance of accuracy and actuality. For predictions of behavior, however, only experiment can finally validate the results; that is, direct experiment, simulation, or previous experience. More on this subject later.

As has been said, checking must not only be done at the end of the engineer's work, but also as it progresses; because the earliest steps have the broadest consequences. Everybody deplores all the wasted work that follows an early error; every reader can recall his own experience of crumpling up sheets of paper in disgust after discovering an instance of this sad fact. The sustained effort and concentration needed to carry through a calculation does not justify our delaying or omitting a running check on validity. The only useful solution is an accurate one, accurate to the appropriate degree. This accuracy always arises out of checking.

It is time to point out that the classroom situation, with its examinations, multiple-choice questions, and pressures for good grades, all encourage the student to rely on guesswork, or on hasty or transcribed answers where he has gaps in his knowledge. This "conditioned" reliance on guessing must be studiously avoided by the would-be engineers. Estimation is all right (Chapter 7) but not guessing, ever. The engineer or student must take an attitude of "no-answer-at-all" until time has been taken to think it out adequately. Fortunately, checking always serves the learning process, frequently merging with it, and speeding the progress to a conclusion.

Many of the formal methods of checking, which we will shortly describe, merely detect error, as indicated earlier, and this is far from revealing the full truth. Each engineer must fortify his checking with a mental framework, with some checking tools and with some outside resources which will guide him to the correct, or satisfactory, answer. Part of the essential mental attitude is willingness to repeat steps; part is constructive discontent; part is imagination; part is a continuing exposure to, and application of, new experiences.

In general, checking can cover the individual steps of a problem, the entire problem, or broad classes of problems. The checking may be done 1) by logic, 2) by mathematical rules, 3) by units and dimensions, 4) by reference to physical or "natural" law, 5) by experimentation, 6) by past personal experience, or 7) by reference to experts.

Checking by Logic: Assumptions

Part of the initial checking is done through reviewing the guiding or governing assumptions. A syllogism in logic is typically symbolized:

If A is true, and B is true, then C must be true.

In other words, the whole fabric of a conclusion *C* is made of two threads of assumptions, *A* and *B*. Both assumptions must be justifiable. Certain assumptions can be validated at the outset of a design; others must await the final numerical answer for their justification.

Special questions must be directed toward any set of assumptions:

Is the assumption necessary?
Does the assumption hide or divert us from an important effect?
Does the assumption apply over the full range of the variables?
Is the assumption stated accurately and unambiguously?
Does the assumption over-simplify the situation?

As an illustration of contrasted assumptions, friction is frequently neglected in problems of motion. The typical textbook inertia problem permits our assuming the absence of friction. However, to find the horsepower required to propel an actual auto on a level road at 60 mph, the windage friction, rolling friction, and transmission friction would have to be considered. Even the air friction of the radiator fan would require almost two horsepower at this speed. If the same car were going up a 10 per cent grade, however, one could probably assume the rolling and transmission friction to be negligible because there is now an added requirement of 40 hp just for lifting the car itself up the hill.

Questions to ask after the numerical solution has been reached are:

Were any assumptions violated, directly or marginally?
Is it now possible to make more or better assumptions?
Do statistical variations in the assumptions affect the answer?
Does the result really answer the original question or need?
Is the result good enough to act upon, or is it necessary to refine the assumptions and rework the solution?

To consider the effects of the numerical solution upon an assumption, suppose the problem was to find the motor power needed to lift to the surface a mine hoist car and load weighing $7\frac{1}{2}$ tons. The mine is a half-mile deep and the hoist car is not counterbalanced. Is it satisfactory to neglect the cable weight? Assuming a 4-to-1 safety factor and standard steel wire cable, one-half mile of cable would weigh about 4000 lb.* At the top of the mineshaft, the cable being wrapped around the drum adds nothing to the hoist weight. However, at the bottom the cable weighs 25 per cent of the hoist car and payload weight. Thus, our original simple assumption won't work at both top and bottom of the shaft.

If we apply the latter checklist of questions to the mine hoist example, we see that the "assumption violated" was discovered by computing the cable weight.

* A 1 in. cable weighs $1\frac{1}{2}$ lb/ft and breaks at 30 tons.

Looking at the next question, a "better assumption" would be the addition of cable weight to the car in proportion to the distance down, but to neglect the inertial effect of that portion of the cable around the drum.

"Statistical variations" along the cable length would not be significant in this case. Statistics would enter, however, if we considered the various dump car weights of coal on successive liftings.

The result seems "good enough" for our purpose if we proceed by assuming the heaviest coal load and by proportioning the cable weight as a function of depth.

Checking by Mathematical Rules

Under the words mathematical rules, we include all the accepted processes and operations: Multiplication, addition, logarithms, square roots, etc.

Most of the applicable rules are so familiar that there is no need for comment. There are a few areas of uncertainty that occasionally make for confusion. A few illustrations will be offered simply as refreshers. Each person has his own Achilles heel of personal uncertainty. Consult a mathematics text or handbook for full treatment when a puzzler arises.

Mixed fractional and negative exponents:

$$(A^{-\frac{1}{3}})^2 = A^{-\frac{1}{3}} \times A^{-\frac{1}{3}} = A^{-\frac{2}{3}} = (\sqrt[3]{A})^{-2} = \left(\frac{1}{\sqrt[3]{A}}\right)^2 = \frac{1}{\sqrt[3]{A^2}}$$

$$A^{-\frac{2}{3}} = A^{-1} \times A^{\frac{1}{3}} = \frac{A^{\frac{1}{3}}}{A}$$

Additions of negative logarithms:

$$\begin{aligned}\log X = 37.12345 - 40 &= 7.12345 - 10 \\ &= -3. + .12345 \\ &= -2.87655\end{aligned}$$

Fractions:

$$\frac{1}{\frac{1}{A} + \frac{1}{B}} = \frac{1}{\left(\frac{A + B}{AB}\right)} = \frac{B}{\frac{1}{A}(A + B)} = \frac{AB}{(A + B)} = \frac{A}{\left(\frac{A}{B} + 1\right)}, \text{ etc.}$$

Inverse functions:

$$\text{arc sin } (\sin A) = A, \quad \log_a a^x = x$$

Any mathematical manipulation should generally be carried out to the end in symbolic notation before numbers are substituted to obtain the final result. The following reasons are important:

One symbol replaces several digits (e.g., x vs. 12.34);
Symbols can be cancelled and reduced to their simplest form;
Limiting cases can best be checked in literal form;
Fewer errors occur with less numerical work.

Finally, in using the mathematical rules, remember *significant figures.* Computations should be made only as accurately as that of the expected measurements or needs. Engineering measurements are rarely made more accurately than 1:1000 or 0.1 per cent. Many classes of problems are very satisfactorily solved with 2 or 3 per cent accuracy. On a 10-in. slide rule, this corresponds to an accuracy of two or three significant figures. Of course, between the numbers 1 and 2 on a slide rule, one can read to three or four figures. Time is best used by working the problem to three significant figures and checking it by approximations to one or two significant figures.

More significant figures may be required in cases of small differences between large numbers; such as,

$$\begin{aligned} x &= \tan 2° - \sin 2° \\ &= .03492 - .03490 \\ &= .00002 \end{aligned}$$

This requires *four* significant figures to produce *one.*
Ask yourself: Are the significant figures really significant, really meaningful? Can I believe them? Can I use all of them?

Checking *decimal points* deserves a little comment. Computation by both slide rule and logarithm requires independent determination of the decimal point. Re-computation of the decimal point for checking usually can be done by sight, or at any rate on some scratch paper. For example, we go through an original expression and come out,

$$\frac{3.14 \times (29.2)^2 \times 1.12 \times 10^3}{238} \stackrel{?}{=} 10.6 \times 10^3$$

We check by approximations, but do not repeat the original steps in the same way:

$$\frac{3.14 \times (29.2)^2 \times 1.12 \times 10^3}{2.38 \times 10^2} \cong 3 \times \frac{(3)^2 \times 10^2}{(3) \times 10^2} \times (1) \times 10^3 = 9000$$

This is a fair check on the 10,600, confirming the decimal point. Converting the numbers to powers of 10 makes it easy to handle very large or very small numbers.

Care in arithmetic is a particularly useful caution. We all experience arithmetic errors and each of us knows there is no excuse. One should form the habit of re-checking every computation immediately, whether a sum, a product, a table look-up, or other. Always do it *another way* the second time, even if it is only slightly different. Add the column from the top down; look up the place in the table from the bottom up; exchange multipliers, and so on. Poor arithmetic will make the bridge fall down just as surely as poor physics, poor metallurgy, or poor logic will.

Checking by Units and Dimensions

Identities. When all algebraically additive terms of an equation are expressed in the same units, the equation is called *dimensionally homogeneous.* Practically all equations of interest to engineers are of this kind. For instance:

$$7 \text{ oranges} + 4 \text{ oranges} = 11 \text{ units}, \quad \text{where the } \textit{units} \text{ must be } \textit{oranges}$$

It is possible to state

$$7 \text{ oranges} + 4 \text{ trees} = 11 \text{ oranges and trees}$$

but this generally is nonsense to an engineer, because he cannot use the mixed term "oranges and trees" in subsequent work. Any "use" stops right there, at that compound term.

Thus, the units or *dimensions* of quantities on both sides of any equation must be identical. These units may not appear to be identical as the equation is written, but they must always combine or cancel to result in identity.

$$100 \frac{\text{oranges}}{\cancel{\text{basket}}} \times 10 \frac{\cancel{\text{baskets}}}{\cancel{\text{tree}}} \times 15 \cancel{\text{trees}} = 15{,}000 \text{ oranges}$$

Or, another example:

$$\left(5 \frac{\cancel{\text{baskets}}}{\cancel{\text{day}}} \times 100 \frac{\text{oranges}}{\cancel{\text{basket}}} \times 2 \cancel{\text{days}}\right) - \left(25 \frac{\text{oranges}}{\cancel{\text{customer}}} \times 20 \cancel{\text{customers}}\right) = 500 \text{ oranges}$$

or still another:

$$\frac{1000 \dfrac{\cancel{\text{oranges}}}{\cancel{\text{tree}}} \times 1 \dfrac{\cancel{\text{tree}}}{\cancel{\text{rod long} \times \text{rod wide}}} \times 160 \dfrac{\cancel{\text{sq rods}}}{\cancel{\text{acre}}} \times \dfrac{1}{4} \dfrac{\text{lb}}{\cancel{\text{orange}}}}{\dfrac{43{,}560 \text{ sq ft}}{\cancel{\text{acre}}}} \cong 1 \text{ lb/sq ft}$$

This matching of units and cancelling them is one of the most necessary and useful methods of checking. No significant physical equation should ever be assembled without writing, collecting and proving out all the units or dimensions. The equation must not be left until the units on both sides agree identically, both with each other and with the purpose of the equation.

Fundamental units: Length, Mass, Time. As we have illustrated in the orchard examples above, we encounter a wide variety of units: Oranges, trees, customers, dollars, days, rods, acres, etc. Whereas oranges, trees, customers, and dollars are unique units, it has been found that many units can be considered composites of a set of fundamental units. For instance, area, acres, or square feet, is length squared (L^2); volume in bushels or cubic feet is length cubed (L^3). Days or seconds are time (T); but frequency is number per unit time ($1/T$). Mass enters into weight, force and inertia (M). It has been found that these three are fundamental units (L, M, T) and are the essential building blocks in any abstract physical system. One or two auxiliary* units are necessary to complete the thermal, electrical, or magnetic systems.

Remembering that velocity is distance per unit time (LT^{-1}) and acceleration correspondingly is LT^{-2}, we can find that the several kinds of force are all the same in what is called the *dimensional* expression.

Linear Force	*Centrifugal Force*	*Weight*
$F = ma$	$F = mr\omega^2$	$w = mg$
$= M(L/T^2)$	$(\omega = 2\pi f;\ f = 1/T)$	$= M(L/T^2)$
$= MLT^{-2}$	$F = ML(1/T^2)$	$= MLT^{-2}$
	$= MLT^{-2}$	

The identity of these expressions should be no surprise, however, because each of the coefficients of the mass m is an acceleration, whether a, $r\omega^2$ or g.

Similarly, all forms of energy (work, W) are found to be ML^2T^{-2}, which is equivalent to force $\times$ distance:

Friction or Elastic Energy		*Kinetic Energy*	
Linear	*Rotating*	*Linear*	*Rotating*
$W = fd$	$W = t\alpha$	$W = \frac{1}{2}mv^2$	$W = \frac{1}{2}I\omega^2$
$= MLT^{-2}(L)$	$= ML^2T^{-2}(\text{No.})$	$= M(L^2T^{-2})$	$= ML^2(T^{-2})$
$= ML^2T^{-2}$	$= ML^2T^{-2}$	$= ML^2T^{-2}$	$= ML^2T^{-2}$

* The unit symbols are as follows: *Metric system* (MKS): L = meter; M = kilogram; T = second; plus the auxiliary units: t = temperature μ = magnetic permeability, k = electric permeability. *English system*: L = ft; M = slug; T = sec; F = lb.

See Appendix D and Reference 4 at the end of this chapter.

where

f = force	α = angle	I = moment of inertia
d = distance	m = mass	ω = angular velocity
t = torque	v = velocity	No. = number

and L, M, T are the fundamental dimensions: mass, length and time.

If confusion arises between lb mass, lb weight and lb force, decide whether it is used as inertia (true anywhere in the universe), weight (true only at the surface of the earth) or force (from whatever source derived: Inertia, gravitational attraction, friction, or elasticity). Then be sure the dimensions express the nature decided upon. (Mass, M; Weight, W; Force, MLT^{-2}).

Thus a convenient tool has emerged, both simplifying the manipulation of units and unifying the concepts across the several fields.

A table included at the end of this book carries all the usual mechanical dimensional units and equivalents (see Appendix D). This table can be used as a source in checking complex expressions for dimensional consistency. The table also lists the dimensions appropriate for 1) heat systems, and 2) electrical systems.

In concluding this portion, note that an expression can be completely reduced to fundamental dimensions, as described, or it may prove simpler to leave it in hybrid or mixed form, as we shall proceed to discuss.

Other units. Although the fundamental dimensions should be used in any new or uncertain case, checking is perfectly satisfactory in any consistent set of units. As in the earlier illustration using oranges and square feet, just be sure the final result gives exactly the same units on both sides of the equation.

As an illustration of using other units to check, let us consider the circuit of a condenser and two resistors as shown in Figure 14.1. The fundamental electrical equations are

$$V_R = IR \tag{1}$$

$$V_C = Q/C \tag{2}$$

$$I = Q/T \tag{3}$$

where

V_R = voltage across a resistance
V_C = voltage across a condenser
I = current flowing (charge per unit time)
R = electrical resistance
Q = electric charge
C = capacitance of the condenser
T = time

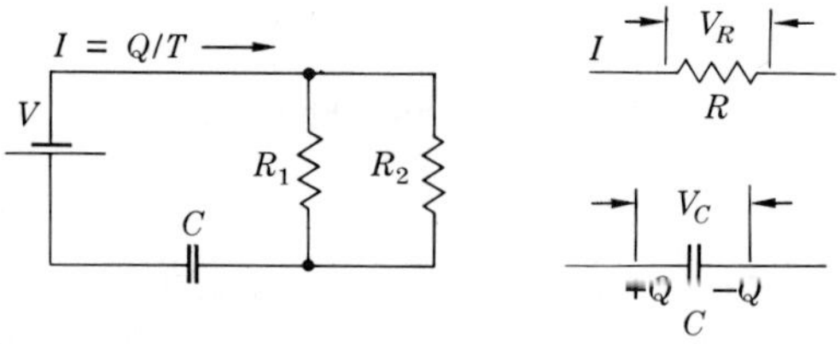

Fig. 14.1 Electric circuit for checking by units, showing resistance and capacitance circuit elements.

We have been given the following relation for the above circuit and we now want to check it dimensionally.

$$V \stackrel{?}{=} I \frac{R_1 R_2}{R_1 + R_2} + \frac{1}{C} \int I dt \tag{4}$$

If we substitute into Equation (4) all the corresponding quantities from the three fundamental equations (1, 2, 3), we find the units all cancel down to V, voltage, units on both sides.

$$V = \cancel{I} \frac{\frac{V}{\cancel{I}} \cdot \frac{V}{\cancel{I}}}{\frac{V}{\cancel{I}} + \frac{V}{\cancel{I}}} + \frac{1}{\frac{\cancel{Q}}{V}} \int \frac{\cancel{Q}}{\cancel{T}} (\cancel{T}) = V + V \tag{5}$$

Thus, Equation (4) checks dimensionally.

Another illustration is taken from the field of heat transfer. The heat loss from one medium equals the heat gained by another, and this is stated for a given geometry,

$$\text{natural logarithm } (q - hAt) - \text{natural logarithm } q = \frac{hT}{c\rho L} \tag{6}$$

or

$$\ln \frac{q - hAt}{q} = \frac{hT}{c\rho L} \tag{7}$$

where

q =	heat flow	=	Btu/hr (Btu hr^{-1})
h =	heat transfer coefficient	=	Btu/hr sq ft °F (Btu $\text{hr}^{-1}L^{-2}t^{-1}$)
A =	area	=	Sq ft (L^2)
t =	temperature	=	°F (t)
c =	specific heat	=	Btu/lb°F (Btu $\text{lb}^{-1}t^{-1}$)
ρ =	density	=	lb/cu ft (lb L^{-3})
L =	length	=	L
T =	time	=	hr

We substitute the units above into Equation (7) and obtain

$$\ln\left(\frac{\text{Btu Hr}^{-1} - (\text{Btu Hr}^{-1}L^{-2}t^{-1} \times L^2 \times t)}{\text{Btu Hr}^{-1}}\right) = \frac{\text{Btu Hr}^{-1}L^{-2}t^{-1} \times \text{Hr}}{\text{Btu lb}^{-1}t^{-1} \times \text{lb } L^{-3} \times L}$$

or

$$\ln\ (\text{number}) = \text{number}$$

hence

$$\text{number} = \text{number}$$

Notice that a very mixed set of units was handled here: Btu, hr, lb, L, t; and yet they cancelled out to unity, or a number.

The introduction of logarithms reminds us that certain quantities are *dimensionless;* their only unit is "number." Those dimensionless quantities which we normally encounter are

Angles (radians are pure numbers)
Trigonometric functions
Exponential functions
Logarithms

Also, the "arguments" of all of these functions must be dimensionless. If, for example, an argument does have dimensions, there must be a counterpart term somewhere else in the equation with the opposite or cancelling dimension. See Equations (6) and (7) above; when natural logarithm q of (6) is changed by putting "q" in the denominator of (7), the argument of the logarithm is then dimensionless.

In the case of trigonometrics or exponentials, check both the argument and the function: That is,

$$\sin\ (\text{argument}) = \sin\ (\text{number}) = \text{number}$$

$$\epsilon^{\text{argument}} = \epsilon^{\text{number}} = \text{number}$$

Checking by Symmetry

One of the applications of logic yields a very useful tool in practical checking. Terms that have analogous symmetrical locations in a system should have analogous or symmetrical locations in the equation. If the actual components of a system may be interchanged without effect, the symbols in the equations must be interchangeable. The converse is true:

Interchangeable symbols in an equation must correspond to interchangeable elements in the physical system.

A sample illustration is in the expansion of $(x + y)^n$. For example, if we saw:

$$(x + y)^4 \stackrel{?}{=} x^4 + 4x^3y + 6x^2y^2 + 2xy^3 + y^4, \tag{8}$$

we would immediately know either the 4 or the 2 was wrong. Since x and y are interchangeable, the expression should be symmetrical, and hence the

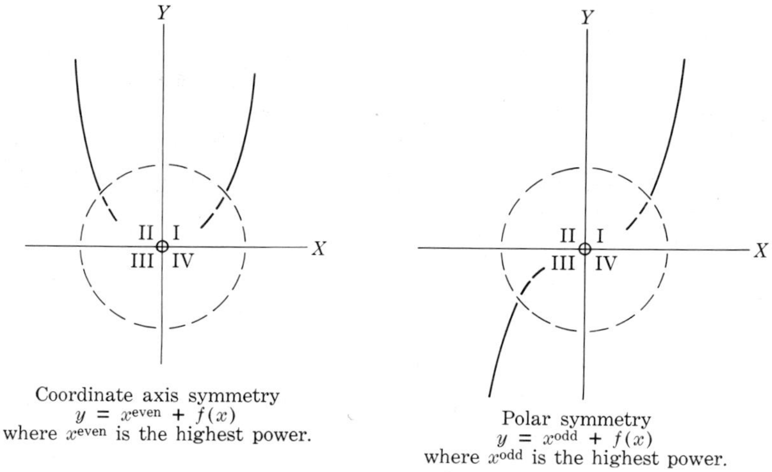

Fig. 14.2 Checking by symmetry. Coordinate axis symmetry vs. polar symmetry for the terminal quadrant of the even and odd exponent polynomials.

coefficients of the x^3y and the xy^3 terms should be identical. What should be the coefficient of the x^3y and xy^3 terms in Equation (8)?

Symmetry also applies graphically. The curve for any expression in which the highest power of the variable is even (x^2, x^4, x^6, . . .) must be symmetrical to the y axis at large values of x. This disregards both the rest of the expression and the shape of the curve nearer the origin, as shown in Figure 14.2.

For odd powers of the variable (x, x^3, x^5, . . .) the curve is similarly symmetrical with the origin (polar symmetry) for large values of the variable.

The next illustrations take us into the systems. Consider series and parallel resistances in an electrical circuit. The correct statement of the total equivalent resistance R_e for the circuit of Figure 14.3 is

$$R_e = R_1 + R_2 + \frac{R_3 R_4}{R_3 + R_4} \tag{9}$$

Normally, a "nonelectrical" person could not vouch for the correctness of Equation (9), but any thinking person would shortly see an error if the following statement was proposed:

$$R_e \stackrel{?}{=} R_1 + R_3 + \frac{R_2 R_4}{R_2 + R_4} \tag{10}$$

Equation (10) must be wrong because R_1 and R_2 are similarly placed in the circuit, as are R_3 and R_4. Thus, logically they must be symmetrically (or

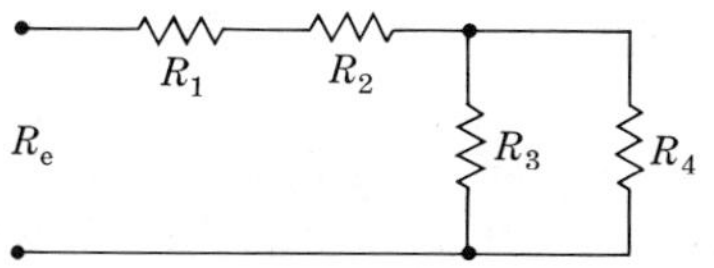

Fig. 14.3 Checking by symmetry. An electric circuit composed of resistances.

interchangeably) placed in the expression. Physically interchanging R_1 and R_2 in the circuit and mathematically in Equation (9) makes no difference in either case. However, making such exchanges between R_3 and R_1 makes considerable difference in the circuit, but no mathematical difference in Equation (10).

However, only certain errors are discernible by symmetry. The error in the following expression cannot be seen through the eyes of symmetry:

$$R_e \stackrel{?}{=} R_3 + R_4 + \frac{R_1 R_2}{R_1 + R_2} \tag{11}$$

Errors which are themselves symmetrical, or double, cannot be discerned by looking for symmetry alone. Other checks must be used as well.

Checking for symmetry also includes algebraic signs and directions of rotation, etc. That is, all the forces along one axis and all the torques in one direction should have the same algebraic sign.

Such checking as we have described in this section applies primarily to the quick, intuitive review, because symmetry analysis cannot compose the correct expression; it only points to areas of certain error. There are problems where inadvertent interchanges of terms leave the expression symmetrical, but now, of course, in error. Figure 14.4 shows an example.

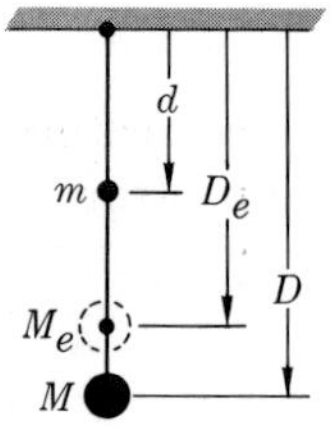

Fig. 14.4 Checking by symmetry and by limits. Double pendulum.

Consider a simple pendulum made up of two masses, m and M. The period of a pendulum is correctly presented as

$$P = 2\pi \sqrt{\text{length}/g} = 2\pi \sqrt{D_e/g}$$

where D_e is the effective length of the pendulum, with the system's center of gravity at M_e. You have been given the equivalent length as $D_e \overset{?}{=} (dM + Dm)/(m + M)$ and wish to check the expression. Although m and M are in symmetrical positions in both the system and the equation, we cannot know so far whether or not D_e is correctly stated. There are other checks that can be applied, and one of them is "limiting-value."

Limiting-Value Checking

Limiting-value checking consists of causing the variables and parameters to take on their limiting or boundary values. Generally these values will be either very small or very large (0 or ∞); but they may also be finite or asymptotic values. With the limiting value substituted, one asks whether the expression meets the known requirements of the problem or system. If the equation doesn't meet these requirements, the equation must be wrong.

For a quick example, the distance dropped by a falling ball is given to you as $S \overset{?}{=} V_0t - \frac{1}{2}gt^2$. Is it correct? At time $t = 0$, the distance S correctly equals zero; but at $t =$ very large, S is negative; although, notice that S is positive at some very small values of t. Since the ball can't *reverse* itself, falling a positive and then a negative distance, the equation must be in error. (Of course, the equation should read $S = V_0t + \frac{1}{2}gt^2$.) The same reasoning would apply if you were given $S \overset{?}{=} V_0t + \dfrac{1}{2g}t^2$. Substitute large and small values of gravity g.

Returning to our problem of the pendulum, which introduced this topic, let us substitute some limiting values in the equation.

The equivalent distance for the equivalent mass was given

$$D_e \stackrel{?}{=} (dM + Dm)/(m + M)$$

In the pendulum, when the two masses lie at the same point, $d = D$; and the equivalent should be $D_e = D = d$; and this proves to be so. It checks.

If $m = M$, the equivalent distance D_e should be halfway between; $D_e = (D + d)/2$ and this checks.

However, when $m = 0$, or $M = 0$, the equivalent should be $D_e = D$ and $D_e = d$, respectively; but this *does not check*. With our above equation, $D_e = d$, $D_e = D$, exactly the opposite.

Further investigation (summing torques about the point) would show that M and m had been erroneously interchanged in the numerator; the correct statement is $D_e = (dm + DM)/(m + M)$.

In limiting-value checking, it is usually most convenient to check the "zero case" first. When checking involves ∞, it is frequently necessary to convert the terms to ratios, with ∞ in the denominator to obtain meaningful results. In the pendulum problem substituting $M = \infty$ (or very large) should result in $D_e = D$ because the effect of m would be negligible. However, the original expression becomes indeterminate:

$$D_e = \frac{dm + DM}{m + M} = \frac{\infty}{\infty} \qquad \text{(indeterminate at } M = \infty)$$

If ratios are obtained, by dividing M, the result is found to be, correctly

$$D_e = \frac{d(m/M) + D}{(m/M) + 1} = D \qquad (\text{at } M = \infty)$$

Similarly, for $m = \infty$

$$D_e = \frac{d + D(M/m)}{1 + (M/m)} = d \qquad (\text{at } m = \infty)$$

A final illustration on limiting-value checking will show how changing the problem-solving methods or using another physical law may be used.

An automobile, weighing 3000 lb, has been moving at 30 mph along the road and the brakes are normally applied to bring the car to a stop. You have worked through the brake friction and the pedal forces and concluded that the decelerating force on the car is $F \stackrel{?}{=} 20$ lb. You can now make an energy check based upon your knowledge of automobiles. The kinetic energy, $\frac{1}{2}MV^2$, must equal the decelerating force times the distance, $F \times D$. Recalling that $M = w/g$, and $g = 32$ ft/sec², we have

$$\text{KE} = \tfrac{1}{2}MV^2 = \tfrac{1}{2} \times \frac{3000 \text{ lb}}{32 \text{ ft/sec}^2} \times (44 \text{ ft/sec})^2 \cong 90{,}000 \text{ ft-lb}$$

or

$$F \times D = 90{,}000 \text{ ft-lb}$$

and thus D should be

$$D = \frac{90{,}000 \text{ ft lb}}{20 \text{ lb}} = 4500 \text{ ft, or approximately 1 mile}$$

Knowing that the normal braking distance is less than 100 ft, you can be sure that the above $F = 20$ lb answer is wrong; the force should be closer to $F = 1000$ lb. This is typical of the use of energy methods to check force problems. This particular problem depended upon the reader's familiarity with the subject. Other uses of energy-method checking depend upon comparison with the boundary conditions and with other knowns in the problem.

Checking Against Experience

The automobile illustration just given is based upon the common knowledge of reasonable values of weights, velocities, and distances which the engineer normally accumulates.

All persons bring into the problem situation much experience about what happens and what can be expected to happen . . . what the probabilities and possibilities are. Incidentally, to repeat an old refrain, the more the young engineer exposes himself to technical and social life experiences, the more frequently and more capably he can apply these valuable tests of personal experience.

When one "sees" water running uphill, he knows he is being fooled by an illusion. There is never any question about whether it is *really* going uphill (resultant of applied forces). The perpetual motion machine is in the same category (conservation of energy). Finally, what has gone into a closed system must either still be there or have come out (conservation of matter). We have these certainties to fall back upon.

The experience of others, experts, associates, and specialists, is not to be disdained. No engineer is "sufficient unto himself," and this holds true particularly in the activity of checking. Solving the problem is what the engineer should do himself; he knows most about it. However, checking can be speeded and put into perspective by short conversations with others. Their observations and evaluations orient the engineer as well as supplying him with concrete numbers, or yes's and no's.

Checking also applies in the social and psychological fields. People do not freely continue a practice or continue to buy a product unless it provides something of value for them.

The basic store of our personal experiences and our conclusions from them are a reservoir for checking that, first, is little enough used, and second, is extremely important as a backstop. After all, the layman, the ordinary person on the street, is the final judge of engineering effectiveness. All that the layman can bring to bear for his judgment is a wide-ranging composite of everyday experiences. These experiences generate challenges and questions, though these may be unvoiced or not even verbalized mentally. Engineers must ask the same questions of themselves.

After the engineer has the sought-for answer, the next most important outcome of any design problem is a separately-checked answer, secured by other thinking. The engineer must check his own work, and he must develop shortcuts and rules of thumb to check the work of others including, particularly, those who work for him. He must expect to be challenged to support any particular conclusion that comes out of his design solutions.

In the typical engineering department, it will be noticed that the drafting and specification checkers are the most experienced and highest paid people in their work groups. This fact points up the importance of those people who are finally charged with the accuracy of the work in its detailed stages.

SUMMARY

Satisfactory operation is the engineer's design goal; but up to the time of such experimental proof, checking is his best reassurance.

1. Running checks should be made, from start to finish, to be sure both that the thinking is rational and pertinent, and that the numbers involved are within expected bounds.
   Plot running curves on sequential computations, either on scratch paper or on the permanent data sheet. If the orderly sequence of points is interrupted, the last point can be validated, or the new direction of the curve can be re-interpreted.
2. The several techniques of checking should be kept available and used as they are appropriate. Only practice will make these fully useful:
   (a) Verifying consistency of dimensions or units.
   (b) Verifying corresponding symmetries between system and equation.
   (c) Validating all the limiting-case situations.
   (d) Applying common sense or reasonableness tests.
   (e) Comparing notes with experts and specialists.
3. The newness of the problems and the single-handedness of the engineer leave him with no alternative but careful checking, in the face of the extreme risks to both himself and his client. The checking

must cover all phases: The *reasoning* or selection of the principles, the *application* of the principles, the arithmetic or *routines* employed, and all the copying or *typography* involved.

4. Alternate methods must be used for checking the original procedures. Technical problems generally are solved by:

   Summing forces or torques about a free body;
   Summing electric currents or fluid flows into a junction;
   Equating inputs to outputs plus storage.

   Alternate methods employ working through:

   Conservation of matter or energy
   Change of energy (force $\times$ distance = energy)
   Change of momentum (force $\times$ time = momentum)
   Thermodynamics:

   First law: Heat and work are interchangeable.

   Second law: Conversion of heat to work is dependent on the temperature at which conversion occurs.

The engineer's real need in the checking process is to walk down a different path to the solution. Always make certain that it *is* different before feeling confident about the agreement. Look for the simplest expression of the essential principles. A simple method, carefully checked and double-checked, is better than a sophisticated check sloppily executed.

EXERCISES

14.1 A handbook gives the following formula for the diameter D of a round shaft to transmit P hp at N rpm with a shaft stress of S lb/sq in. Does the formula make sense?

$$D^3 = 321{,}000P/NS$$

(*Hint:* Assume all the torque is transmitted by uniform shearing forces in a band within the outer one-third of the radius of the shaft.)

14.2 A handbook gives the moment of inertia of a hollow circular cylinder about its axis as

$$I = \frac{\pi ma}{2}(r_1^4 - r_2^4)$$

where r_1, r_2 are outer and inner radii, respectively, in feet.
a = length (feet)
m = mass per unit volume of body, lb/ft^3
What are the dimensions of I?

14.3 In converting from one circuit to another (a delta to a star, Figure 14.5) the following equivalents were given to you:

$$R_a \stackrel{?}{=} \frac{R_1R_3}{R_1 + R_2 + R_3} \qquad R_b \stackrel{?}{=} \frac{R_3R_2}{R_1 + R_2 + R_3}$$

Check these equations for suspected error. What checking technique applies?

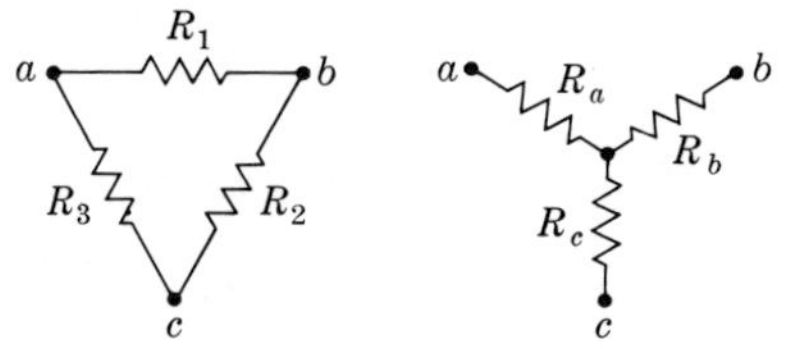

Fig. 14.5 Checking exercise. Electric circuit "star-delta" transformation.

14.4 The volume of a spherical zone (Figure 14.6) is given as:

$$V = \frac{\pi h}{24}(3a^2 + 3c^2 + 4h^2)$$

Check this equation at all the limits of the values of a, c and h.

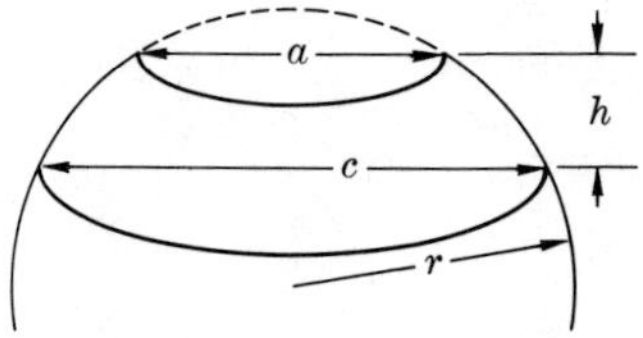

Fig. 14.6 Checking exercise. Volume of a spherical zone.

REFERENCES

1. Ver Planck, D. W. and B. R. Teare, Jr.: "Engineering Analysis," chap. 6, John Wiley & Sons, Inc., New York, 1954.
2. Smith, R. J.: "Engineering as a Career," 2nd ed., p. 268, McGraw-Hill Book Company, New York, 1962.
3. Wilson, E. B., Jr.: "Introduction to Scientific Research," pp. 308–313, 343–344, McGraw-Hill Book Company, New York, 1952.
4. "Marks Mechanical Engineers Handbook," 6th ed., Sec. 1-73, McGraw-Hill Book Company, New York, 1964.
5. Ipsen, D. C., "Units, Dimensions and Dimensionless Numbers," McGraw-Hill Book Company, New York, 1960.

Optimization Fifteen

"Knowledge and timber shouldn't
be used much until they are seasoned."
Oliver Wendell Holmes

Meanings

Throughout the book we have endeavored to promote, in fact we have insisted upon, the broadest professional attitudes in practicing engineering design. This breadth embraces the whole range of technical *and* nontechnical concerns that the engineer must have, both as an individual and as a technologist. His problems plead for breadth at all stages, from the needs analysis onward; but probably the keenest demand comes with the decisions of *optimization.*

Optimization, as a word, has come into more popular use in recent years because of the newly available optimum-seeking techniques and facilities and because of the growing recognition of the necessity for optimization. We *do not have* all the minerals, all the water, even all the fresh air we need. But there is a difference between optimum and optimal, between seeking and attaining, and we shall discuss this difference in a moment.

Two other words must always be implied and understood when optimization is mentioned, for without these two words, "optimizing" is meaningless. These two words are *system* and *criteria*. System defines the boundaries of our concern: The components, the interrelations, the extent of our optimization. The criteria define our objectives and measures of value; hopefully, the criteria tell us how to make decisions. The more clearly the system and the criteria are set forth, the more appropriate and the more effective will be our optimizing. Even with the system and the criteria given, we must yet discover *how* to optimize.

One other factor, never neglected in practice but frequently not recognized in theory, is the optimization of the engineer's time and effort, or that of his group or employer. Each project, as well as each problem, deserves attention to optimality. In the project, the probability of error or loss is balanced against the cost of additional development time, considering the expected increased confidence and probability of final success.

Systems vs. Subsystems

Optimization applies only to entire systems in any real sense. Subsystems and components are always sub-optimal or sub-optimized. We define a *system* as a set of relevant design interrelationships which for a given discussion are relatively isolated and independent in purpose and use; whereas *subsystems* may be any aggregate of relationships within a system. Notice that a system may be arbitrarily identified and bounded. In an earlier chapter we mentioned that the systems of engineering concern were machines, processes, structures, information systems, and living-systems. Let us illustrate how any individual system fits into a hierarchy of systems.

A telephone system illustrates many of the various shades and flavors of the optimizing problem. But how much do we include in a telephone system? Many descriptions fit the various telephone systems in operation now. Starting at the world level, a telephone system may be:

1. Intercontinental Radio and Satellite Communications system.
2. American Telephone and Telegraph Company.
3. Local regional telephone company.
4. Independent village telephone exchange, with its switchboard and customers.

Of course, there are systems and subsystems within all the categories, even within the smallest exchange just mentioned:

1. Ringing system.
2. Power supply system.
3. Lightning protection system.
4. Voice to current transformation (speaker-receiver).

Arbitrarily, in a given problem, each of these "systems" might be considered the total concern and yet none, even the largest, is a fully self-sufficient system in the complete sense of the word. Even the world satellite system involves rocketry and launching bases, two subjects never normally of the remotest interest to "telephone engineers" doing their day-to-day job. When one attempts to optimize his particular system, just how much does he include? When he cuts out a small portion of the total to sub-optimize, what is he overlooking? Does he know the whole system well enough to select the portion to optimize? Of course, almost any object of engineering interest can be improved, but our present purpose is to distinguish between improvement and optimization.

Optimum vs. Optimal

The criteria for optimality are seldom, if ever, pure and exclusive. They usually conflict with the facts of life. As we have said before, we cannot have at the same time the most economical *and* the safest device. We cannot have the greatest number of features *and* the fewest moving parts, or the longest life *and* the highest efficiency. We cannot have the maximum chemical reaction-yield *and* the widest process tolerances. Only a few of the "mosts" do realistically go together:

Smallest size generally goes with lightest weight.

Fewest parts generally mean greatest reliability, lowest cost, and simplest manufacture.

In general, the desirables do not harmonize, if in fact they don't actually conflict. This is illustrated by the following specific examples taken at random:

In laundering, the degree of cleanliness must be balanced against textile wear.

In mining, the purity of the ore is balanced against cost of beneficiating.

In paper making, increasing the traveling speed of the newly-formed sheet must be balanced against the risk of clean-out and restart if the sheet tears and piles up.

In construction, the desirable reduction of water ratio in concrete must be balanced against difficulty in filling the form.

In desalination of water, increasing the purity of the water output diminishes the process flow and increases the maintenance.

In an electric motor, higher starting torque means higher current in-rush and larger size feeder wires or an undesirable dip in the line voltage.

In all these cases, the actual operating value chosen by the engineer (he must eventually do the choosing) is a compromise among the several alternatives. In the electric motor example, the engineer must increase the size of the motor's iron core or the size of the wire if the starting current is dimming the lights too much, or blowing a fuse too frequently. His optimum is seldom the technical *best* he knows, only the better of conflicting alternatives. It is never optimum, only optimal. He seeks the best but never attains it.

The act of choosing is called the "trade-off" in present parlance. Our motor engineer in adding iron to the core has just traded weight and cost for reduced starting current. If he had the choice, he might have traded outside diameter (size) for starting torque. In any case, he did not achieve all he could have technically. We are defining *optimum* as the technical best considering the state of the art, and *optimal* as best within the total constraints of the accepted system. Of course, not even the optimum achieves the conceptual "ideal", so engineers still try to improve the "optimum."

Criteria

The word criterion was used in Chapters 12 and 13 in connection with symbolic modeling, where the "Criterion Function" was called the fundamental design equation and was mathematically expressed as

$$\text{Criterion Function,} \quad CF = \sum_i a_i x_i \quad \text{(to be maximized or minimized)}$$

In general, "criteria" cover all the influences and values involved in measuring optimization. Many of the important criteria, however, are not technical at all; very real issues arise because of the law, or because of psychological, political, artistic, or humanitarian considerations.

The engineer uses the various criteria as any professional does. In other words, he must be well enough educated and willing enough to accept responsibility, that he will share the burden of decision with his superior or client or customer. It may be he must accept all the responsibility himself, since his customer is not qualified or refuses to do so. A frequent illustration involves the safety of citizens using a product, or employees using a machine or tool. The engineer decides that it is safe (enough) and recommends construction. The client or customer can only accept or reject that recommendation. (Safety is measured by reduced probability of failure.)

Where possible, it is naturally preferable for the engineer to quantify the criteria, as discussed in Chapter 13. But, whether they are quantified are not, the engineer cannot escape consciously applying the criteria, in whatever form they appear. He can never make excuses: "I didn't know it infringed the patent," or "How did I know the operator would do that stupid thing?" or "It seems artistic enough for me!" The engineer searches for and tests with all the criteria issuing from the project and from the culture itself.

The value system of the culture presumably has been absorbed by the engineer working in that culture. The value system ranges broadly, from interpreting the "sacredness" of human life to deciding what constitutes personal privacy under various living conditions. How much armor is added to protect a fighter-plane pilot, or how far apart are the fireplace grills placed in a public park? An individual's value system covers relative compensations, class and racial distinctions, if any, accepted behaviors and protocol (morals), standards of beauty, equivalents of exchange and the place and kind of religion and philosophy in his society. From these examples, it is seen that the American engineer working in Indonesia or Nigeria or Taiwan must first clearly recognize that he is a foreigner, and that he will generally make assumptions and decide questions on bases different from those of his hosts.

When to Optimize[4]

Although optimization sounds like a desirable procedure at all the stages of design, don't expect much during the exploratory or developmental stages of an art. At that time, the engineer just cannot know enough of the whole field to do much optimizing. He must be satisfied if "the wheels just turn over." Optimization then is far down the road. The cause and effect mathematical relationships must be known and agreed upon before anything like real optimization is possible. As a rule of thumb, perhaps 90 per cent of the design principles and laws must be established and familiar, and a considerable period of practice must already be history. That is, it would have been foolish to have attempted to optimize the internal combustion engine in 1900, except on a restricted or component basis. Fuels, materials, combustion, and mass-machining operations were all in their relative infancy. Optimization really came 20 or 30 years later. Certainly the engineer does the best he can, but we are referring to the optimization of a system by an intensive thorough-going attack. In the early development stages, order-of-magnitude errors and yet-to-come experimental information tie the hands of the would-be optimizer, or "pull the rug from under" his careful predictions.

In summary, we return then to the several general fundamentals of optimization:

1. The engineer's objective is always to seek an optimum solution, to best meet the identified needs, using resources which are always limited, both for the project and the product.

2. Only independent systems can be optimized, and the chosen solution is never really optimum (the best technically conceivable); it is only optimal (the best compromise).
3. All subsystems are sub-optimized, due to known, handicapping but unavoidable system constraints.
4. Criteria for measuring the results of optimizing cover the full spectrum of the client's interest, and all must be applied in due priority.
5. Optimize only when the technical field is well enough developed, and you know enough about it. Optimize what you can; settle for "satisfactory operation" for the balance.

Methods of Optimization: Subjective

The first, most frequent and broadest method of optimization is the subjective kind. This is used intuitively by every person making the ordinary decisions of living. What is the shortest way to the store? How best should the furniture be arranged? What is the best way to allocate my income? These decisions all consist simply of mental survey and appraisal of the alternatives. This method of optimization takes several forms in design:

1. *Mental juggling of the parameters.* Whether on quantitative or qualitative matters, the engineer continually performs the mental scanning and sorting of choices.
2. *"One Hoss Shay."** Where the choice exists, all parts should be made of equal strength, equal durability, equal life. Ideally, this yields maximum economy, maximum efficiency.[4]
3. *Collect several functions into one part.*[3,4] These results come only with conscious effort from the start. Examples are: Using the skin of the airplane for structural strength; modifying a decorative trim strip to serve also as the required handle; using an auto engine's cam shaft also to drive the fuel pump and the ignition distributor; using a helicopter blade to carry a jet propulsion engine at its tip; using the lighting system to supply the heat in an office building; making the water tank legs also the supply and delivery pipes.

Methods of Optimization: The Use of General Principles

Another category which all experienced engineers use consciously or unconsciously consists of their collection of general principles from the expertise of different fields. A few examples are listed on the next page:

* Oliver Wendell Holmes' legendary carriage that was built so well that it lasted "one hundred years to the day," then fell apart all at once.

1. *In the use of materials for structure:*[3] Design to use tension members, instead of compression or bending members (reduce the number of compression members to a minimum); avoid stress raisers, such as notches, sharp fillet corners, perforations; spread material as far out as possible from the axis of the structure (i.e., use thin-walled tubes, wide-flanged beams); in dynamics, keep the major forces between stationary interfaces instead of moving interfaces, etc.
2. *In fluid circuits:* Modulate or control at low pressures and/or low flow rates if possible, etc.; avoid sources of sudden shocks or shock waves, such as snap-acting valves, drastic changes of section, etc.; on rotating seals, make any leakage flow toward the center, *against* centrifugal force.
3. *In electric circuits:* Switching, modulation and control should be done at the lowest energy levels (volts and amperes); in communication, fidelity is gained by frequency modulation at the cost of increased bandwidth, whereas bandwidth reduction is gained by amplitude modulation at the cost of noise; in inductive power devices, one can generally trade copper for iron and vice versa (e.g., larger conductors for less iron in the magnetic path); transistors can replace vacuum tubes except under high-frequency, high-power applications, and that application limit has been rising steadily over recent years.
4. *In chemical processes:* While exposed reactant area is normally maximized in order to maximize rate of reaction, heat transfer or separation of products may impose a different area limit. Temperature or pressure changes normally control separation of products; but under certain critical conditions, a third component is necessary to accomplish the process. At a mathematical optimum of distillation-column geometry, most designs are dynamically unstable.
5. *In energy conversion:* The greater number of cylinders in an engine, the greater the power output per cubic inch of displacement and per pound of weight; because of smaller size, the speed may be greater also. In ship propulsion, the design objective is to minimize the production of waves in the wake, to put as little energy as possible into the water.

The preceeding examples are only a sampling to demonstrate the many aspects of optimization that rest largely upon experience, and yet contribute vitally to the results.

Methods of Optimization: Graphics

There are many ways in which graphics serves optimization, some of the ways having been referred to in Chapter 11.

1. *Conceptualizing and visualizing:* The graphic visual form, and our freedom to change it, help in optimizing size, shape, relationship, proportion, etc. We can easily experiment by sketching "better" forms, "better" use of the given space, etc.

2. *Rapid generalization:* Envelopes of motion, loci, or successive approximations help us find the "best."
3. *Mathematical equivalents:* Differentiation, curve plotting, and nomographs all can graphically speed mathematical maximization, etc.

The examples shown in Figures 15.1 and 15.2 are simple illustrations that include several of the above graphical suggestions. Figure 15.1 shows how a prompt and relatively accurate answer can be obtained graphically for a maximizing problem that would require much more time to solve conventionally by differential calculus. The question is: What is the longest

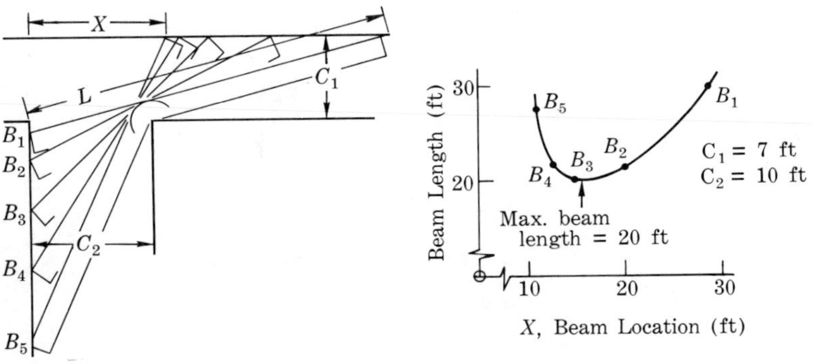

Fig. 15.1 Graphical optimization. Maximum length of a beam which can be moved around a corner in corridors.

beam of given width that can be inched around a corner between perpendicular corridors of widths C_1 and C_2?

The steps are:

1. Draw several locations of the beam: B_1, B_2, B_3 over a range.
2. Plot the length L vs. a coordinate X.
3. Find *minimum point* on the curve of L vs. X.

Answer: This minimum point is the maximum beam length sought.

Another example of optimization by graphics is shown in Figure 15.2, wherein the maximum angular velocities of an oscillating shaft are to be equalized between the forward and return strokes. Although these equal maximum angular velocities can be obtained analytically as well, this combination of velocities with visual lobe symmetry is virtually impossible except by graphics.

A rotating crank-arm CA in Figure 15.2a drives the oscillating shaft S by means of lever L through connecting rod CR. This is a "four-bar linkage." In many applications where space is at a premium, the crank CA must be

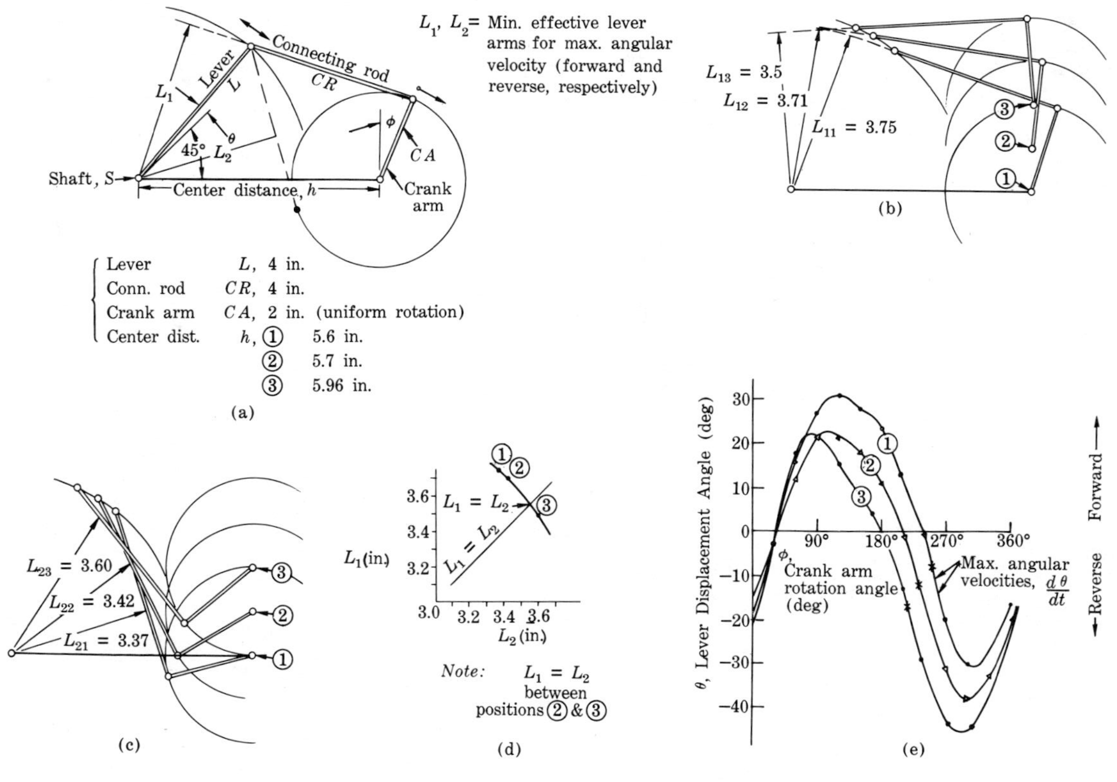

Fig. 15.2 Graphical optimization. Equalizing the maximum shaft velocities and improving the displacement angle symmetry for a four-bar linkage.

The problem is to find the center distance h which equalizes the lengths L_1 and L_2 of the effective lever arms.

placed close to the lever L and the resulting oscillatory motion of S is non-symmetrical and nonsinusoidal, producing very different forward and return angular velocity curves. In certain product applications, these velocities should be equalized, and this can be approximated by adjusting the center distance h, in Figure 15.2a, until the minimum effective lever arms L_1 and L_2 are equal.

The method shown uses interpolation by plotting these effective lever arms, L_1 and L_2, forward and reverse (Figure 15.2b,c,d), and finding the value of h where $L_1 = L_2$. The plot of displacement (Figure 15.2e) shows how distorted the wave is for other center positions, such as No. 1, and how relatively symmetrical the wave is at position No. 2. Thus, by graphics, a center distance near position No. 2 is chosen because the forward and reverse angular velocities are relatively equal, and the displacement angle θ is most symmetrical and uniform. The center of symmetry would be moved from the present value of $\theta = 0°$ to $\theta = -7°$ to balance the lobes, in Figure 15.2e.

Methods of Optimization: Analytical

True engineering optimization is such a complex, demanding, and sophisticated process that the usual terms and concepts grossly over-simplify the realities.

When we use the words "analytical optimization," we imply that we can express all the parameters, variables, and constraints in explicit mathematical terms, and if not all of them, at least an adequate number and variety. Strenuous as *this* is, it still meets only half the difficulty, because we must know what to suggest, what to consider, what to analyze for our given case. The selection of the alternatives is a crucial aspect of optimization, as it is of any important stage of design. Even if we can express many of the conditions mathematically, let us agree that most of the mathematics is only capable of maximizing or minimizing a limited portion of an engineering problem. Of course, with extensive modeling talent and effort available, and with the use of computers to process vast quantities of data, more and more "optimizing" can be done analytically.

The General Analytically Optimized Problem

Where analytical procedures are available, and when the criterion function[1] can be adequately expressed quantitatively, the optimizing problem takes the form of the following three sets of equations. The first is the value expression to be maximized (or the cost or penalty to be minimized); the second set constitutes the physical principles of operation; and the third represents the boundary conditions or limits imposed by the problem situation. Setting up and processing these equations or their successors comprises what is usually called *mathematical optimization.*

1. Value expression: $CF = f(u, v, w, x, \ldots)$ (to be maximized) (1)

2. Functional constraints: $FC_1 = F_1(u, v, w, x, \ldots) = 0$ (2)
$FC_2 = F_2(u, v, w, x, \ldots) = 0$
$FC_3 \ldots$ etc.

3. Regional constraints: $\text{lim}_1 \leq R_1(u, v, w, x, \ldots) \leq \text{LIM}_1$ (3)
$\text{lim}_2 \leq R_2(u, v, w, x, \ldots) \leq \text{LIM}_2$
$\text{lim}_3 \ldots$ etc.

We shall illustrate these conditions in three-dimensional space, first graphically and then algebraically. Even though we cannot visualize surfaces in four (or more) dimensional space, remember that mathematics does not have such limitations; it can deal with n-dimensions (hyperspace) and generally arrive at needed solutions.

The criterion function equation (1) is a curved surface or family of surfaces, in simple cases having some sort of a domed shape, as shown in Figure 15.3. The peak of one dome (for two variables u, v) or the peak of the highest dome (for three variables u, v, w) is the desired maximum of the criterion function CF. The location of the peak defines the values of u, v, and w at which the optimum occurs. In general the locus of the maxima wanders through space, so no simple section cut through the surfaces necessarily leads to the maximum quickly.

The criterion function is shown as a simple surface; and it is correspondingly simple to maximize with so few variables. Usually, however, there are many variables acting in physically constrained relationships, such as forces, sizes, and weights in a bridge truss or the currents and resistances in electric circuits. The functional constraints, Equations (2), relate variables, and generally allow the elimination of surplus variables, reducing the criterion function to the smallest possible number. We will illustrate this shortly.

The last set, Equations (3), provide boundaries within which the solution must lie. In the general case, the expressions $R(u, v, w, x)$ describe curved surfaces; but for our simple illustration we are showing them as planes for maximum and minimum values of u, presumably taken from the problem situation. These limits, lim_1 and LIM_1, are generally man-made or practical limits, although nonetheless real and necessary. Notice in our illustration, Figure 15.3, the left-hand limit forbids the higher available maximum value of CF_{4m} on w_4.

One other comment before we illustrate algebraically. Even though the *CF* surfaces are shown as simple domes, in general they may be saddle-shaped or waved with higher adjacent peaks or shaped otherwise. More

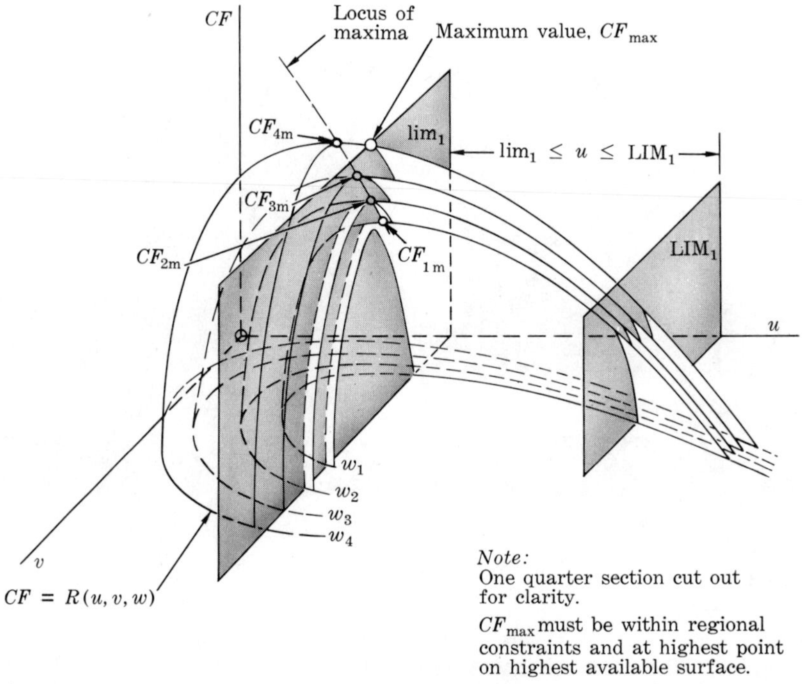

Fig. 15.3 Family of generalized criterion function surfaces. General variables u, v, w are shown to form maximum values of *CF* along the "locus of maxima." One regional constraint is shown as two planes $\text{lim} \leq u \leq \text{LIM}$ within which the maximum value is shown as CF_{max} along the crest of w_4.

advanced mathematics, beyond our present scope, enables the engineer to find the absolute maximum or supremum. A number of sophisticated optimum-seeking mathematical techniques are now adapted to computer use; but the purpose is still the same in all cases: Find the peak value of a surface within the permitted regions.

BOOK PROJECT EXAMPLE: Bucket Conveyor Optimization

The bucket type of conveyor is preferred when sand or gravel is to be raised an abrupt vertical angle as in Figure 15.4. If we are given an approximate flow in pounds per hour, we can use a rapidly moving belt

(a) (b)

Fig. 15.4 Transit-mix concrete loading plants, showing the bucket conveyors for lifting the aggregates. (a) With buckets spaced $1\frac{1}{2}$ ft; (b) With buckets adjacent.

with few buckets or a slowly moving belt with many buckets. The former consumes more power and the latter requires more capital equipment. Let us find a geometry that minimizes the total annual cost.

This cost, of course, is the sum of the energy cost C_p and the interest cost C_i on the capital investment, and can be expressed as

$$C = C_p + C_i \qquad (\$/\text{yr}) \quad (\text{to be minimized}) \tag{4}$$

We will use the following nomenclature to expand this equation into an expression of useful variables:

Let

K = energy cost, \$/hp-hr

H = time in operation, hr/yr

Q = flow rate, lb/sec

A = annual output, lb/yr ($= 3600\ HQ$)

V = belt velocity, ft/sec ($= bQ/m$)

h = lift height, ft

I = interest rate, \$/\$/yr or %/yr

C_F = cost of basic frame and belt, \$

P = power required, hp

$P = Q(h + V^2/g)/550$, hp, from Chapter 9

C_B = cost of buckets, \$ ($= mNC_b$)

C_b = cost of buckets per unit capacity, \$/lb

N = number of buckets, bkt, ($= 2h/b$)

b = bucket spacing, ft/bkt

m = capacity per bucket, lb/bkt

g = gravity, 32 ft/sec² List (5)

The following auxiliary equations bring in the parameters for velocity, frame cost, number of buckets, and annual output, respectively:

$$V = bQ/m \quad \text{(ft/sec)} \quad C_B = mNC_b \quad (\$)$$
$$N = 2h/b \quad \text{(bkt)} \quad A = 3600\ HQ \quad \text{(lb/yr)} \tag{6}$$

Now, using Equations (6) the power term C_p in (4) can be written in progressive forms leading to an expression of useful variables. To begin with, the cost C_p is the product of the unit energy cost K and the amount of energy used per year HP.

$$C_p = KHP \qquad (\$/\text{yr}) \tag{7}$$

Substituting for H, P, Q, and V from above, we find the power cost to be

$$C_p = KA(h + b^2Q^2/m^2g) \times .505 \times 10^{-6} \qquad (\$/\text{yr}) \tag{8}$$

Now, proceeding on to the interest portion, we write the product of the interest rate and the total investment,

$$C_i = I(C_F + C_B) \qquad (\$/\text{yr}) \tag{9}$$

Then, using $C_B = mNC_b$ and $N = 2h/b$, we now have

$$C_i = I(C_F + 2mhC_B/b) \qquad (\$/\text{yr}) \tag{10}$$

Thus the cost equation (4) becomes the following, using the expanded forms of C_p and C_i, from (8) and (10) respectively,

$$C = KA(h + b^2Q^2/m^2g) \times .505 \times 10^{-6} + I(C_F + 2mhC_b/b) \qquad (\$/\text{yr}) \tag{11}$$

In its direct algebraic form, Equation (11) holds true with any one or more of its symbols as variables, or for that matter, as constants. In the design situation, however, certain quantities are relatively fixed, while others naturally may vary. For instance, we know the cost of electric power, the interest rate on borrowed money, the planned annual output, and we may choose reasonable values for height to lift and unit cost per bucket, etc. The natural variables for this equation seem to be the spacing b, the bucket capacity m, and possibly the flow Q or hours per year H. After we fix the design, these are no longer variables, but single values, or parameters. Now, treating them as variables, we will find their values for minimum cost by differentiating and equating the derivative to zero.

$$\frac{dC}{db} = KA(0 + 2bQ^2/m^2g) \times .505 \times 10^{-6} \times I(0 - 2mhC_b/b^2) = 0$$

Thus

$$b^3 = 1.98 \times 10^6 m^3 ghIC_b/KAQ^2$$

or

$$b = m(1.98 \times 10^6 ghIC_b/KAQ^2)^{1/3} \quad \text{(for min. } C) \tag{12}$$

This, then, is the value of b for minimum total cost C. Notice that we could have converted b/m into a single variable, in the original equation (11), in which case one can show that the minimum cost C would have occurred at the same value as that deduced from the Equation (12), namely

$$b/m = (1.98 \times 10^6 ghIC_b/KAQ^2)^{1/3} \quad \text{(for min. } C) \tag{13}$$

The graphical equivalent of our cost equation (11) is shown in Figure 15.5, where both the criterion function surfaces $C_0C_1C_2$ and the regional constraints $1 < b < 10$ and $20 < m < 100$ are depicted, respectively. The several surfaces nesting upward are those with increasing values of Q. Notice that we have solved a minimizing problem, so the surfaces are troughs, instead of domes as shown in Figure 15.3, although the purposes and processes are similar. Small chart inserts show the C–b plane profile and the C–m plane profile, both taken through a common minimum point C_{min}. The locus of the minima, $b/m =$ constant, is a straight line through the origin, at a ratio b/m, which is proportional to $Q^{-2/3}$, as seen in Equation (12). Equation (11) also shows that the cost C rises with increasing values of Q, although not proportionally.

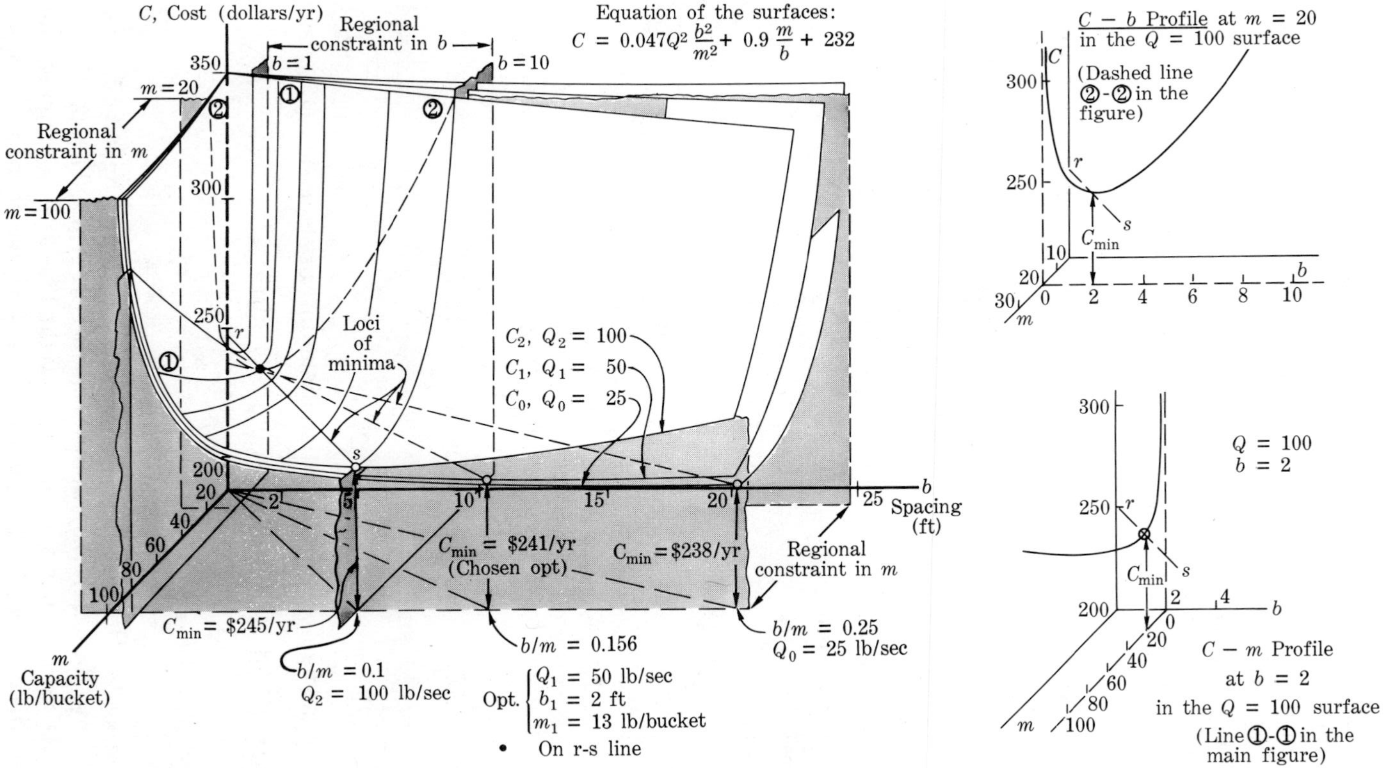

Fig. 15.5 Family of criterion function surfaces for a conveyor problem. Minimum value of cost shown at $C_{\min} = \$241/\text{yr}$. Optimum values of bucket spacing, bucket capacity and flow rate shown as b, m, and Q respectively, chosen $b = 3.1$, $n = 20$, $Q = 50$.

We may find out how well we check with current practice, by substituting realistic numbers into the two equations, (11) for the general cost, and (13) for the minimum cost. At the same time we can suggest regional constraints, showing them in parentheses after the given number, in List (14) below.

Referring back to our nomenclature table, List (5), we estimate as follows:

K = energy cost:	0.015	\$/hp hr	
H = time in operation: ?		m/yr	$(200 < H < 8760)$
P = power required: ?		hp	
Q = flow rate:	50	lb/sec	$(15 < Q < 75)$
A = annual output:	200,000,000	lb/yr	
V = belt velocity: ?		ft/sec	
h = lift height:	75	ft	$(25 < h < 125)$
I = interest rate:	0.06	%/yr	
C_F = frame cost:	\$2000		
C_b = bucket cost:	0.1	\$/lb	$(0.1 < C_b < .5)$
N = number of buckets: ?			
b = bucket spacing: ?		ft	$(1 < b < 10 < 2h)$
m = bucket capacity: ?		lb	$(20 < m < 100)$

List (14)

Substituting these numbers into the general cost equation (11), we find

$$C = 1.5(75 + 78.b^2/m^2) + .06(2000 + 15m/b)$$

or

$$C = 117.b^2/m^2 + .9m/b + 232 \quad (\$/\text{yr}) \tag{15}$$

The same figures applied to the minimum cost equation (13) yield

$$b/m = [(1.98 \times 10^6)(32)(75)(.06)(0.1)/(.015)(2 \times 10^8)(50)^2]^{1/3}$$

or

$$b/m = 0.156 \text{ ft/lb} \quad (\text{for } C_{\min}) \tag{16}$$

The cost for this $C_{\min}$ value of b/m is then found by substituting (16) back into Equation (15), and the result is \$241/yr. This particular value holds for any point along the straight line $b/m = 0.156$ at the bottom of the surface C_1. Our above choice of Q determined $C_{\min}$.

For other values of Q, Q_0 lower and Q_2 higher, corresponding surfaces C_0 and C_2 define the resulting lower and higher annual costs, respectively.

As a check on practicality, if we examine Figure 15.4, we see the spacing in these particular cases is 1 or 2 ft, and each bucket holds about $\frac{1}{3}$ cu ft or 33 lb of sand or gravel. Thus $b/m = \frac{2}{33} = .061$ ft/lb, or about $\frac{2}{5}$ of the amount we deduced through Equation (16). The difference probably is due both to different assumptions and to an optimization which includes the storage hoppers, making the conveyor larger so the hoppers could be smaller. For whatever reason, the difference emphasizes the necessity of choosing a sufficiently inclusive system to justify the time and effort of optimizing.

To conclude this example, let us return to the original optimizing equations (1), (2), (3), and identify the corresponding equations in the example we have just finished. We will use the same nomenclature:

1. Value expression:

$$C = KA(h + b^2Q^2/m^2g) \times 550 \times 10^{-6} + I(C_F + 2mhC_b/b) \qquad (11)$$

or

$$C = f(b, m, Q, \text{etc.}) \quad \text{(to be minimized)}$$

2. Functional constraints:

$P = Q(h + V^2/g)550$	or	$F_1(P, Q, h, V) = 0$		List (5)
$V = bQ/m$		$F_2(V, b, Q, m) = 0$		
$C_B = mNC_b$		$F_3(m, N) = 0$	and	
$N = 2h/b$		$F_4(N, h, b) = 0$		
$A = 3600\ HQ$		$F_5(H, Q) = 0$		(6)

3. Regional constraints:

$200 < H < 8760$	or	$\text{lim}_1 < H < \text{LIM}_1$	
$15 < Q < 75$		$\text{lim}_2 < Q < \text{LIM}_2$	
$25 < h < 125$		$\text{lim}_3 < h < \text{LIM}_3$	
$.1 < C_b < .5$		$\text{lim}_4 < C_b < \text{LIM}_4$	
$1. < b < 10$		$\text{lim}_5 < b < \text{LIM}_5$	
$20 < m < 100$		$\text{lim}_6 < m < \text{LIM}_6$	List (14)

The rest of the symbols were constants, of course.

Our choice of the exact optimum point on the "cost-surfaces" depends upon our knowing reasons for particular values between the regional limits. In the real case, we would search for those reasons and finally settle upon a particular set of values. With our present limited knowledge, we will simply tentatively choose mid-range values and, using our equations, close with these numbers:

$b = 3.1$ ft/bucket	hence,	$V = 7.7$ ft/sec
$m = 20$ lb/bucket		$N = 40$ buckets
$Q = 50$ lb/sec		$H = 1100$ hr/yr
		$P = 7$ hp
		$C = \$241$/yr

This point occurs at the intersection of $m = 20$, $b = 3.1$ in Figure 15.5.

Invention for Optimizing[2,3,4]

Thus far in this chapter we have stressed systems, criteria, and analytical methods for optimizing, and have said little about the creating the configuration which we expect to optimize. All optimizations depend upon an initial configuration of elements, and only the most rare and sophisticated optimizing techniques will add or subtract elements; most techniques simply proportion the elements or assign values to the parameters. It may seem obvious advice but we desire to emphasize that the engineer must know enough about the field to choose correctly his primitive form or shape in accordance with the fundamental principles at work. Otherwise his careful optimization of the parameters simply might be a waste of time. The invention of the initial configuration is critical to useful optimizing.

In Chapter 9 we worked through an illustration of finding the best angle for a guy-wire pole support (Figure 9.1). There we allowed only one variable, the angle of the guy wire to the pole. From the angle we computed the minimum weight of wire. Now, we will add another variable, the pole itself. For simplicity, we will assume the pole is a steel pipe* able to withstand compressive loads, weight for weight, like a cable supports tension or stretching loads.

Having now both wire and pole as variables, we must propose a configuration or shape to optimize. Knowing only a modest amount about structures, we choose four shapes that seem reasonable, and show them in Figure 15.6. We are sure that we must keep the number of pieces to a minimum generally, but we are not sure whether or not the lighter loads on the

* There is not a great difference in cost between a steel pipe, a wooden pole, or a concrete pole. Also the pipe is assumed not to buckle.

form in Figure 15.6c would offset the added brace P_2. We don't know whether it will be one of these shapes or another which will prove to be the lightest. Remember that we are trying to minimize the weight of material, and are measuring the structure by the equation, as in Chapter 9,

$$\text{Weight} = \sum_i \frac{\text{Force}_i\ (\text{lb}) \times \text{Length}_i\ (\text{ft})}{6000} \quad \text{lb} \tag{17}$$

We computed the original shape, Figure 9.1, by applying Newton's laws, summing forces at the top point, where the cable, pole, and guy wire

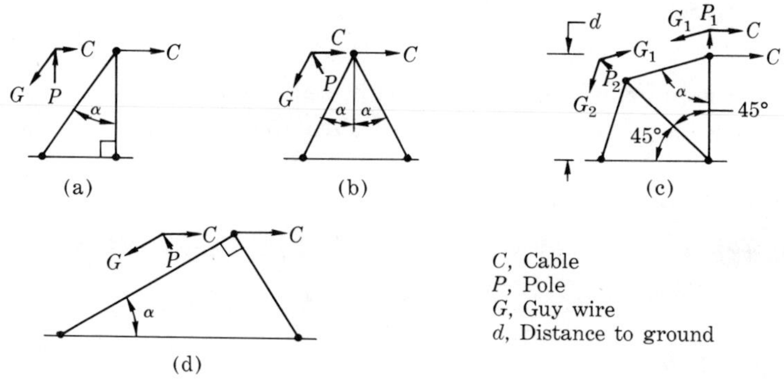

Fig. 15.6 Optimization of pole-guy structure. Configurations to be analyzed. (Reference Figure 9.1). Cable to ground distance = d.

meet. Using Chapter 9, the forces and lengths for Figure 15.6a are found to be, respectively,

$$\text{Forces:}\quad G = C/\sin\alpha,\ P = G\cos\alpha = \frac{C\cos\alpha}{\sin\alpha},\ C = C \quad \text{lb}$$

$$\text{Lengths:}\quad g = p/\cos\alpha,\ p = p\ (=d) \qquad \text{ft}$$

The total weight of guy and pole, using Equation (17) above, hence is

$$W_t = \frac{1}{6000}(Gg + Pp) = \frac{1}{6000}\left(\frac{C}{\sin\alpha} \times \frac{p}{\cos\alpha} + \frac{C\cos\alpha}{\sin\alpha} \times p\right)$$

$$W_t = \frac{pC}{6000}\left(\frac{1}{\sin\alpha\cos\alpha} + \frac{\cos\alpha}{\sin\alpha}\right) \tag{18}$$

The minimum weight occurs at $dW_t/d\alpha = 0$. Thus, differentiating, we have

$$\frac{dW_t}{d\alpha} = \frac{pC}{6000}\left[\left(\frac{-\cos\alpha}{\sin^2\alpha}\right)\left(\frac{1}{\cos\alpha}\right) + \left(\frac{\sin\alpha}{\cos^2\alpha}\right)\left(\frac{1}{\sin\alpha}\right) - \left(\frac{1}{\sin^2\alpha}\right)\right] = 0$$

By clearing fractions, we obtain

$$\frac{1 - 3\cos^2\alpha}{\sin^2\alpha\cos^2\alpha} = 0$$

or

$$1 - 3\cos^2\alpha = 0; \quad \cos^2\alpha = \tfrac{1}{3}; \quad \alpha = 54.8° \tag{19}$$

Thus the angle for minimum weight, considering both pole and wire, is $\alpha = 54.8°$, versus the original 45° obtained by considering the wire alone. A measure of value is the trigonometric portion of the weight equation. Let us call this the index I_a; so that for Equation (18) we have

$$I_a = \frac{1 + \cos^2\alpha}{\sin\alpha\cos\alpha} = 2.83 \qquad (\text{for } \alpha = 54.8°) \tag{20}$$

The lower the index, the lower the weight and the better the design. We can proceed with the three other forms, Figure 15.6b,c,d, and we will find their indices and optimum angles to be

$$\left.\begin{aligned} I_b &= \frac{2}{\sin 2\alpha} = 2.0 && (\text{for } \alpha = 45°) \\ I_c &= \frac{\sin^2\alpha - \sin\alpha\cos\alpha + 3}{\sin\alpha\,(\sin\alpha + \cos\alpha)} = 2.81 && (\text{for } \alpha = 57.7°) \\ I_d &= \frac{2}{\sin 2\alpha} = 2.0 && (\text{for } \alpha = 45°) \end{aligned}\right\} \tag{21}$$

Thus, as we look back over the results we see that the 90° angle at the top correlates forms b with d, but as yet we have no real clue why the simple form of a was not as light in weight. In fact, form c was lighter than a. We simply picked out geometries intuitively, hoping to find a true optimum. Even now we don't know whether or not a lighter form (lower index) can be assembled. This brings out the point of this whole exercise, namely:

> *All optimization proceeds from a form or a configuration, and progress is generally random unless you have discovered the ruling principles.*

In this case, all practicing structural engineers know the main principle, and you may have deduced it by now. It is this:

> *Locate the structural members at right angles to each other, so that each does not add any load to the other.*

Perpendicularity is the desired geometry (not permitted in form a) and it must be drawn in place or permitted, before the problem can be analyzed. Only two of the versions we tried could be varied so as to bring the members perpendicular to each other. Only by creating or inventing suitable forms can any optimization procedure be expected to produce the *optimum*.

We will summarize optimization at the end of the chapter; but before doing so, we will illustrate one more useful example: Maximizing the criterion function with regional constraints.

Optimizing with Regional Constraints: Linear Programming

As we mentioned earlier in this chapter, functional constraints are imposed on the present or prospective system by the operation of natural or physical laws, such as those of physics, geology, economics, etc. The regional constraints, on the other hand, are usually discrete bands of inherent or even man-made restrictions and comprise limits between which the variables or functions must exist and operate.

There are many illustrations of distinct upper and lower limits, in contrast to unrestricted variation. One example is the production rate of a punch press in a factory; it can turn out a maximum of x parts per hour. Another example is the automobile speed limit on throughways: No less than y or more than z mph. Or, a ship must carry a minimum of b tons of ballast or not more than c tons of cargo. A new product going into production should yield no less than u \$/unit profit or more than v per cent of the market available. An airplane has upper and lower limits of air speed. All of these limits are to be taken seriously but the ship's and plane's limits are the most dramatic; neither of them is stable at the lower limits, and it is dangerous for them to travel at the higher limits. Such limiting capability of a process or a machine occasionally impacts on a design, so our last illustration of an optimizing technique will demonstrate a method, known as *linear programming*, for dealing with simple regional constraints.

Linear programming was developed during World War II to best schedule the use of limited production facilities, where the relations between variables could be expressed in linear (first-power) terms. It is a method of analysis and not synthesis. Its use in design is simply to test the adequacy or to propose the best use or operation of a system which has been planned and developed by other means. It is an after-the-fact technique, as are most analytical tools, but still a valuable concept and a useful tool with which to test a design before release.

In our book project, the transit-mix company, we shall assume we have designed the plant and are now testing its operating effectiveness. At the plants's planned first stage, we find several limitations on the two most frequent concrete mixes which we will sell, No. 1 for regular commercial foundations and walls, and No. 2 for a state highway project. The limits are:

1. The composition tolerances (per cent sand, gravel, cement) require batch checking, which delays the weighing and loading of No. 2 by a ratio of 2:1; and we can supply only 16 loads/hr of No. 1 mix.

2. Because of distance, we can deliver $\frac{5}{3}$ as much No. 2 mix as No. 1 and the limit is 9 loads/hr for No. 1.
3. Two truck loading bays are planned, but only one will be in operation, and it can deliver only 10 loads/hr total.

One additional factor is the relative profitability of the two different mixes. We find the profit is 50 per cent higher on No. 2 than No. 1 (the ratio is 3/2).

Our operating question is, "What combination of No. 1 and No. 2 production rates will give us the maximum profit?" (Assuming we are free to make and sell in the market as much or little of each mix as we choose.)

We assign variables,

x = number of loads/hr for No. 1 mix

y = number of loads/hr for No. 2 mix

and write the expressions for the conditions described above.

$$\text{Tolerances:} \quad x + 2y \leq 16 \quad (x,\ y \text{ nonnegative}) \tag{22}$$

$$\text{Delivery:} \quad 5x + 3y \leq 45 \tag{23}$$

$$\text{Loading bays:} \quad x + y \leq 10 \tag{24}$$

$$\text{Profit:} \quad CF \sim 2x + 3y \quad (\text{to be maximum}) \tag{25}$$

These are not like the usual algebraic equations met in the past because (1) these are inequalities, (2) there are more expressions than unknowns and (3) our object is to maximize an extra expression, *CF*. However, graphical (or analytic geometrical) methods both help us solve the problem quickly and provide a useful visualization of the process of solution.

For instance, in Figure 15.7a, the shaded area under the line in the first quadrant meets the conditions of (22) above:

The maximum x is 16.

The maximum y is 8.

For intermediate x, y, the sum $x + 2y$ doesn't exceed 16. x, y are nonnegative.

Similarly, the conditions of (23) and (24) are met by the corresponding lines and areas in Figure 15.7b, except that the shading includes *both* requirements by lying under whichever line is closest to the origin. Thus the net permitted region for the three limits is that lying under all three lines as shown shaded in Figure 15.7d.

Now, how do we deal with the profit relation, Equation (25)? The question is: Which x, y, of the permitted set that we have found will give the largest value of *CF* when measured by $2x + 3y$?

Looking at it graphically, $CF = 2x + 3y$ is a line of fixed slope,

$$m = -\tfrac{2}{3}$$

but variable position. It ranges from a line intersecting the origin, where $CF = 0$, $(x = 0 = y)$ to higher and higher values of CF, moving parallel

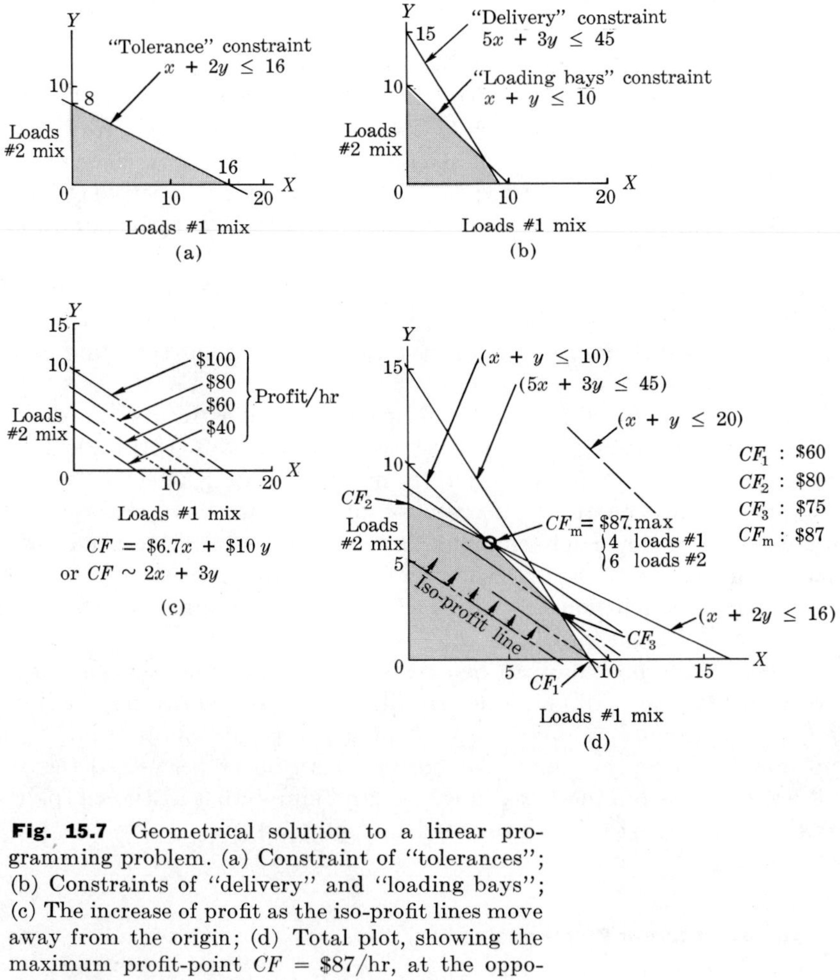

Fig. 15.7 Geometrical solution to a linear programming problem. (a) Constraint of "tolerances"; (b) Constraints of "delivery" and "loading bays"; (c) The increase of profit as the iso-profit lines move away from the origin; (d) Total plot, showing the maximum profit-point CF = \$87/hr, at the opposite corner of the shaded permitted area.

to itself as shown in Figure 15.7c. The various lines are called iso-profit lines because CF is a constant along each one. This family of parallel lines has been given dollar values by assuming No. 1 profit at \$6.67/load/hr and No. 2 at \$10.00/load/hr ($CF = 6.67x + 10y$, \$/hr). The proportionality sign, $\sim$, has been converted to an equal sign, $=$, by multiplying the right-hand side of the relation by the constant \$3.33/load. Thus, where the line crosses $x = 6$, $y = 0$, $CF = \$40/\text{hr}$, and CF becomes proportionally higher as it moves out.

Thus, intuitively we can deduce what has been proven to be a general principle,* namely, the maximum CF occurs at the farthest distant corner point in the permitted shaded area which represents Equations (22), (23), and (24).

To summarize, all these conditions have been put together in the graph of Figure 15.7d, showing the permitted-area-corner which yields the highest value: $CF_{\max} = \$87/\text{hr}$, at $x = 4$, $y = 6$ loads/hr. Any other combination of x, y yields a lower CF. To illustrate the latter, the other corners of the permitted area have been shown, together with their respective CF's, 1, 2, 3 (\$60, \$80, \$75, respectively). These lower values of CF correspond to intermediate iso-profit lines as the maximum was being searched for. Each CF_i corresponded to a particular mix of (x_i, y_i) ranging widely from (9, 0) to (0, 8). But there is only one maximum point, unless a constraint line happens to be parallel to the CF line, in which case any (x, y) along that line gives the same $CF_{\max}$.

To show the influence of the particular constraints, another line, $x + y \leq 20$ has been drawn in (i.e., two loading bays working). This new constraint is now so far out that it has no limiting effect. The solution $CF_{\max}$ would now be at the intersection of the other two lines $x + 2y = 16$ and $5x + 3y = 45$. Such changed or changing situations occur often in practice.

Thus the conclusion to our project example is to mix and sell 4 loads/hr of No. 1 and 6 loads/hr of No. 2.

Clearly this problem can be extended to three dimensions analogously, and intuitively to n dimensions. In addition, the process may be inverted and the criterion CF *minimized* by starting at a point remote from the origin and moving in to the innermost corner of an outer permitted region. We will leave the technique here, however, and simply list at the chapter's end some references for those interested in further study.

Review of Linear Programming

For our interest, the linear programming problem represents optimization with only regional, and no functional, constraints. The geometric

* The general problem, even with large numbers of variables and inequalities, has been solved algebraically by mathematician George Dantzig, who developed the Simplex Method for linear programming.

solution we have demonstrated has the value of picturing the successive steps taken toward the optimum as well as visualizing where the optimum must be. The significance or indifference of the various constraining limits shows clearly in the geometrical depiction. The line $x + y \leq 10$ was limiting in our example, whereas $x + y \leq 20$ would have been utterly unrelated. Our geometrical method was shown in two dimensions; it could have been extended to a three-dimension case, but for four dimensions or more we would need the so-called "Simplex" algorithm to solve the problem.

The general types of optimizing problems which linear programming solves are:

1. *Mixing.* To obtain lowest batch cost incorporating several ingredients having different properties and costs (i.e., gasoline blends, cattle feeds, ore combinations).
2. *Capacity allocation.* To select best use of production facilities and the various optional products that can be manufactured (i.e., factory machines and processes).
3. *Production scheduling.* To arrange the best schedules in face of fluctuating sales demand, backorder, inventory, and storage costs.
4. *Transportation sequence.* To lay out the best routes and schedules with given trucks, ships, distances, and needs.

SUMMARY

1. Optimization in the use of both the client's and society's resources, while meeting the consumer's need, is the *basic objective* of the engineer.
2. Optimization is a selection process, and the attendant engineering decisions rely upon criteria specifically or generally agreed upon in advance. All types of quantitative and qualitative criteria may be encountered by the engineer.
3. Optimization is relevant only for independent systems; subsystems must remain sub-optimal to favor the system. (No optimized system is ever composed of optimized subsystems.) "Freedom to choose the best" in the system becomes "constraint within limits" in the subsystem.
4. Practically all desirable optimizing objectives conflict with some other objective, so "trade off" decisions are necessary.
5. Optimizing procedures come in many guises: Subjective, graphical, analytical (explicit or iterative), algorithmic, and empirical.

6. The general analytic form of the optimizing problem is as follows:

The value expression: $CF = f(u, v, w, \ldots)$ (to be maximized or minimized)

The functional constraints: $FC_i = F_i(u, v, w, \ldots) = 0$

The regional constraints: $\lim_i \leq R_i\ (u, v, w, \ldots) \leq \mathrm{LIM}_i$

EXERCISES

15.1 A high-voltage power transmission line is being rerouted from a "straight-through" configuration to a horizontal angle of $A°$ or $B°$ on one side, as shown in Figure 15.8. The same tower will be re-used but new "dead-end" insulators will be connected to the same support points and the original vertical string of insulators continued in use for the tower clearance. The problem is to keep the spark distance, line to tower, to a minimum of 5 ft. See the top view in Figure 15.8. What is the maximum horizontal angle, and is it A or B, to which the

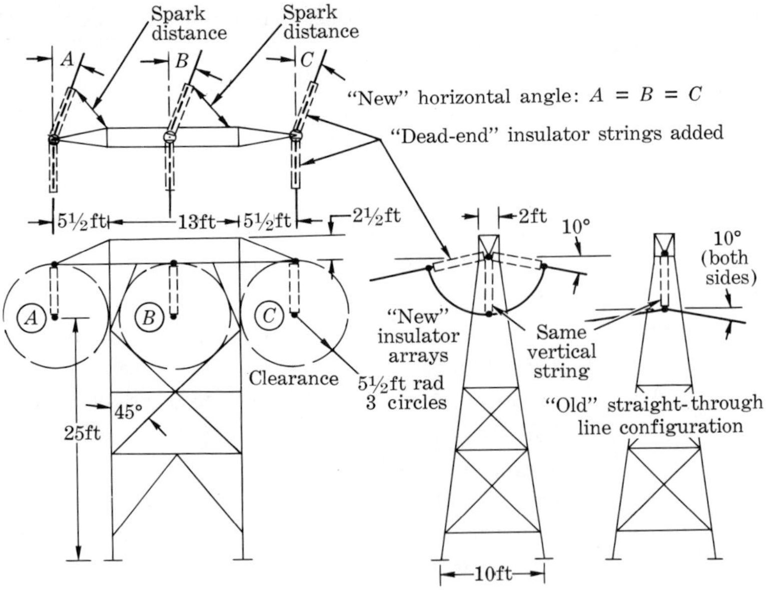

Fig. 15.8 Graphical optimization exercise. High-voltage transmission tower with the line being changed from "straight-through" to a new horizontal angle. (Reference Figure 19.4)

line can be relocated, and still maintain a sparking clearance of at least 5 ft to the nearest tower member? (Disregard the change of forces on the tower.)

15.2 A high temperature furnace F is at the focus of a parabolic mirror which is lined up with the sun's rays by a plane tracking mirror, called a heliostat. See Figure 15.9 for the dimensions. 1) What is the optimum size of this plane-mirror, given the sun-angle data in the referenced figure? Use the heights H_1, H_2 and widths W_1, W_2 measured from the pivot-point perpendicular P.P. 2) How much would H_1, H_2, W_1, and W_2 be if the parabolic mirror were 50 ft square?

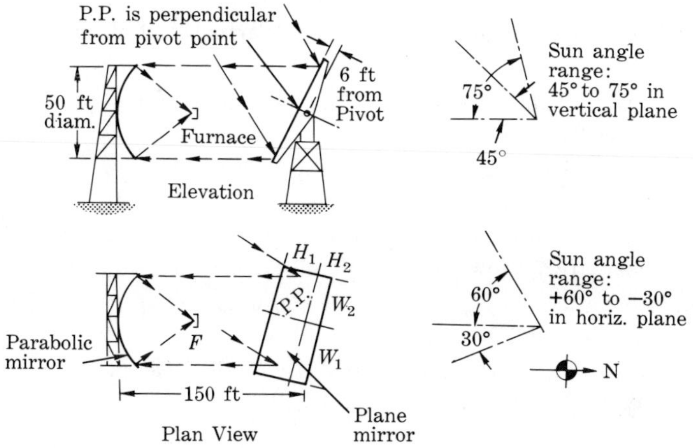

Fig. 15.9 Graphical optimization exercise. High-temperature solar furnace using a parabolic reflector supplied by a plane mirror heliostat.

15.3 Talk with some professional engineers in one specialty, such as acoustic engineering, or aeronautical engineering, and learn a few more generalized principles (like those on page 264), which are important in practice, but do not immediately result from theoretical considerations.

15.4 What are the conflicting criteria that were likely to have been important in arriving at the decisions for optimizing a) a portable tape recorder dictating machine; b) a city-wide waste collection and disposal system?

15.5 Go through the analysis, in Figure 15.6d, for minimizing the weight. Prove that $\alpha = 45°$ at minimum.

15.6 Draw the lines and determine graphically the minimum value of C for:

$i_1 + i_2 \leq 5$ (i_1 and i_2 nonnegative)
$i_1 - 2i_2 \leq 2$
$i_2 - 2i_1 \leq 2$
$C = i_2 - i_1$ (to be minimum)

REFERENCES

1. Asimow, M.: "Introduction to Design," chap. 11, Prentice-Hall, Inc., Englewood Cliffs, N.J., 1962.
2. Shanley, F. R.: "Strength of Materials," McGraw-Hill Book Company, New York, 1957.
3. Shanley, F. R.: "Weight-Strength Analysis of Aircraft Structures," Dover Publishing Company, Inc., New York, 1960.
4. Shanley, F. R.: "Design Notebook, Elementary Mechanics and Strength of Materials," EDP Report No. 1-62, UCLA, Department of Engineering, Los Angeles, 1962. (Plus personal communications.)
5. VerPlanck, D. W. and B. R. Teare, Jr.: "Engineering Analysis," John Wiley & Sons, Inc., New York, 1954.
6. Krick, Edward: "Engineering and Engineering Design," chap. 5, John Wiley & Sons, Inc., New York, 1965.
7. Johnson, R. C.: "Optimum Design of Mechanical Elements," chaps. 1, 3, 4, 7, John Wiley & Sons, Inc., New York, 1961.
8. Hall, A. D.: "A Methodology of Systems Engineering," chap. 10, D. Van Nostrand Co., Inc., Princeton, N.J., 1962.
9. Ackoff, R. L.: "Scientific Method, Optimizing Applied Research Decisions," chap. 2, John Wiley & Sons, Inc., New York, 1962.

Linear Programming references

Charnes, A. A., W. W. Cooper and A. Henderson: "Introduction to Linear Programming," John Wiley & Sons, Inc., New York, 1953.

Gass, S. I.: "Linear Programming," McGraw-Hill Book Company, New York, 1958.

Garvin, W. W.: "Introduction to Linear Programming," McGraw-Hill Book Company, New York, 1960.

Digital Computers and Design

Sixteen

"Give me a lever long enough,
and a prop strong enough, and
I can move the world singlehanded."
Archimedes

Introduction

The digital computer has leap-frogged in a period of twenty years from a remote laboratory at Harvard into business offices, factories, power plants, physical and life science laboratories, and into every engineering field in the profession. For example,

A digital computer now controls a Port Arthur (Texas) oil refinery, bringing an efficiency increase of almost ten per cent over manual control.

Airline reservations are checked, requested, and recorded on nationally interconnected computerized systems.

Electric power transformers are now designed and optimized, and bills of material are printed out with existing computerized programs. (Thus, one company was able to dispense with the services of one hundred engineers.)

The Project Ranger flights to the moon were impossible to predict and control without the high-speed response of the Jet Propulsion Laboratories' digital computers.*

Numerical control of machine-tool and manufacturing operations is now applied through computers to job shop factories.

The computer makes possible rapid reduction of vast accumulation of data and the retrieval of desired information, whether for a business or an engineering purpose.

* The error of a single character among the millions of symbols in a space-maneuver computer program cost NASA an entire moon-shot mission.

This unprecedented advance in computing and control capability puts new demands upon engineering education, and requires all engineers to defend their professional standings by some familiarity with the computer and its uses. Properly used, the computer (and its already large library of prepared programs) greatly extends the engineer's effectiveness. The whirlwind speed of the computer permits the engineer to examine many more alternatives before being forced to come to a decision. Even at a cost of $50 to $300 per hour (for large commercial systems), a few minutes or hours of computer time will yield substantial design economies. It has been estimated that for a moderate-sized computer, one machine-second equals a man-month.

It seems that the digital computer may well be Archimedes' long lever. The prop that we still need is the imagination and the logic to balance that lever's strength.

In the near future the engineer may expect to have access to large high-speed computers through decentralized satellite consoles located conveniently nearby. He may put his own problem on the local console or work through a programmer-operator. In any case, he will scarcely be conscious that he is time-sharing the large computer with 5 or 50 other similar consoles operating simultaneously.

The Antecedents

The compact, reliable electrical-electronic machine of today traces its ancestors back at least to the abacus invented by the Chinese 3000 years ago. This is the primitive digital calculator, surprisingly prescient, having counters for "fives" as well as counters for "ones."

Blaise Pascal, the French philosopher-scientist, invented in 1671 a calculating machine still on display in a French museum. The next major step came in 1804 when another Frenchman, J. M. Jacquard, developed an automatic punched-card device for controlling looms. Thus, intricately patterned cloth could be made as easily on his loom as plain cloth could be made on others. In 1833, this punched-card idea was adapted by Charles Babbage, an English mathematician, to program arithmetic calculations in his "analytic engine," really the first digital computer. Babbage created an extensive theoretical design, but died in 1871 without building any more than small sections of his invention. He planned for all calculations to be done mechanically, storing the numbers as counter-wheel locations.

Little further was done until the pressures of World War II stimulated projectile-table calculating devices for the military services. In 1944, Howard Aiken at Harvard completed the Mark I, Automatic Sequence Controlled Calculator. This electro-mechanical device could operate upon and store 72 numbers of 24 digits each, and was controlled by punched tape. Another more electronically controlled version, ENIAC, was finished in 1946 at the University of Pennsylvania, this time using IBM punched cards as input and output.

The giant step forward, eclipsing all others, occurred in 1945 when von Neumann, Burks, and Goldstine* invented and developed the "stored-program" concept, coding the arithmetic operations so the computer could summon up commands that were numerically stored in its memory exactly as numerical data were stored. This concept permitted the computer to operate at its inherent high speed, not held back by the response limitations of its human operator. At this time it became possible to "loop," or repeat arithmetic operations in controlled succession upon an array of data entries. Also it was now possible to "branch" out to other steps in a computation, relying upon computer-made decisions relative to algebraic sign and zero value. This decision-making ability now is used by the computer to change its instructions, to *write* commands as well as *execute* them. It can operate on commands as numbers and then later execute those changed commands.

Thus, to use the computer directly, the engineer today must write his problem in a numerically coded sequence of commands, detailing each step necessary, from beginning to end. These commands together with the data are fed into the computer in IBM cards or magnetic tape (*input*), processed through the arithmetic sequence, and the answers are then delivered, either typed on a page, or put on new IBM cards or magnetic tape (*output*).

The demand now for modern computers has far outstripped early expectations, so that literally thousands of computers are sold and installed each year in the United States alone.†

Essential Blocks and Operation

The block diagram for the desk-size general purpose computer is shown in Figure 16.1. In this computer, the commands and data are fed into the INPUT block on IBM cards; the numerical information is transferred to MEMORY, processed through the ARITHMETIC unit, back to MEMORY and thence to OUTPUT. As output, the answers are automatically typed on a page or punched into IBM cards. For the rest of the chapter we will continue to use CAPITALS for the computer functional components.

The effective processing speed of this computer is generally limited by the INPUT/OUTPUT speed. Although the contents of 100,000 IBM cards per minute can be processed between MEMORY and ARITHMETIC, generally only 250 to 500 IBM cards per minute can be mechanically fed

* Reference 13, Datamation, p. 24.

† More than 4000 each year since 1960.

through the INPUT or OUTPUT equipment. Typically, on a "small" design problem, the input/output may take two or three minutes, whereas computation takes five to thirty seconds. Incidentally, cards are used for the input in order to maintain the freedom to insert or rearrange commands

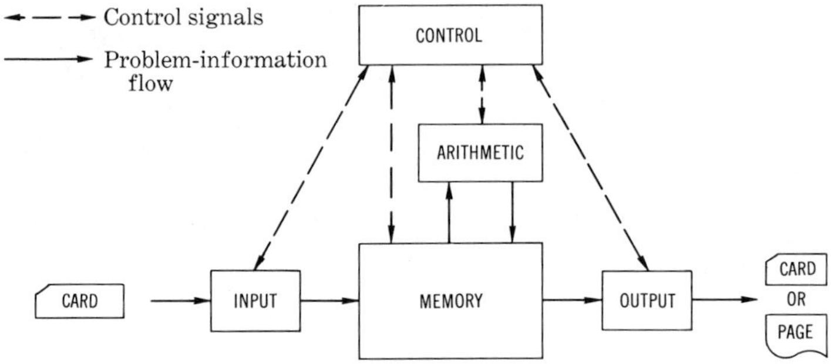

Fig. 16.1 General-purpose computer basic elements.

while the program is being perfected. The program must be tried out on the computer with familiar data and results, before the engineer can be sure of the answers when he uses unfamiliar data.

Another alternative is employed on larger, faster computers as shown in Figure 16.2. Here the card input is converted in advance to magnetic tape,

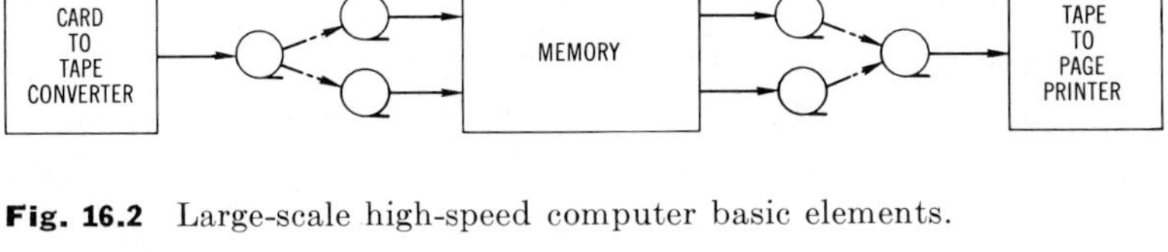

Fig. 16.2 Large-scale high-speed computer basic elements.

which in turn is fed into the larger MEMORY of this system, running several tapes at once if desired. Thus the input and output limitations on computing speed are removed. This large processor computes one problem after another rapidly, working out of and into its MEMORY, producing successive answers which are later transferred to tape, and still later typed out as a printed page. In the latter case, going from tape-to-computer-to-tape may require only a few seconds. On problems with large quantities of data, the computing time may grow into hours.

In both of these computers, the INPUT is the device which accepts the IBM cards and converts the pattern of punched holes into electric impulses, so that, coded, timed, and guided by CONTROL, the corresponding numbers are deposited in assigned "addresses" in MEMORY. Since a computer can work on a maximum of only two numbers at one time, the ARITHMETIC unit must receive these in the intended order, operate on them (add, divide, compare) and return at least one of them to MEMORY before going on for more. Instructions for these arithmetic operations have been coded as numbers and also stored in particular locations in MEMORY. The sequence proceeds as these commands are identified, decoded, and executed by the CONTROL unit in the prepared order: Command No. 1, No. 2, No. 3, and so on all the way to "Halt."

Upon completion of the entire sequence of commands, the CONTROL causes the OUTPUT to punch the answers from MEMORY on new IBM cards or to type out the answers on paper, depending on the equipment in use. Or, the results may be transferred from MEMORY to magnetic tape, for other kinds of processing later.

The sequence of instructions that causes numbers to be fetched, operated upon, interpreted, executed, and re-stored, in order to carry out a given assignment, is called a *program*. Generic programs have been written to handle classes of problems, such as solving any quadratic equation,* or designing an electric solenoid with given constraints,† etc. Of course, if a specific problem doesn't happen to fall into a class existing in present "program libraries,"‡ a specific program must be written for that problem.

Considerable skill must be developed to write efficient programs; and engineers who knew about a problem but not about programming had difficulty in communicating with trained programmers who knew about computers but not about the engineer's problem. As a result, computer specialists developed a program to write programs. Several have been put in use, the one of most direct engineering interest being FORTRAN§ (FORmula TRANslation). Others are ALGOL, SOAP, MAD, COBOL, etc. It is beyond our scope to offer here more than a brief comment about FORTRAN, although at the end of the chapter we will illustrate several solutions using FORTRAN. FORTRAN is a problem-oriented language (coding), relatively easy for the engineer to learn and apply. The computer makes two runs with

* Reference 1, p. 15.

† Reference 1, p. 53.

‡ Each major manufacturer has a published listing of its "program library," usually cross-referenced to programs of other computing systems.

§ References 1, 2, 3.

a FORTRAN program written by the engineer: 1) to convert the original (source deck) into a "machine language" program, now punched into a new deck of IBM cards, (the object deck), and 2) to run this new deck through in solving the original problem.

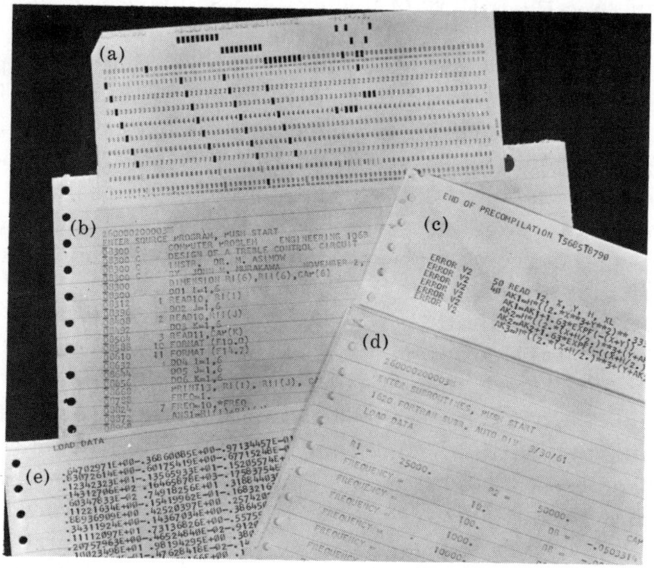

Fig. 16.3 Computer-punched card and type-out examples. (a) IBM card showing punchings for all numbers and letters; (b) FORTRAN program listing; (c) Error code listing during pre-compiling; (d) Answers typed out for program (b) above; (e) Typical mass data input or output in floating point notation (i.e., $E + 01 = 10^1$).

Figure 16.3 shows several computer-typed pages; (a) an IBM card punched to identify the hole-coding of the various characters, (b) a FORTRAN program, (c) a computer listing of certain errors the computer found in the program, (d) the typed answers, and the last, (e) a massive data listing.

How Numbers Are Set to Work between INPUT, OUTPUT, and MEMORY

Let us expand on the foregoing generalities and go into a little more detail about the way these various equipment blocks operate. At a constant rate, the INPUT* unit mechanically feeds the IBM cards singly off the

* Chapters 1 and 2 of Reference 2.

bottom of the stack and runs them (or the magnetic tape) past reading heads which convert the holes in the cards (or the magnetized spots on the tape) into synchronized sequences of coded electrical impulses. These trains of pulses are delivered to MEMORY in the same order; they are deposited in assigned MEMORY locations, or addresses, and remain there until intentionally changed.

Practically all computers are equipped with typewriter keyboard inputs also. Simultaneously with the pressing of the key, the character is typed for record and converted into electric pulses which are sent into the machine for MEMORY or CONTROL.

Let us look more closely at MEMORY. This is a numbered sequence of storage locations, physically located in the computer. The cell at each "address" contains one number, whose constituent digits must be within the cell digit-capacity for the particular machine. The maximum number of digits storable in each cell ranges from 1 to 7, depending upon the computer. If only one digit is stored per cell, a number having several digits is stored in cells of adjacent "addresses." The whole group is called a "word." For example, the number 236918 is stored below in addresses 0905 through 0910:

Address in memory:	0911	0910	0909	0908	0907	0906	0905	0904
Digit in the cell:		$\bar{2}$	3	6	9	1	8	

This is a *variable-word-length* memory, the bar over the 2 being called a "flag" and identifying the left-hand terminus. With this flag in place one can address 236918 simply at 0905.

Certain computers store the entire number in one address (*fixed-word-length*), six digits being the limit in the case below:

Address in memory:	0909	0908	0907
Number in the cell:		236918	

Floating-point numbers. So far we have not considered the decimal-point problem. Decimal points are usually located by converting the given number 236918 into floating-point or "excess-50" exponential form for actual storage in memory and internal use in arithmetic. Thus, with the block frame indicating the memory cell, our number in floating-point notation would be

$$23691.8 = .236918 \times 10^5 = \boxed{5523691}$$

This converted number signifies that an exponent of 5 had been added to the median value 50 (to avoid negative exponents) and the number 23691 had been converted to a "tenths" decimal. Since the computer allowed only 5 places, the last digit 8 was truncated. Thus,

23.6918 in floating point is	5223691
.236918 in floating point is	5023691
.000236918 in floating point is	4723691

The computer "knows" that the first two numbers are the exponent and treats the group accordingly. The algebraic sign is provided for also, the computer simply accepting the plus or minus put in front of the number.

Data or commands stored in MEMORY. During the program, when we want to find a number in a computer memory, we simply call for it by its address. For fixed-word-length memories, "0908" means the contents of cell 0908, whatever digits or number may be in there at the time. For variable-word-length memories, "0908" means the digits of the number starting at cell 0908 and going left to the digit with the "flag." Also, these numbers either may be data from the problem or they may be commands from the program, If the same number 236918 had been identified as commands, the significance to the computer would have been:

O	P		A	D	D	R
2	3		6	9	1	8

where the 23 was an operation code (i.e., add, subtract, etc.) and the 6918 was the address of the number upon which to execute that operation. Let us continue with the construction features of a memory.

Forms of MEMORY. Practically all machines in production now use magnetic memories, employing either magnetic discs, drums, or miniature cores for continuous internal active use, or tapes for intermittent and external use.

Disc memories (stacked like phonograph records) carry a thin film of magnetic material on both sides, and the addresses consist of imaginary divisions of numbered concentric rings, each ring further subdivided angularly into small numbered segments. The drum memory similarly has a thin magnetizable layer; except in its case, the working surface is wrapped around the cylinder. In both cases, the disc and drum are continually rotating at relatively high speed during computation, the magnetic heads riding close on a thin cushion of air. Similar to the conventional tape recorder, the magnetic "writing" heads convert the high-frequency electrical current impulses

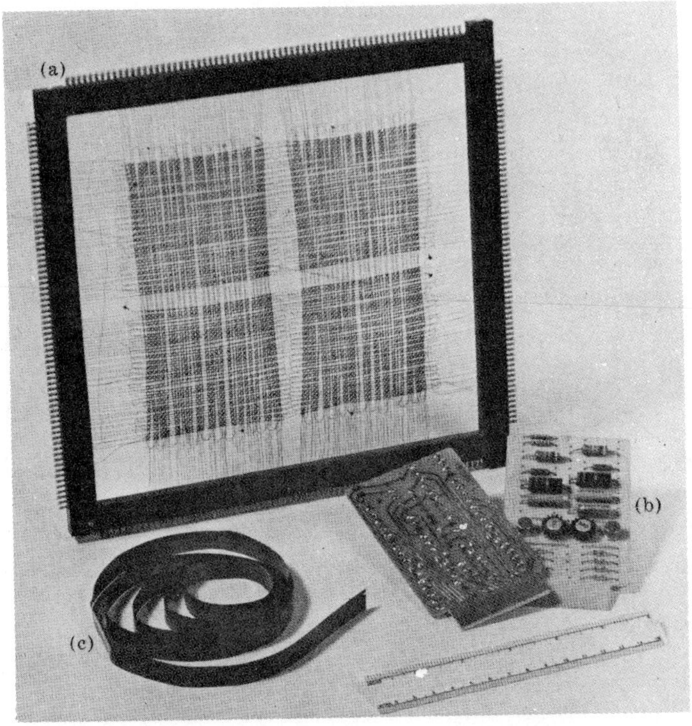

Fig. 16.4 Computer components. (a) Magnetic core memory plane for IBM 1620 showing the 100×100 $X - Y$ grid of wires. One small core is at each intersection; (b) Front and rear views of a modular plug-in circuit board, showing transistors, diodes, resistors, condensers, and the etched connector strips dip-soldered on the rear side; (c) Curled portion of standard six-channel magnetic memory tape.

into closed-spaced zones of surface magnetism. In fact, the zones are crowded together at over 500 lines/in. "Reading" heads perform the inverse.

The other preferred form of magnetic memory is the *core-plane*. In this configuration, minute iron cores, $\frac{3}{32}$ of an inch in diameter (about the size of this "O") are strung on an X-Y coordinate grid of wires as shown in Figure 16.4. Signal current pulses are sent simultaneously through one X

and one Y wire and the reponse is picked up by a third wire which threads *all* cores. The magnitude of the pulse in each wire, X_1, Y_1 is one-half that necessary to magnetize a core. Thus, only at the intersection X_1Y_1 does one core become magnetized, or retain the desired signal. Several parallel planes acting in conjunction provide corresponding XY cores to store the array of signals which represent one *alphanumeric* character (alphanumeric = letter or digit). For example, six of the planes shown in Figure 16.4, each containing 100 X-wires and 100 Y-wires each, combine to store 10,000 characters.

Magnetic memory is employed because of its high speed of magnetization and demagnetization and because of the ease in location of the given zone or core. Other memory principles have been, and continue to be, under development, but magnetic storage has proved most advantageous and most economical for large memories so far.

Memory access time. The time required to locate a desired cell is called access time. The search for this cell must be done *serially* for the rotating disc or drum, but it is possible to go to it *directly* (random access) in the case of the core plane MEMORY. This access time is measured in microseconds (μ sec) for each digit. In random access memory, 5 to 10 μ sec may be required to locate a ten-digit number and to transfer it into other parts of the circuit for airthmetic use. When a number is extracted from a cell for use, it is immediately and identically replaced. When another number is inserted, the original is erased. This is called "nondestructive read-out" and "destructive read-in." The MEMORY is also used, of course, to store intermediate results secured during the execution of a program. The magnetic tape shown in Figure 16.4 employs a powdered iron-oxide magnetic coating applied to very thin Mylar polyester tape.*

Other operating components. The ARITHMETIC unit consists of the electronic circuitry necessary to perform various normal arithmetic and logical operations. The speed of a typical *add* operation for two numbers of ten decimal digits each might be 2 μ sec. The memory access time generally limits computing speed, rather than the arithmetic process time. The etched circuit boards shown in Figure 16.4 are typical of the modular electronic assemblies of transistors, diodes, resistors, etc, used in the ARITHMETIC section.

The CONTROL is the power supply and the synchronizer for the high-frequency† trains of pulses that are the message units flowing over the large complex of switching circuits comprising ARITHMETIC and MEMORY. The CONTROL also produces the two machine cycles called I-time (Instruction) and E-time (Execution). During the former, the next command

* Reference 4, page 732.

† 100,000 cps to 1 million or more cps.

is obtained from MEMORY and decoded; the contents of the data address are brought in for use, and a counter prepared to bring in the next command after execution. During the E-time, the operation is performed.

The OUTPUT[5] is similar to the INPUT in reverse, both being under the supervision of CONTROL and both working between MEMORY and the outside world or the waiting operator. The output data can be typed directly, but this is slowest of all; only ten digits can be delivered per second because of mechanical limitations. Using punched cards, this number becomes 100 to 200 digits per second, and using magnetic tape and line printer, 300 to 500 digits per second.

In review, the computer can perform arithmetic operations at very high speeds and with very high reliability. At one addition per 10μ sec, the computer can carry out over 100,000 additions per second, a prodigious performance. But this is the job it is designed to do.

Special Uses and Varieties of Computers

The basic ability of the digital computer we have described is high-speed addition. The computer has absolutely no inherent capability of producing a closed-form solution for problems in calculus, differential equations, trigonometry, or even geometry. All such engineering problems must be converted by numerical methods into equivalent simple problems in arithmetic and algebra. The computer can only slavishly follow arithmetic instructions. Integration, for example, must be done by summing rectangles, triangles, trapezoids, etc. Differential equations must be solved by iterative successive approximations. Trigonometric functions and power or root functions must be "looked-up-in-a-table-in-memory" or found by converging series approximations, called sub-routines. New procedures* or specific algorithms become necessary with new types of problems.

The high-speed and many-digit operation and memory capacity of the computer prevents the need for these substitute methods from being a handicap. But new methods or conversions must be found for new cases, and the engineer now has a new need to become familiar with numerical methods, *per se*.

Generally, the computer is useful to the engineer where the following conditions characterize his problem situation:

1. The problem solving is reducible to a concrete series of logical steps;
2. Many repetitions are needed for a given series of arithmetic steps, usually repeating the operations on a large number of successive or incremental data entries.

* *Heuristics* is the name given to those methods not having rigorous analytical proof, yet capable of seeking or discovering solutions. (Greek: *heuriskein*, to discover.)

The computer's greatest functional power is in "looping" (item 2 above), wherein the commands are a discrete series of arithmetic operations performed on successive addresses in MEMORY. There can be successions of loops, and loops-within-loops, each working with different decision choices determining when to come out of the loop and continue the program.

Each computer has an array of "legal" commands or instructions built into its hardware by the manufacturer. These commands form the vocabulary by which we "talk to" the computer. They consist of such commands as:

Add, multiply, etc.
Read in data, read in commands, etc.
Store.
Transfer control if negative, if zero, etc.
Type, punch cards, etc.
Perform sub-routine, etc.
Permit manual operation.

Thus the engineer, having decided that his problem is appropriate for the computer, must communicate either by learning the operation codes (or FORTRAN) himself, or by working through a trained programmer. Contrarily, if the engineer's problem is short, or uses only a few data items, or is long without iterations, it probably is not worth the time or money to use the computer.

There are many types, sizes and prices of computers that meet the differing needs of business, industry, education, and government:

Inventory, payroll, insurance, banking.
Medical, sociological, governmental statistics.
On-line plant and process controls.
Military, land, sea and air, missile warning and controls.
Engineering and scientific problem solving.

Each of these applications requires different types of input and output, different operating speeds and different memory capacities. Each presents different space, power and operator limitations. They range in size from a shoe-box to a room-full . . . but all of them are electronic and electro-mechanical.

We are generally concerned here only with the engineering problem computer, the last type listed above. It usually takes the form of a desk-size integral processer and memory, or else a floor console with separate units for input-output. A summary of the characteristics of several of the computer models useful to engineers is given in Table 16.2 at the end of this chapter. The IBM 1620, shown in Figure 10.8c, is a typical desk-size engineering problem computer.

Use of Computers in Design

There are many ways in which computers may be used in design, as we shall see in a moment, but most of them come under the heading of labor- or time-saving. Important as these considerations are, they do not directly help in the thinking, the creating, the solution-synthesizing effort that is the core of design. With the exception of a few self-adaptive or simple "learning" routines developed, the computer cannot be said to "think." Thus the engineer is still left with this strictly cerebral responsibility. The machine is yet neither a replacement for, nor much of an aid to, creativity. As design and the decision-making in design become more fully understood, some of the searching for alternative solutions will certainly be relegated to the computer. It will be necessary however to bring into play the full range of the value systems of the designer, his client, his organization and the culture before the decisions issuing from such computer use can become meaningful.

The variety of design process and the means of solutions of a problem within that process, are shown in the matrix[6] in Table 16.1. The letters in the squares refer to the relative number of references found in a recent literature search, and indicate the relative development activity along the respective lines. The absence of activity measures the challenge and possibly the need and difficulty. There are many other instances of design by computers which are not published because of their proprietary competitive value. Their presence is alluded to only in conversation with engineers in industry.

The terms listed down the left side of the Table under "Nature of the Process," are amplified below, simply presented in the form chosen by the author of the respective article.

1. *Decision:* Comparison of alternatives by economic choice, looking as far into the alternates as possible.
2. *Optimization:* Computing many thousands of solutions for shortest-wire circuitry, minimum-interference telephone cables, etc.
3. *Analysis:* Solutions in greater detail and accuracy; Monte Carlo solutions heretofore not possible; discipline-oriented programs to solve linear systems.
4. *Synthesis:* Given the circuit elements and input/output requirements, the components were assembled.
5. *Process control:* Refinery, power plant, and chemical process "real time" applications.
6. *Simulation:* Digital equivalent of an analog model in mechanics, or of business systems, etc.
7. *Data reduction:* Transcribing, scaling, computing, editing.
8. *Information retrieval:* Using classification systems to broaden the base and to speed retrieval.

Computerized Solutions to Problems in Design

The matrix in Table 16.1 illustrates a number of kinds of computer-amenable problems encountered in design. Those problems most economically solved by hand or desk calculator should not even reach for an IBM card, much less come face to face with a computer console. However, critical problems which are laborious or sophisticated or otherwise unsolvable

Table 16.1 Application Matrix. Engineering Design with Computer. (From R. L. Strom, Reference 6.)

Means of Solution / *Nature of Process*	A. *Massive Computation**	B. *Iterative Solutions*	C. *Algorithmic Solutions*	D. *Monte Carlo*	E. *Heuristic Solutions*	F. *Boolean*	G. *Other*
1. Decision processes		*L*	*L*		*L*		
2. Optimization	*L*	*VH*	*H*			*L*	
3. Analysis	*L*	*H*	*VH*	*L*		*L*	
4. Synthesis	*L*	*M*	*H*		*L*	*L*	
5. Process control	*L*	*L*	*M*		*L*	*L*	
6. Simulation	*L*	*H*	*VH*	*L*		*L*	*L*
7. Data reduction and analysis	*L*		*M*				
8. Information retrieval			*L*				
9. Other		*L*	*L*				*M*

Legend: Number of references found: $L = 1–4$, $M = 5–9$, $H = 10–14$, $VH = 15+$.
* Unsolvable or impractical manually.

should be taken to the "machine with the flashing lights." More and more, technology is compelling the engineer to increase his effectiveness by enlisting all feasible help; and the computer has now become accessible and its cost within reason.

The problem types listed in Table 16.1 as A, B, and C come within our scope and will be illustrated here.

Massive Computation

Type A, Massive Computation, is illustrated by a criterion function equation which is to be maximized by permutations over a large number of alternatives. A typical form is

$$CF = af_1(x) + bf_2(y) + cf_3(z) \qquad \text{(to be maximized)} \tag{1}$$

where a, b, c are constant coefficients and

x takes the values successively $m_1, m_2, \ldots, m_m$
y takes the values successively $n_1, n_2, \ldots, n_n$
z takes the values successively $p_1, p_2, \ldots, p_p$

The computation is simple; but even if m, n, p were just five alternatives each, 5^3 or 125 different CF values would have to be computed to find

```
Ø83ØØ C       FIND MAX(CF=AX + BY + CZ)
Ø83ØØ         DIMENSION X(5),Y(5),Z(5)
Ø83ØØ         READ,A,B,C
Ø8348         DO 1Ø I=1,5
Ø836Ø 1Ø      READ,X(I),Y(I),Z(I)
Ø8516         CF1=Ø.Ø
Ø854Ø         DO 2Ø I=1,5
Ø8552         DO 2ØJ=1,5
Ø8564         DO 2Ø N=1,5
Ø8576         CF=A*X(I)+B*Y(J)+C*Z(N)
Ø878Ø         IF(CF-CF1) 2Ø,2Ø,11
Ø8844 11      CF1=CF
Ø8868 2Ø      CONTINUE
Ø8976         PRINT 3Ø,CF1
Ø9ØØØ 3Ø      FORMAT(E14.8)
Ø9Ø22         STOP
Ø9Ø3Ø         END
END OF COMPILATION Ø9Ø29T95ØØ
T981Ø X       T977Ø
T976Ø Y       T972Ø
T971Ø Z       T967Ø
T966Ø A
T965Ø B
T964Ø C
T962Ø I
T961Ø CF1
T96ØØ ØØØØØØØØ99
T958Ø J
T957Ø N
T956Ø CF
```

Fig. 16.5 FORTRAN program for the equation $CF = ax + by + cz$, to be maximized with five permutations each of x, y, and z respectively.

the largest. The problem snowballs immoderately if we must add even a few more terms or more alternatives. The program to compute and type out all of the values is simple: A loop within a loop, and both within a third loop.

The FORTRAN statements to perform this are shown in Figure 16.5, where

$$f_1(x) = x; \quad f_2(y) = y; \quad f_1(z) = z$$

An added sub-program has been included, ranking all the answers by size and typing out only the largest. This sub-program starts at line 08780.

If the reader is not familiar with problem flow-charting and computer program writing, he is referred to Appendix G, where an illustration is worked out and coding conventions for FORTRAN are briefly listed.

```
Ø83ØØ C        COMPUTE C =AD**2 + BD + C/D + D/D**2
Ø83ØØ 2Ø       FORMAT(2F1Ø.4)
Ø8328          READ,A,B,C,D,D1
Ø84ØØ 1Ø       DF1=A*D1*D1
Ø8448          DF2=B*D1
Ø8484          DF3=C/D1
Ø852Ø          DF4=D/(D1*D1)
Ø8568          CF=DF1+DF2+DF3+DF4
Ø8628          PRINT 2Ø,CF,D1
Ø8664          D1=D1+Ø.Ø1
Ø87ØØ          GO TO 1Ø
Ø87Ø8          END
END OF COMPILATION Ø87Ø7T966Ø
T979Ø A
T978Ø B
T977Ø C
T976Ø D
T975Ø D1
T973Ø DF1
T971Ø DF2
T97ØØ DF3
T969Ø DF4
T967Ø CF
T966Ø TØØØØØØØØT
```

Fig. 16.6 FORTRAN program for the equation $C = aD^2 + bD + cD^{-1} + dD^{-2}$, from $D_1 = 1$ to $D_1 = 3$, by steps $\Delta D = 0.01$. (Termination is manual.)

Another type A design problem is the explicit equation which is to be maximized or minimized (or simply evaluated) over a large number of values of a variable. In Chapter 12, the water tank problem yielded a fourth power equation for cost which has the form

$$C = aD^2 + bD + cD^{-1} + dD^{-2} \tag{2}$$

If this equation needed more careful evaluation than that of the original graphical approximation, many values of D could be substituted in, starting say, with $D = 1$ and going up to $D = 3$ by increments of $\Delta D = 0.01$. Such a procedure is an economical and typical, although inelegant, method of solution for even very complex equations. It is particularly straightforward when the limits of the variables are known, but it also can be used exploratorily to find those limits. The FORTRAN statement for finding the solution to Equation (2) is shown in Figure 16.6.

Iterative Solution

The type B problem, Iterative Solution, will be illustrated here by equations of three primary types, although others also arise in practice.

1. Equations containing transcendental functions.
2. Simultaneous equations.
3. Differential equations, not amenable to known methods.

In all of these cases, there is presumably no rational direct method of solution, so the basic approach of numerical methods and computerized iteration is convenient or necessary.

Iteration consists of finding and using a numerical relation between proximate values of a variable so as to converge upon the desired solution value. The relation is called a *recursion* expression or formula and has the general form

$$x_{i+1} = f(x_i) \qquad \text{or} \qquad x + \Delta x = f(x) \tag{3}$$

The function $f(x_i)$ takes many forms; but the object, of course, is to converge upon the solution with as few iterations as possible. The principle is shown graphically in Figure 16.7a, wherein successive values of $f(x)$, or y, approach zero as x is given values $x_0, x_1, x_2, \ldots, x_n$. There is always a residual error in iteration, and this is shown as ϵ, between the solution value x_s and the last computed value x_n. This chart doesn't show how the successive x values are found; but Figure 16.7b shows one method (Newton-Raphson) simply using the tangent to the curve (dy/dx) to locate the next x_i. We will amplify on this shortly. In each case, however, the engineer must choose the first value x_0.

If the equation can be broken into useful halves $(f_1(x) - f_2(x) = 0)$, the solution x_s can be found at their intersection as in Figure 16.7c. Here the iteration consists of going from one curve to the other, closing in at the crossing point. Starting at x_0, $f_1(x_0)$ is found and equated to $f_2(x_1)$. This leads to the value of x_1, from which $f_1(x_1)$ is computed and the process continued until x_{i+1} becomes very nearly equal to x_i. At this point $x_{i+1} \cong x_i \cong x_s$. Thus, in general,

$$f_2(x_{i+1}) = f_1(x_i) \tag{4}$$

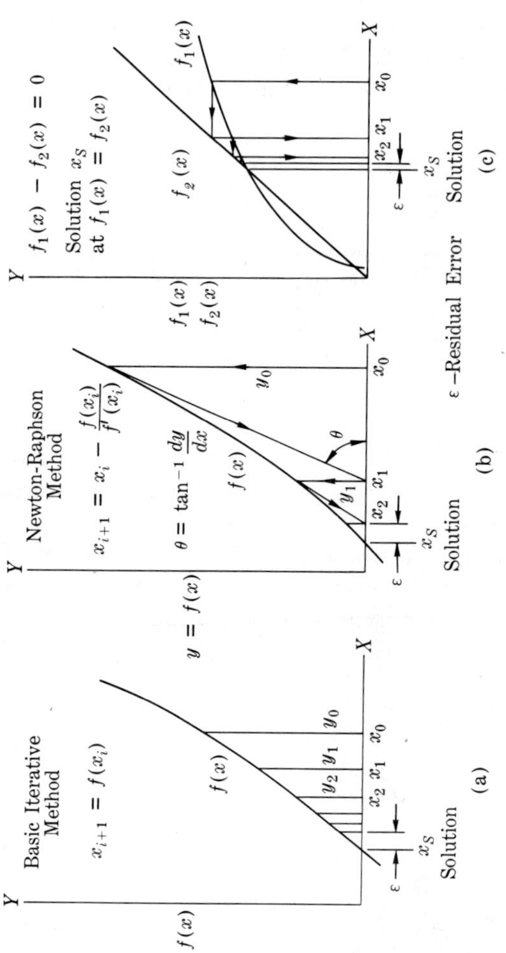

Fig. 16.7 Iterative method of solution of equations.

leading to the explicit value of x_{i+1}. Then

$$f_2(x_{i+2}) = f_1(x_{i+1}), \text{ etc.}$$

Transcendental equations. These will be illustrated by the following examples of those two kinds of transcendental equations which are within our scope: trigonometric and exponential.

Trigonometric equations (wheel counterweight). The usual types of trigonometric equations which are not directly solvable are those which combine the following terms:

a. x and $\sin x$

b. $\sin x$ and $\sin 2x$

c. $\sin x$ and $\sin^2 x$

We shall illustrate with an example of the first type. Each new equation must be approached in the light of its individual peculiarities, but the following example is instructive.

An equation involving angle and sine of the angle arises when both arc and chord enter into a shape. Such a case occurs when computing the sector-shaped counterweight typically used on locomotive driver wheels or oil well reciprocating pumps. See Figure 16.8a. The side area of the whole counterweight is shown as $2A$, with a lever arm C measured to the center of gravity of the counterweight. One half-area A lies in toward the center of the wheel; but all we need to work with is the outer half, sector A.

Subtracting the triangles from the segment area $r^2\theta$, we find the area A to be

$$A = r^2\theta - (r^2 \sin 2\theta)/2 = r^2(\theta - \sin 2\theta/2) \qquad \text{(sq ft)} \tag{5}$$

where

r = radius of wheel (ft)
θ = arc, half angle (radian)

also,

M = unbalanced moment, ft lb

t = thickness of counterweight, ft

ρ = density of counterweight, lb/cu ft

C = lever arm, wheel center to center of gravity, ft

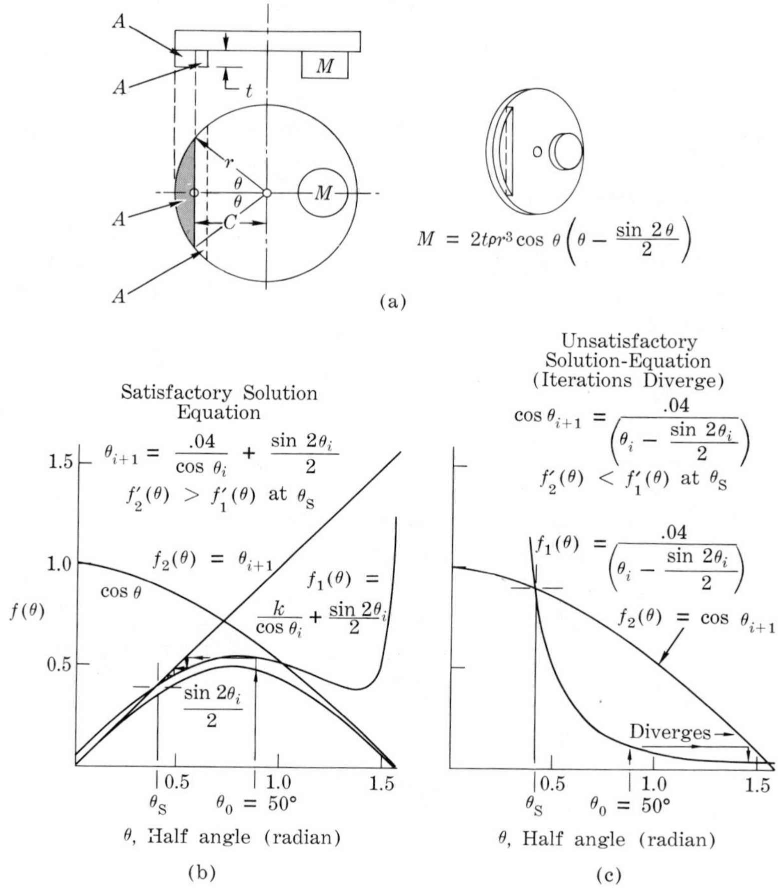

Fig. 16.8 Iterative solutions; counterweight example. (a) Counterweight geometry; (b) Converging solution, $\theta_s \cong 25° \cong 0.43$ radian; (c) Solution not obtainable iteratively, even using identical original equation.

The moment of the whole counterweight must equal the otherwise unbalanced moment M on the opposite side; so, equating moments, we have

$$M = \text{weight} \times \text{lever arm} = 2 \times At\rho \times C \tag{6}$$

or

$$M = 2t\rho r^2(\theta - \sin 2\theta/2)(r \cos \theta)$$

$$M = 2t\rho r^3 \cos \theta(\theta - \sin 2\theta/2) \quad \text{(ft lb)} \tag{7}$$

Thus

$$\cos \theta(\theta - \sin 2\theta/2) = K = M/2t\rho r^3 \tag{8}$$

In Equation (7) we would normally be given the unbalanced moment M, the density ρ and radius r. This leaves the two variables t and θ; and of course, thickness t would be limited to a small size relative to radius r, and θ is limited to that angle (about 60°) which provides area A equal to one half of the half circle ($A = \pi r^2/4$ is the maximum counterweight).

To reduce Equation (7) to the one variable θ, we may assume a reasonable value for the thickness t and combine all into one constant K, as in Equation (8).

$$(M = 500 \text{ ft lb}; \quad t = \tfrac{1}{2} \text{ ft}; \quad \rho = 450 \text{ lb/cu ft}; \quad r = 3 \text{ ft})$$

We can develop our iteration or recursion equation simply by rearranging Equation (8), so that

$$\theta = K/\cos \theta + \sin 2\theta/2 \tag{9}$$

from which

$$\theta_{i+1} = K/\cos \theta_i + \sin 2\theta_i/2 \tag{10}$$

Equation (10) is read as follows:

1. Assume a value θ_0, within acceptable limits (0–60°).
2. Compute $K/\cos \theta_0 + \sin 2\theta_0/2$. This equals θ_1.
3. Compute $K/\cos \theta_1 + \sin 2\theta_1/2$. This equals θ_2.
4. Continue until $\theta_n \cong \theta_{n-1}$, to the allowable degree of difference.

The curves of Figure 16.8b shows how this converges in relatively few steps on the solution $\theta_s \cong 25° = 0.43$ from an initial value $\theta_0 = 50°$.

The computer program in FORTRAN, together with the typed out iterations, are shown in Figure 16.9a,b.

In setting up the recursion equation, notice that the variable must be easily deducible from the $i + 1$ term. Otherwise the next value is not available. That is, rearranging Equation (8) in this form,

$$\theta \cos \theta = K + \cos \theta \sin 2\theta/2$$

would have been no help, because θ_{i+1} could not be directly obtained from either side.

One other alternative seems possible, in the form $f_2(x_{i+1}) = f_1(x_i)$:

$$\cos \theta = K/(\theta - \sin 2\theta/2) \tag{11}$$

or

$$\sec \theta = (\theta - \sin 2\theta/2)/K \tag{12}$$

```
Ø83ØØ C       THETA = K/COS(THETA) + SIN(2*THETA)/2
Ø83ØØ         READ,AK,THETA
Ø8336         PRINT 1Ø
Ø836Ø         AN=1.
Ø8384         THET1=Ø.
Ø84Ø8 21       A=COSF(THETA)
Ø8432         B=2.*THETA
Ø8468         C=SINF(B)
Ø8492         AK1=AK/A
Ø8528         AK2=C/2.
Ø8564         THETA=AK1+AK2
Ø86ØØ         IF((THETA-THET1)-.ØØØØ1)31,31,3Ø
Ø8676 3Ø      THET1=THETA
Ø87ØØ         K=1
Ø8724         GO TO 4Ø
Ø8732 31      K=2
Ø8756 4Ø      PRINT 11,AN,THETA
Ø8792 1Ø      FORMAT(19HITERATION     THETA)
Ø8854 11      FORMAT(F1Ø.Ø,F1Ø.5)
Ø8882         AN=AN+1.Ø
Ø8918         GO TO (21,22),K
Ø8994 22      STOP
Ø9ØØ2         END
END OF COMPILATION Ø97Ø31955Ø
1981Ø AK
198ØØ THETA
1977Ø AN
1976Ø 1ØØØØØØØØ1
1975Ø THET1
1974Ø ØØØØØØØØ99
1972Ø A
197ØØ B
1969Ø 2ØØØØØØØØ1
1968Ø C
1967Ø AK1
1966Ø AK2
1964Ø 1ØØØØØØØØ4
1961Ø K
196ØØ ØØØ1
1958Ø ØØØ2
```

(a)

Fig. 16.9 (a) FORTRAN program for counterweight example. $\theta_{i+1} = .04/\cos\theta_i + \sin 2\theta_i/2$, where $\theta_0 = 50°$.

but neither of these will work because of another and more important reason: The iteration leads *away* from the solution when going from f_1 to f_2. In order for a sequence to converge, the first expression $f_1(x_i)$ must have a lower absolute value of slope at and near the intersection than the second, $f_2(x_{i+1})$, from which the next value of x is obtained. Figure 16.8c shows the problem

```
ITERATION      THETA
       1.      .23552
       2.      .26804
       3.      .29687
       4.      .32156
       5.      .34201
       6.      .35841
       7.      .37121
       8.      .38096
       9.      .38824
      10.      .39360
      11.      .39750
      12.      .40031
      13.      .40233
      14.      .40377
      15.      .40479
      16.      .40551
      17.      .40603
      18.      .40639
      19.      .40664
      20.      .40683
      21.      .40695
      22.      .40704
      23.      .40711
      24.      .40715
      25.      .40718
      26.      .40721
      27.      .40722
      28.      .40723
      29.      .40724
STOP
```

(b)

Fig. 16.9 (b) Solution iterations for counterweight example.

with Equation (11). The companion equation (12) is similarly handicapped. Figure 16.10 shows the various successful configurations from which can be deduced the general geometric requirements for convergence.

Exponential equations (electric cable insulation).[1] The one other transcendental equation which we will illustrate involves exponentials; and this particular case comes from a problem in electric cable insulation. We will develop it very briefly because it is solved in much the same way as the last example.

When a voltage is applied between conductor and sheath in the typical power cable whose cross-section is shown in Figure 16.11, the highest electrical stress appears at radius r. The ratio between line voltage and this

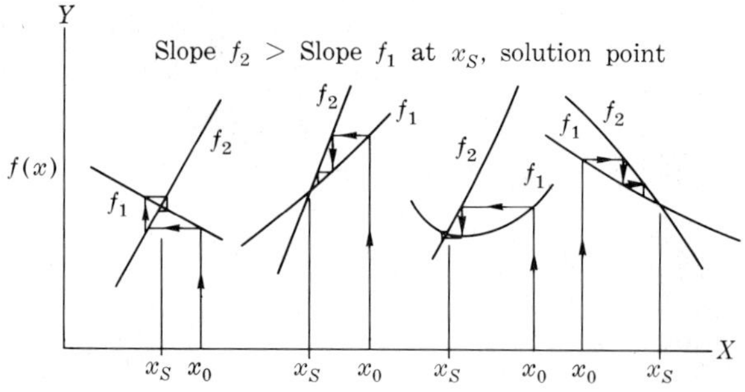

Fig. 16.10 Types of iterative solutions, based on slope f_2 > slope f_1.

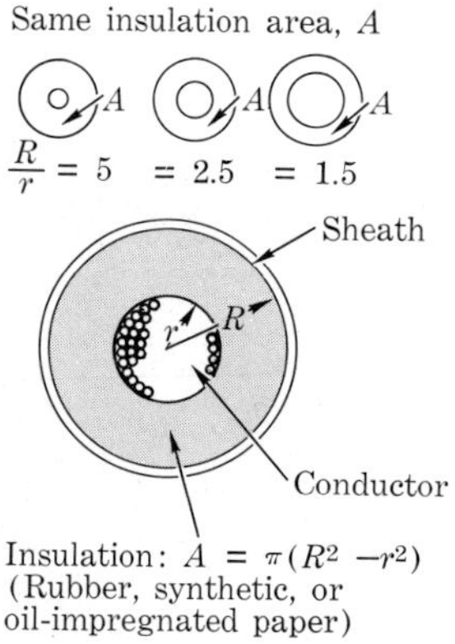

Fig. 16.11 Iterative solution. Cable example, $x_{i+1} = \epsilon^{(1-1/x_i^2)}$, where ($x = R/r$). Showing three different cables with same area of insulation A, corresponding to three values of R/r.

maximum electrical stress gives us one relation,* which we will call utilization, u, a constant for a given r and R.

$$u = (\text{line voltage/max. electrical stress}) = r \ln (R/r) \qquad (13)$$

* Derivation:

$$\text{voltage} = (\sigma \ln R/r)/2\pi\epsilon$$

$$\text{Max. stress} = \sigma/2\pi\epsilon r$$

where

σ = charge per unit length of cable
ϵ = dielectric constant of insulation

Thus,

$$(\text{voltage/max. stress}) = r \ln R/r$$

The problem is to proportion R with r so as to best use a given volume (section area) of insulation. The designer is interested, because this would allow such a cable to carry the highest safe voltage, and hence presumably the most power. Thus we want to minimize area A with respect to the ratio R/r, which we shall call x.

$$A = \pi(R^2 - r^2) = \pi r^2(R^2/r^2 - 1) = \pi r^2(x^2 - 1) \tag{14}$$

From Equation (13) we substitute $(u/\ln x)$ for r, and the expression for area becomes

$$A = \pi u^2(x^2 - 1)/\ln^2 x$$

When we set $dA/dx = 0$, we find the minimum value of area occurs at a value of x which solves this equation:

$$x^2 \ln x - x^2 + 1 = 0$$

or

$$\ln x = 1 - 1/x^2$$

or, exponentiating,

$$x = \epsilon^{(1-1/x^2)}. \tag{15}$$

This is a transcendental equation which can be solved by an iterative method similar to the trigonometric equation (9); otherwise the equation can most easily be solved only graphically.

We will leave this example now, simply stating the equation's *recursive expression*

$$x_{i+1} = \epsilon^{(1-1/x_i^2)} \tag{16}$$

from which $x = 2.21846$ in a few iterations; using a permissible error of $x_{i+1} - x_i = .000001$.

Simultaneous equations (heat-transfer).[1] The second illustration of the iterative method is the application to simultaneous equations. We shall use the example of a relatively simple heat-transfer problem, in which the computation proceeds from equation to equation. Having started with assumed values of two of the variables, the computations lead to closer values which are used for the next round.

The problem situation is the wall of a furnace, using a layer of high temperature firebrick on the inside plus a layer of more common brick on the outside. We know the inside and outside temperatures, the internal heat supply, and the properties of the brick. The question is what wall thicknesses should be specified. Figure 16.12 shows the situation diagrammatically. We will have three equations and three unknowns, x, y, T, arising from the usual design question of "How big must it be to meet the constraints?"

The heat flow through a plane is $Q = K(T_f - T_r)/x$, Btu/hr/sq ft, where K is the conductivity, T_f and T_r are the furnace and room temperatures and x is the wall thickness. If the temperature of the interface plane between the two kinds of bricks is called T, the above heat flow equation

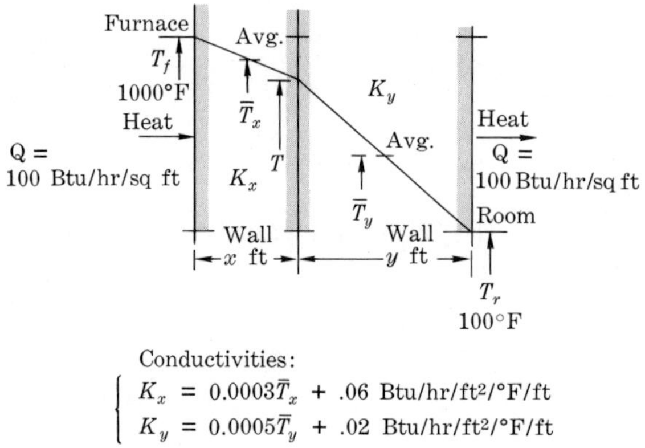

Fig. 16.12 Iterative solution. Simultaneous equations; double brick wall example. Temperature is shown graphically along the vertical dimension, starting with room temperature T_r at the bottom and rising to furnace temperature T_f at the top.

takes the following sequential form, equating heat flow through successive walls and their combination,

$$Q = \frac{T_f - T}{\dfrac{x}{K_x}} = \frac{T - T_r}{\dfrac{y}{K_y}} = \frac{T_f - T_r}{\dfrac{x}{K_x} + \dfrac{y}{K_y}} \tag{17}$$

The Ks vary with the average temperature of the bricks,* $(T_f + T)/2$, $(T + T_r)/2$, but they introduce no new variables. Equations (17) represent only two equations because any third can be derived from the other two. Thus, we will formulate a needed third equation by assuming that the cost difference between types of brick makes it desirable, as a first approximation, for us to use twice as much of the cheaper.

$$y = 2x \tag{18}$$

In an other problem situation, some other relation might be appropriate, but a suitable third equation must be supplied. (Other relations might be: $x + y = c_1$; or $x = c_2$; or x, y given, Q, T_f, T unknown.)

* See any recent mechanical engineering handbook.

Our nomenclature and given values are:

Q = heat flow through a unit area,	100 Btu/hr/sq ft	
T_f = furnace temperature,	1000 °F	
T_r = room temperature,	100 °F	
$\bar{T}_x$ = average temperature in wall x, $(T_f + T)/2$, °F.	—	
$\bar{T}_y$ = average temperature in wall y, $(T + T_r)/2$, °F	—	
T = temperature at interface x-y, °F	unknown	
T_i = temperature at interface, at *ith* iteration	unknown	
x, y = wall thickness, ft		
K = wall conductivity, (Btu/hr/sq ft)(ft)/°F	—	
$K_x = .0003\bar{T}_x + .06$, (Btu/hr/sq ft)(ft)/°F	—	(19)
$K_y = .0005\bar{T}_y + .02$, (Btu/hr/sq ft)(ft)/°F	—	(20)

The easiest iterative method consists of assuming K_x, K_y in their operating neighborhoods, and then using the following equations, derived from Equations (17), (18), (19), and (20).

$$x_i = \frac{(K_x + K_y)(T_f - T_r)}{Q(K_y + 2K_x)} \qquad \text{[from Equations (17) and (18)]} \tag{21}$$

$$T_i = T_f - Qx_i/K_x \qquad \text{[from Equation (17)]} \tag{22}$$

The value of T_i above leads to new values for $\bar{T}_x$ and $\bar{T}_y$ and consequently to a new K_x and K_y; so the next steps are:

$$\bar{T}_{xi} = (T_f + T_i)/2 \tag{23}$$

$$\bar{T}_{yi} = (T_i + T_r)/2 \tag{24}$$

$$K_{xi} = .0003\bar{T}_{xi} + .06 \qquad \text{[from Equation (19)]} \tag{25}$$

$$K_{yi} = .0005\bar{T}_{yi} + .02 \qquad \text{[from Equation (20)]} \tag{26}$$

$$x_{i+1} = \frac{(K_{xi} + K_{yi})(T_f - T_r)}{Q(K_{yi} + 2K_{xi})} \tag{27}$$

$$T_{i+1} = T_f - Qx_{i+1}/K_{xi} \tag{28}$$

Thus, the set of Equations (23) through (28) are the simultaneous equations leading to a solution,

$$T_{i+1} = f_1(T_i) = f_2(x_i)$$

When the sequence is started at $K_x = .07$, $K_y = .03$, the solution converges in a few steps to $x = 0.155$ ft; $y = 0.311$ ft; $T = 822$ °F. The same solutions result in a few steps also if K_x and K_y are assumed twice their values to begin with.

Since there are other ways of setting up simultaneous equations iteratively, one other method of solving the above problems will be shown. The same basic equations (17), (18), (19), and (20) still apply; but now the value of T_i is assumed first and the other quantities calculated from it:

T_i assumed

$$\bar{T}_{xi} = (1000 + T_i)/2$$

$$K_{xi} = .0003\bar{T}_{xi} + .06$$

$$x_i = K_{xi}(1000 - T_i)/100$$

$$y_i = 2x_i$$

$$T_j = -400 \pm 200(6.25 + 100 \cdot y_i)^{1/2} \text{ (the quadratic roots for } T)^*$$

$$T_{i+1} = (T_i + T_j)/2$$

The expression for T_j does not converge alone; however if it is averaged with T_i, the expression does converge on the same answers as those of the previous method: $x = 0.155$ ft; $y = 0.311$ ft; $T = 822$ °F.

In conclusion, the iterative solution of simultaneous equations is extremely useful, but also uncertain. The method must be proved by sample calculations and convergence must be demonstrated. One other hazard of course is the possibility that the equations may have no real solution. The objective to remember in setting up recursive expressions is to find a new expression for the variable first estimated; and the several equations must lead to it:

$$x_{i+1} = f_1(x_i)$$

Differential Equations

Another iterative use of the computer in design is in the integration of otherwise unsolvable differential equations. Velocities, accelerations, slopes, and flows all involve rates of change, and these are frequently set up in the dy/dx or d^2y/dt^2 derivative forms. Standard mathematical procedures permit the solution of many types. But, there are many which cannot be directly integrated. For instance, the following expression† rigorously requires an infinite series for integration:

$$\frac{dy}{dx} = \frac{\log y}{x} \quad \text{or} \quad dy = \log y \, dx/x \tag{29}$$

$$* \; Q = \frac{K_y(T - T_r)}{y_i} = \frac{[.0005(T + T_r)/2] + .02(T - 100)}{2x_i} = 100$$

† ($\int dx/\log x = \log(\log x) + \log x + (\log x)^2/4 + \cdots$)

If Equation (29) is changed to differential form, and we choose a constant small value for Δx, an iterative expression quickly appears. We will change x_i first, each time, to x_{i+1}:

$$x_{i+1} = x_i + \Delta x$$

$$\Delta y_i = \frac{\log y_i \, \Delta x}{x_{i+1}}$$

$$y_{i+1} = y_i + \Delta y_i = y_i + \frac{\log y_i \, \Delta x}{x_{i+1}}$$

or

$$y_{i+1} = y_i + \frac{\log y_i \, \Delta x}{(x_i + \Delta x)} \tag{30}$$

Equation (30) can be directly evaluated starting at initial values of $y_i = 10.0$ and $x_i = 1.0$, and $\Delta x = .02$, for instance, these x, y values coming from the boundary conditions of the problem. The number of repetitions would depend upon the need, but the number might be as many as 100 to cover the required region. The print-out would proceed as follows:

$\Delta x = .02$	x_i	y_i	$\log y_i$	$x_i + \Delta x$	$\log y_i \, \Delta x/(x_i + \Delta x)$	y_{i+1}
	1.00	10.0000	1.00000	1.02	.01960	10.01960
	1.02	10.0196	1.00087	1.04	.01922	10.03882
	1.04	10.03882	1.00170	1.06	.01890	10.05772
	1.06	10.05772 . . . . .etc.				

The plot of x_i and y_i constitutes the solution to the original differential equation (29).

The type C problem, *Algorithm*, consists of an iterative procedure (or theorem) which is developed for convenience, economy of operations, and simplicity in solving any amenable problem, which is not more easily solved otherwise. The categories range widely, from the long-division square-root and polynomial algorithms to the elaborate "Simplex Algorithm"* developed by Dantzig for linear programming problems.

The simplest algorithm is that for raising an expression to a power.

A^n is found by $A_{i+1} = A \times A_i$, taken n times.

The same algorithm applies to the computation of a polynomial,†

$$f(x) = a_1x^3 + a_2x^2 + a_3x + a_4)$$

or,

$$f(x) = ((((a_1)x + a_2)x + a_3)\; x + a_4)$$

* See Linear Programming References, Chapter Fifteen.

† From Reference 1.

This is obtained by the algorithm, "Multiply by x and add the next constant," three times. In general, this is shown as

$$f_n(x) \overset{i=n}{\underset{i=1}{}} = [f_i(x)]x + a_{i+1} \qquad \text{where } f_1(x) = a_1 \tag{31}$$

The form of Equation (31) could have been applied to the optimizing equation developed in Chapter 12.

$$CF = aD^2 + bD + d/D + e/D^2$$

or

$$CF = (aD^4 + bD^3 + cD^2 + dD + e)/D^2 \qquad \text{where } c = 0$$

Thus

$$CF = (((((a)D + b)D + 0)D + d)D + e)/D^2$$

In general,

$$CF \overset{i=4}{\underset{i=1}{}} = [\{f_i(D)\}D + a_{i+1}]D^{-2} \qquad \text{where } f_1(D) = a \tag{32}$$

The square root of a number may be found by a very simple algorithm,* wherein the first value x_1 of the variable is initially estimated.

$$x_{i+1} = 0.5(x_i + A/x_i) \tag{33}$$

when

$$x_{i+1} - x_i = \text{very small}, \; x_i = (A)^{\frac{1}{2}}$$

For example, if $A = 0.49$, and the initial x_1 were chosen 0.5,

$$x_2 = 0.5\left(0.5 + \frac{0.49}{0.5}\right) = 0.74$$

$$x_3 = 0.5\left(0.74 + \frac{0.49}{.074}\right) = 0.7015$$

$$x_4 = 0.5\left(0.7015 + \frac{0.49}{0.7015}\right) = 0.7002 \;.\;.\;.\; \text{etc.}$$

Such an algorithm is used by computers as a subroutine generally, when the square root command is called for.

All common arithmetic operations use algorithms, simply routinizing the operations of multiply, carry, add, shift, borrow, etc. Algorithms are used to organize the routines of accounting: Payment of wages, overtime, deductions, and taxes. The algorithm may be called a "formula," but its prime characteristic is its repeated use to explore, to evaluate, or to sum up a large number of repeated similar situations.

As with any tool in design, algorithms are developed and used when appropriate, in other words, when convenient, competitive, and productive. We will close this section with a summary of the uses of the computer in design and then present a project illustration of a simple algorithm.

* From References 1,10

SUMMARY

The digital computer is now used in all the stages and phases of design. Its greatest aid is in rapid computation for evaluation of alternatives, in its high speed of information retrieval (when the information is in the computer), and in its simulation of system operation. The computer is of least help in synthesis, *de novo* creativity and decisions learned from experience.

Much effort is currently being expended on "software" and shortly more help may be anticipated in the areas of decision-making, graphic design, and automated design.

A wide variety of computers are available at economical costs (direct or shared); and engineers have little excuse to deny themselves the added effectiveness. See Table 16.2 for this array of computers and characteristics.

BOOK PROJECT EXAMPLE: Algorithm

One of the problems of a young company is choosing the most advantageous location of the plant with respect to customers and supplies. The best location of the Econocrete Company is in the "center of gravity" of its transportation, taking into account costs, proportions of sales, and distances.

A map of the neighboring towns has been assembled with colored pins showing the location of various sized customers (or sales regions) and the respective supplies. See Figure 16.13. There are naturally different costs associated with hauling the sand and gravel aggregate to the plant, compared to trucking the mix on to the customers.

Based upon records and cost investigations, the following formula seems to represent fairly the cost per cubic yard mile, both from the supplier to the plant and then to the customer.

$$\text{Cost} = .03 \sum_{s} f_s d_s + .12 \sum_{c} f_c d_c \qquad \text{(dollars/cu yd mile)} \qquad (34)$$

where

f_s = fraction supplied by the given supplier
d_s = distance from given supplier, mile
f_c = fraction bought by given customer
d_c = distance to the given customer, mile

The coefficients represent \$.03/cu yd mile and \$.12/cu yd mile for aggregate and mixed concrete respectively.

The map shows the distances on mile grids for some representative cases, S for supplier, C for customer and P for a tentative plant location. The distances are measured along the coordinates, not as the crow flies. The objective, of course, is to find the above total transportation cost for the given plant location shown, and enough other locations, so as to find the location having minimum cost.

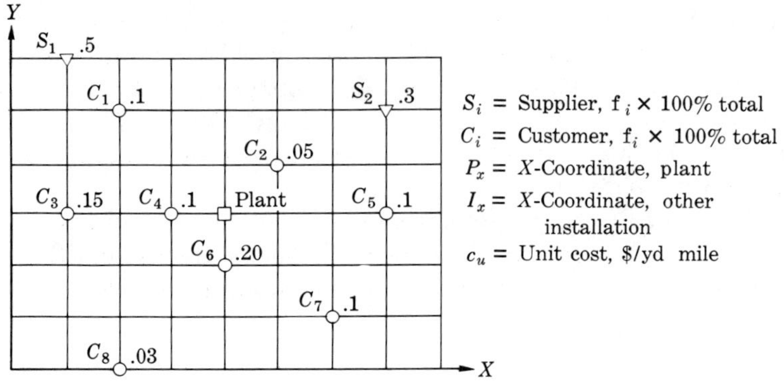

Fig. 16.13 Book Project Example: Algorithm for transportation costs.
Cost $= c_u f_i(|P_x - I_x| + |P_y - I_y|)$.

Equation (34) above converts into the following when the x-y coordinates of each location are used:

$$\text{Item cost} = c_u f_i(|P_x - I_x| + |P_y - I_y|) \tag{35}$$

$$\text{Total cost} = .03 \sum_s f_s(|P_x - S_x| + |P_y - S_y|) + .12 \sum_c f_c(|P_x - C_x| + |P_y - C_y|) \tag{36}$$

where P_x, P_y, S_x, S_y, C_x, C_y are respective coordinates for plant, supplier, and customer.

c_u = unit cost of commodity (.03 or .12)
I_x, I_y = installation coordinates, supplier, or customer
f_i = installation fractional supply or use

The absolute values, $|P_x - S_x|$, are used because plus or minus distances cost equally.

Thus, the algorithm for item cost of transportation is Equation (35). This expression applies to all installations, and serves to construct the total cost equation (36). In either form, (35) or (36), the algorithm can be utilized on a digital computer, totaling the costs for all installations versus each alternative plant location, P_x P_y. The whole array is ranked in value and the minimum P_xP_y quantified and located.

Table 16.2

	Fixed Point Add Time (*μ sec*)	*Word Length*	*Storage Words*	*Memory Access Time* (*μ sec*)	*Type of Input*	*Type of Output*	*Average Monthly Rental* ($)	*Average Price, Basic System* ($)
1. Control Data Corporation CDC 160A.	19	12 bits	8,000 (Core)	6 (Core)	Card, paper tape.	Card, paper tape.	2,250	90,000
2. General Electric GE 225.	36	20 bits	4,096 (Core)	18 (Core)	Card, paper tape, magnetic tape, typewriter.	Card, magnetic tape, printer, typewriter.	8,000	400,000
3. General Precision LGP 30.	260	31 bits	4,096 (Drum)	(Drum) 270 min 7500 max	Paper tape, typewriter.	Paper tape, typewriter.	1,300	50,000
4. International Business Machines IBM 1401.	290	Variable	1,400 to 16,000 (Core)	11 (Core)	Card, magnetic tape, optical character reader.	Card, magnetic tape, printer.	3,000	150,000
5. International Business Machines IBM 1620 II.	200	12 bits/instruction and variable field	20,000 (Char.)	10 (Core)	Card, paper tape, typewriter.	Card, paper tape, typewriter, lineprinter.	1,800	90,000
6. National Cash Register NCR 310.	19	12 bits	4,096 (Core)	6 (Core)	Paper tape, typewriter.	Paper tape, teletypewriter.	2,500	100,000
7. Sperry Rand UNIVAC 1004.	128	Variable field	(Alphanum.) 961 (Char.)	8 (Core)	Card.	Printer, card.	1,700	65,000
8. International Business Machines IBM 7094 II.	4	36 bits	32,768 (Core)	1 (Core)	Card, magnetic tape.	Card, magnetic tape, printer.	70,000	3,000,000

REFERENCES

1. IBM Corporation: "General Information Manual, Engineering Analysis for Computers," #F20-8077, White Plains, New York, 1961.
2. Anderson, D. M.: "Basic Computer Programming, IBM 1620 Fortran," Appleton-Century-Crofts, New York, 1964.
3. Organick, E. I.: "A Fortran Primer," Addison-Wesley Publishing Co., Inc., Reading, Mass., 1963.
4. Ledley, R. S.: "Digital Computer and Control Engineering," McGraw-Hill Book Company, New York, 1960.
5. Smith, G. M., V. C. Wolford, M. L. James: "Analog and Digital Computer Methods in Engineering Analysis," International Textbook Co., Scranton, Penn., 1964.
6. Strom, R. L.: *Role of Computers in Design,* Engineering Research Colloquium, UCLA, February 1962.
7. "Adams Sheets on Computer Characteristics," Charles W. Adams Co., Bedford, Mass.
8. Douthwaite, G. K.: Good Problems for a Course in Introductory Digital Computing for Engineers, *Journal of Engineering Education,* Vol. 55, No. 3, November 1964, p. 83.
9. Westervelt, J. H., R. C. Wilson, et al.: *Use of Computers in Mechanical Engineering Education* and *Use of Computers in Industrial Engineering Education,* etc., Ford Foundation Project on Use of Computers in Engineering Education, University of Michigan Computing Center, 1962–63.
10. Lynch, Charles J., Design by Digital Computer, *Product Engineering,* April 11, 1960, pp. 43–58.
11. Baumann, D. M., S. J. Fenves, L. A. Schmit, Jr.: *Impact of Computer Concepts on Engineering Design,* Report #1, Committee on the Use of Computers and Mathematical Techniques in Engineering Design, Commission on Engineering Education, Washington, D.C., 1965.
12. Computer-Aided Design, *Product Engineering Magazine,* Monograph #R113, McGraw-Hill, Inc., New York, 1960–1965.
13. Burks, A. W., H. H. Goldstine, J. Von Neumann: Preliminary Discussion of the Logical Design of an Electronic Computer, *Datamation,* September 1962.

Communication and the Engineer

Seventeen

"You cannot teach a man anything;
you can only help him find it by himself."
Galileo

Communication and Information

It is a very unusual engineer nowadays who can finish an engineering assignment by saying: "I've put it together personally, and I know it does what it is supposed to do. I'm done." These comments would imply that the engineer had built and tested the product or system himself and that it was not necessary for him to work with, consult, or inform anyone else throughout the process, or afterwards.

This simple working situation may have occurred in earlier days, and with men like Henry Ford and Thomas Edison, but it just doesn't occur any more. Even the lone inventor working in his garage or basement must "sell" his idea to the financing sources or to a customer.

Under our present division of labor, the engineer must communicate with a vast array of people continuously; to get information, identify the problem, prepare the solutions, test them, prepare drawings and prototypes, tell the shop what to build, and finally tell the client what he now has.

The engineer doesn't know enough even to develop the design alone; and once he has a plan, he hasn't enough skills or enough hands to implement it alone. This may be obvious, but one occasionally observes engineers still *trying* to do it all alone.

In short, the engineer must communicate to succeed. And by communicate we mean far more than *speak*, or *instruct*, or even *persuade*. Communication embraces these, but it goes on to include feedback: the *listening* and the *reacting* to other persons. Without the return loop of observing the results, the originator has no way of knowing the effectiveness of his communication attempts, or whether his design really "works" for its purpose.

A new and operational definition of communication* has been suggested, and we paraphrase it as follows:

> "Communication is the process by which one organism (object or machine) discriminatively secures a response from another."†

Communication is thus evaluated by the degree to which the response conforms to the intent of the originator.

In the context of machines, we can readily see this principle quantified. Communication can of course be applied to machines. We "tell" a car where to move or a refrigerator how cold to chill. A machine operator (or machine) sets a dial which "orders" an automatic plastic molding press to produce 10,000 parts. If 9,990 acceptable parts are produced in response, 10 parts of the 10,000 or one-tenth of one per cent of the message did not go through. The response erred by that amount.

The counterpart analogy of information between humans seems at the outset difficult, if not impossible, to quantify. How can one measure human response?

Claude Shannon[5], working in the telephone-system context at the Bell Telephone Laboratories, has developed a mathematical theory of information. This was done first by defining information statistically in terms of what one *could* say, not what one *did* say. Information thus connotes freedom of choice, the number of alternatives possible for the message. If a person could send only one message, there was zero information in it, there was no point in sending it at all.

With limited message possibilities, as with a set of numbered pre-written message-telegrams, the receiver knows the message is one out of twenty (for example). Other factors being equal, that particular quantity of information is represented by a *probability* of 1 in 20, or 0.05.

Now with the possibility of numerical treatment of information established, the transmitting channels can be investigated, along with the various kinds of *noise* which affect information flow.

* The material that follows in this half of the chapter is largely based on Rosenstein et al., Chapters 1, 2, 3, (Reference 1); and Tribus, Chapter 24, Reference 2.

† S. S. Stevens quoted in Reference 1.

Information Theory

We will open the door on information theory only a crack. The various concepts and understandings that come from this theory are important to everyday communication; and this is why we mention the theory at all.

The theory starts with the elements of the communication system. This is shown in its simplest form as Figure 17.1, the physical (mechanical, acoustical, or electrical) system.

In order for the desired message to go from the sender to the destination, the transmitter encodes, the channel carries, and the receiver decodes the message. The coding, of course, is the conversion of numbers, indentifications, and meanings into transmittable form, such as electrical currents,

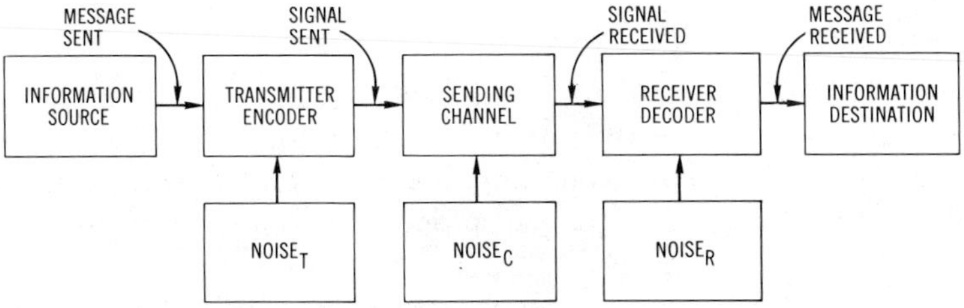

Fig. 17.1 Basic elements of all communication activities or systems.

punched cards, pulses of sound or light, etc. Interference by the intrusion of mechanical or electrical error is inevitable. Stray induced electric currents, backlash in mechanical systems, damaged or bent IBM cards, and defective parts all contribute error, such error being equivalent to the familiar static or *noise* in a telephone or radio system. As you see by the figure, noise or distortion can enter into the coding, into the transmission, or into the decoding in a physical form. We will discuss other kinds of noise in a later section.

The important concepts so far are:

Intended results require communication (Effect comes from cause.)
Source → Flow system → Destination (Communication requires a transmittal agent.)
Coding-decoding (Information changes form.)
Noise (Error in transmission is inevitable.)

Information Measures

It has been found convenient to express information numerically in terms of the binary code, or two digits. For example, if one chose a symbol to represent one of a possible group of 32 messages ($= 2^5$), among which he were equally free to choose, this would be called 5 bits of information (BIT = *BInary digiT*). This conclusion comes from Shannon's equation for information:

$$H = -\sum_i p_i \log_2 p_i \text{ bits/symbol used} \tag{1}$$

where p_i = probability of use of interdependent symbols $1, 2, \ldots, n$, and

$$\sum_i = \sum_1^n = \text{summation over all } n \text{ symbols}$$

In the case of 32 equally probable messages above, $p_i = 1/32$. Changing the sign and inverting the logarithm, the equation becomes:

$$H = 32(\tfrac{1}{32} \log_2 32) = \log_2 32 = 5 \tag{2}$$

Equation (1) can be extended to the case of complete certainty where probability equals 1 ($\log 1 = 0$) and hence the information $H = 0$. Only where there exists choice among possible messages does H have a non-zero value. As the range of choice increases, so does the quantity of information carried.

Channel capacity is defined as the number of bits of information transmitted correctly per unit time. If a computer-to-computer wire connection between two plants had a set of 2^m symbols among which to choose freely, ($\log_2 2^m = m$ bits/symbol), and the channel could transmit n symbols/sec) the channel capacity would be mn bits/sec. If there were 64 symbols ($m = 6$) and the wire could carry 100 symbols/sec ($n = 100$), the capacity would be 600 bits/sec. In the case of black and white television, let us assume $m = 1$ (dot, no dot) and the picture is 500 lines high by 500 dots wide, repeated 30 times/sec. This requires a channel capacity mn of 7.5 million bits/sec.

So far, we have assumed error-free transmission; but of course error creeps in. The only way to assure accuracy is to use special coding with redundancy. The English language itself is a code with considerable redundancy. About half the letters could be garbled or left out of a sentence and the meaning could still be conveyed:

"a–d th– m—n–ng c—ld st—l b– c–nv–y–d"

Redundancy is evident by the occurrence of u following q, by the double consonants like mm, ll, pp, etc. Yet certain redundancy must always exist to guarantee the accurate reception of the message. Important messages must never depend upon the exact transmission of every symbol correctly.

As one final sally into information theory[2], let us illustrate the significance of the binary measure.

How much information is there in 365 equally probable numbers? This is another way of asking how many questions you need to ask to ascertain a person's birthday.

$$H = -\Sigma p_i \log_2 p_i = 365(\tfrac{1}{365} \log_2 365) \tag{3}$$

or

$$H = 8.5 \text{ bits}$$

The implication is that by asking eight or nine yes/no answer questions, you should be able to learn the day and month. Let's try:

Question	*Answer*	*Answers then Possible*
1. Is it on or before June 30?	No	July 1–Dec. 31
2. Is it on or before September 30?	Yes	July 1–Sept. 30
3. Is it on or before August 15?	No	Aug. 15–Sept. 30
4. Is it on or before September 7?	No	Sept. 7–Sept. 30
5. Is it on or after September 18?	Yes	Sept. 18–Sept. 30
6. Is it on or after September 25?	Yes	Sept. 25–Sept. 30
7. Is it on or after September 28?	No	Sept. 25–Sept. 27
8. Is it September 25?	No	Sept. 26 or 27
9. Is it September 26?	Yes	Sept. 26 *Answer**

* If the answer to question 9 had been "no," we would still have had the answer, since there were only two choices.

This gives an interpretation under equal probabilities. If they are not equal, the questions should be divided so as to partition the alternates into "probability halves" each time, in the same way our questions above were divided into simple numerical halves.

We recapitulate this section by the following:

Communication is measured by specific response to the stimulus.

The communication system is complex and requires coding, transmission, and decoding in every case.

Noise interference decreases the accuracy of the message.

Redundancy must be employed to offset the errors introduced by noise.

Information can be measured probabilistically in terms of what *could* be sent.

The information quantified in binary units implies what was actually sent among the known possible alternatives.

Communication and Reports

When communication is extended to include men and women, the diagram of Figure 17.1 becomes much more complex, not only in blocks, but in ramifications. To include not only the codings, but the various meanings and interpretations, the orientations and purposes that persons can put into the coding as well as the decoding, opens a Pandora's box of confusion and potential trouble. Figure 17.2 shows the expanded communications diagram applicable to any combination of persons or machines.

Now "noise" includes *semantic* and *psychological* noise as well as mechanical or physical noise.

To illustrate the semantic and psychological noises, consider the situation of a young engineer discussing his project with an experienced mechanical drafting designer. The engineer is trying to communicate his ideas about

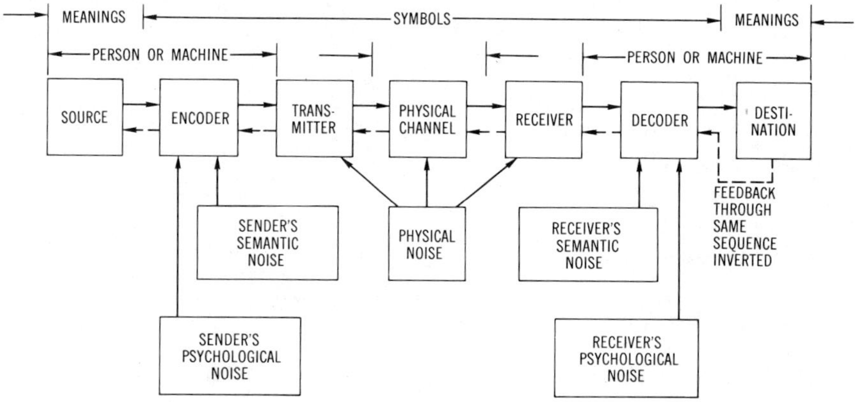

Fig. 17.2 Complete block diagram of a communication system.

a new etched-circuit board to replace a conventional wired panel in a particular application. This is a situation where "etched-circuit-board" could cause considerable confusion to the designer, not being familiar with this new technology. Is the circuit etched? Or is the board etched? Is there actually a circuit on a board? Does this young fellow know what he's talking about? Is the idea practical? Has it been used in production? The semantics of the new meanings and the psychological pressure of defending his job-worthiness among new ideas, combine to make the drafting designer defensive, probably unsympathetic, and certainly not an easy recipient for communication. This illustration can be duplicated in either direction, up or down, and with all levels of communicators, from top management, supervisor, engineer, and sales, on out to the customer.

The importance of appreciating and applying the meanings of the diagram of Figure 17.2 can hardly be over-emphasized. It is a widely published fact that most engineers who have job difficulties, or are discharged, have encountered trouble for *personal* and not *technical* reasons. These perplexed men cannot get along with others—they find communication difficult or impossible with individuals who are important in their circle of operation. The remedies may be easy or difficult, but good communication precedes any significant success.

The Communication Diagram and Engineering Reporting

The engineer can extract many useful cues to help his writing by applying the Figure 17.2 to the "report" situation.

The diagram's first implication seem very clear: There are many influences which impinge on and determine communication and we are *not* aware of either the presence or nature of most of them. Very few of the causes or effects are conscious or verbal, yet they are very real. Consider, for example, the intangibles of personality clash, of getting off on the wrong foot, of the effect of distracting mannerisms or style. These are several psychological noises.

The second implication is that a writer must prepare for all the contingencies of the reader in advance. In a report there is no feedback to suggest the running additions or corrections that a normal speaker makes in face-to-face conversation.

Let us continue to interpret the diagram for the case of communication through technical reports. Although the relation is close, the encoder is not the transmitter; neither is the receiver the decoder. The encoder is the translator from idea to word. The transmitter is the word, not yet on paper or tape. The channel is the physical sheaf of papers.

It doesn't require great imagination to picture the varieties of *noise* indicated in Figure 17.3. Any errors in wording or spelling, any new or unusual symbols, any incomplete diagrams or wrong references to pictures, any detail defects of any kind will inject noise and hence reduce the likelihood that the message will come through clear and strong.

Noise in the channel will consist of incorrect writing mechanics, clumsy typing, disordered pages, soiled or damaged paper, blurred printing reproduction, shipping damage or even inconveniently folded blueprints.

Semantic noise, caused by ambiguous words or by specialized words from an unfamiliar jargon, can be just as distracting or disturbing to the reader.

Even as I write this, I wonder about my own semantics. I just used the words "writing mechanics" above. By this I meant punctuation, spelling, paragraphing, capitalizing, and footnoting. But "mechanics" may well have meant something entirely different to the reader. Perhaps it suggested draftsmanship, penmanship, typing, or even grammar. Erroneous alternate interpretations are thought through and excluded by the skillful writer.

Noise in the transmitter enters primarily as inept logic of assembly, as poor organization of the report, poor sentence structure, or poor charts. These all introduce noise even though the details may be well coded in familiar symbols.

The list goes on and on, but the effect is very important: Communication ability decreases inexorably as the noise introduced into any part of the system increases.*

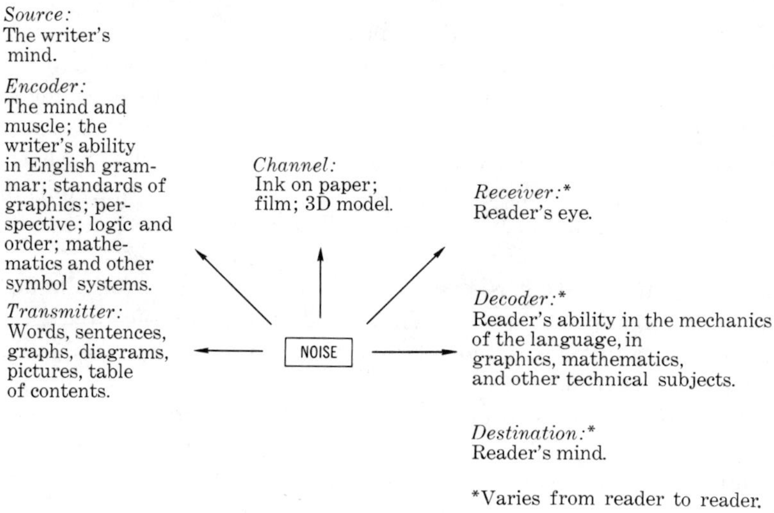

Fig. 17.3 Block diagram interpretation for communication via reports.

Of course, we cannot control the reader and the distractions to which he may be subjected. But, as we shall later explain, the writer must plan for an expected variation in readers. This is why we added "variable" to the receiving-end designations in Figure 17.3.

When the noise, in any stage, is due to carelessness or negligence, the reflection on the writer can only be damning. The reader is likely to project to the writer the trouble he has with the immediate garbled message, and discredit even the writer's serious thinking.

* For those interested in information theory, the influence of noise on C, Channel Capacity (bits/second) is: $C = W \log (P + N)/N$; where P is the power of the message signal, N is the power of the noise. Notice that when $N = P$ or $N > P$, C becomes relatively very small. Thus we come to understand the importance of the frequently heard phrase, signal-to-noise ratio, P/N.

Conversely, a clearly related message, accurate, complete, and understandable, will carry its own persuasion. You, the writer, are at the finishing end of the design; the reader is just at the beginning. You have the perspective and all the arguments. He has only a few superficial ideas because of limited exposure. You know the "answers."

The Nature of the Report

The word "report" means to carry-back, and the two focusing points it suggests are:

To whom do we carry something back?

What do we carry back?

You see, what we carry back depends upon the receiver. Having performed a given engineering assignment, we would write an entirely different report to the sales manager who is to use it with a customer than we would write to our supervisor who reads it only to approve the technology.

The audience and how to reach it. Thus, the very first questions we ask as we consider starting a report are:

Who will read the report? How busy is he? How interested is he? How much does he know about the subject or the specific job?

What variety of readers is expected? What is the range of their technical skill and vocabulary? How important are the various readers?

What will the reader do with the report? Is it for information (file it)? Is it for instruction (use it)? Is it for command (act on it)?

Is the report confidential? How wide is the distribution? Is the report to be published?

Visualize the range of readers possible, regardless of the organization within which the report is written. All of these persons may at some time or another read a report of yours:

Stockholders, owners.
Top management.
Supervisor.
Associate.
Subordinate.
Other department or function.

Vendor, supplier.
Customer, client.
Public official.
Inspection agency.
Professional society.
News agency, public.

Each of these people has his own orientations, vocabulary, range of technical knowledge, intelligence level, and sex, to say nothing of the differences in temperament. Clustering the readers into groups is the first demand upon the writer of reports; then comes identifying their characteristics and deciding whether or not one report will (or can) cover them all.

Having chosen the report form, we must realize we have a semi-captive audience. The reader must at least pick up our report; he generally has paid for it either in total or in part, and he hopes it will be of value to him. At least it is about a subject with which he "should" be concerned.

However, what happens after he starts to read generally depends on the writer. The reader is usually busy, and when necessary, he will look for convenient stopping places, so the report can be picked up again after the interruption. The most important information should be first, tapering down through the document so the least important is last. Thus, the reader can stop anytime he has seen enough to satisfy him. He may read only the summary, or every word. The narrative should be complete, with referenced illustrations integrally enmeshed, so the reader isn't frustrated by implications of later information. He may actually be short-changed by not seeing the too-well-hidden gist of the whole effort.

R. R. Rathbone[1] has well summarized the essentials for coding the information most effectively for the reader:

1. *Make the title meaningful and, if possible, brief.* The reader depends upon the title for an accurate statement of what he is reading about.
2. *Make a summary or abstract.* One of the best ways of gaining the attention of the reader is by presenting a clear concise summary of the high points. Don't hesitate to be specific or quantitative. This settles incipient questions and the reader can then concentrate on your logic.
3. *Provide sufficient background material.* Enough background description must be given to allow the *least informed* reader to understand why the design was undertaken and what the lesser-known or general-purpose words mean in this context. For instance, the "nuclear device" might mean the gamma-ray sensing element that measures the thickness of the plastic sheet being rolled. If properly done, explanation in this step *reduces* semantic noise.
4. *Associate the familiar with the unfamiliar.* Supply starting points and analogies to bring up to your discussion level the reader who is new to the project or to the field. For the technical reader, we could expand upon the above "nuclear device" or "gamma-ray sensing element" by comparing its behavior to the optical case in which the absorption equation holds: $I = I_0e^{-ax}$, where x is the thickness of the sheet. For the less technical or commercial reader, we might simply describe the device as similar to "measuring the absorption of light passing through a translucent sheet." Each case would require appropriate introduction into the actual "gamma-ray" absorption phenomenon.

5. *Describe the whole before the parts.* The background and introduction gives the reader part of the picture; but be sure he has the whole picture before you lead him into the details. The little pieces must fit logically into something bigger or they float around in the reader's mind making noise (and preventing or retarding the decoding). Eliminate the pseudo-choices, the plausible but incorrect alternatives, and lead him down a clearly-marked thought path. This is not a detective story; it is an exposition, a planned elucidation.

6. *Develop a sound structure.* Having the whole picture, the reader is ready for the pieces that make it up. But the way you present them and their organization into the whole is crucial to the reader's comprehension. Follow a recognizable pattern, such as from left to right, inside to outside, or following the line of process or information flow. If one method of development seems the most logical to you, it probably will be best for the reader also. If in doubt, outline several methods and review them (after a waiting period) to choose the best.

7. *Emphasize the important ideas.* You are the one in the best position to separate the important from the unimportant. Let this choice show up in the space and emphasis you give, both to the ideas and advantages as well as to the components and the details of description. No matter how much you have to tell, remember that a long report is usually not read, only skimmed. Secondary information belongs in an appendix.

8. *Be sure to separate fact from opinion.* The reader wants to know clearly what are the *facts*, what are the opinions, and whose opinions they are. In the presentation of a design, the writer will mix many other persons' ideas and opinions in the final composite. This obliges the writer to acknowledge both the originator and the contribution, separating the appropriated extract from the writer's interpretation of it. The reader normally is not much concerned about the originator; but he does want to know the factual origin, so he can make up his own mind independently. The responsibility of the author, of his section (e.g., the Engineering Department), or of the organization (the ABC Co.) are involved in any opinions or recommendations expressed. Remember that the organization is responsible for anything issued in its name or under its letterhead, unless specifically stated otherwise.

9. *Use clean, precise, straightforward language.* The best indication of good writing is unobtrusiveness coupled with maintenance of interest. With good writing, you don't notice the style, yet you want to keep reading on. The ideas are put across without the carrier being seen. All the bluff, the extras, the window-dressing should be pruned from technical report writing, yet without emasculating the writer's personality. That personality subtly comes through even in a simple report. The best way of impressing the reader is with understandability rather than with big words or elaborate phrases, or even with elegant technical statements. Pick the right situation before you decide to use the meticulously precise technical statements. Most of the

time your audience knows less than you do about the subject and simply wants information, not oratory.

Specifically Mr. Rathbone points out some of the traps to avoid:

Fuzzy qualifiers:

"*Large amount* of energy." (How much?)

"*Light-gauge* metal." (How thin?)

Euphemisms:

"Of the *six* men hired, *all were useful* except the first *three*."
(Only the last three were useful.)

Overformal words:

"*Effectuate the procedures leading to the consideration.*" (Get the material ready for the decision.)

Jargon and coined words:

"*We will finalize the cost using available expertise.*" (The best available men will complete the cost.)

Clichés:

"*By and large, it stands to reason, each and every one* will agree." (Everybody generally will agree.)

Dead wood:

"*Due to the necessity of.*" (Because)

"*In the case that.*" (If)

10. *Consider the rapid reader.* The remaining suggestions of Mr. Rathbone are particularly useful in practice because, as we said, the majority of readers do not have time to go through a whole report at one sitting. His recommended practices also lend vividness and sharpness to any writing and for any reader, but they are particularly crucial to help the rapid reader

Use descriptive headings and sub-headings freely.

Place topic sentences near the beginning of paragraphs.

Use simple sentence structure when the thought is complex. Break the thought *and* the sentence into pieces.

Make full use of graphic aids.

Adjust the pace to fit the subject and the reader.

Avoid footnoting secondary or reference material. Footnotes are distracting. Work the material in, leave it out, or put it in the appendix.

Provide white space around the text: Wide margins, good indentation, short paragraphs.

11. *Choose the appropriate report form.* In considering the variety of forms that technical report writing assumes, one naturally questions whether a single set of writing instructions will cover the diversities.

Engineering communication comprises quite an array, ranging in importance from executive decision-making material down to the quick answer for an informal question. Keep in mind that some of the following forms constitute the total engineering output and results of a strenuous group effort on a large project. Such is certainly the case where the communication is an engineering bid for a contract. In general, the following list goes from the simplest communication to the most complex.

Engineer's Writing	*Purpose and Content*
Memo	Informal note asking or answering a specific question or two.
Letter	More formal communication, but similar to a memo.
Informal report*	Structured statement of a problem and a solution, including results and conclusions. It may be interim or final.
Instruction	Specific directions, materials or procedures used in carrying out a technical task.
Specification	Complete detailed technical description, supplementing a set of drawings to permit purchase or construction.
Formal report*	Report on a larger problem or to a more formal audience.
Proposal*	Recommendations for proceeding on a technical alternative, with advantages and disadvantages.
Article	Presentation in a book or periodical of new technical information, or a digest of earlier work, resulting from a study or research.
Technical Paper*	A report on the conduct and results of a major problem, representing usually more than a year's work and constituting a significant contribution to the art.

Of this list, the indicated (*) items fall into the class of on-the-job professional reports with which we are concerned.

Although the audience cannot be truly divided into three types, a reasonable approximation can be made, and this leads us into the respective varieties of report organization shown in Table 17.1.

Notice the differences compared to student reports. Generally the student report carries all the information the student has had time to collect, and is frequently oriented toward impressing the reader rather than accomplishing a task. Our discussion here concerns the professional report, covering the three types listed below. In the next section we will plunge into the actual writing task, just as the engineer must do after he has collected the information, chosen his purpose, organized his material, and decided whom he is addressing.

Table 17.1 Engineering Report Organizations for Different Readers.

Nontechnical Reader; Manager, Client, etc.	*Technical Supervisor or Associates*
1. Letter of transmittal.	1. Letter of transmittal.
2. Title page or heading.	2. Title page or heading.
3. Summary and conclusions.	3. Background of problem.
4. Methods; procedures, logic.	4. Conclusions and recommendations.
5. Verification; extension.	5. Assumptions.
6. Appendix; detail results.	6. Body; argument.
	7. Review of results.
	8. Appendix; data; sample calculations.
	9. References.
	10. Distribution list.

Publication through Professional Channels

1. Title page.
2. Table of contents, illustrations list.
3. Acknowledgements.
4. Preface, foreword.
5. Abstract.
6. Introduction, background.
7. Body, procedure, argument.
8. Results, conclusions.
9. Projection for future work.
10. Summary.
11. Appendix, data.
12. References, bibliography.

The Preparation of the Report

The outline. The well-prepared engineer has several resources he can promptly call upon as he begins to draw up his outline:

A project notebook;
A project file folder;
A card file of references, patents, competitors' equipment, etc.;
Previous interim reports.

These materials, together with his project statement or report objective, start the engineer's thinking as he selects the report form and the internal section headings he is going to use. If he doesn't have all of the desired materials to start with, he may have to pause part way through his writing to get the missing information.

The new writer normally looks upon an outline as medicine: It is a bad-tasting prescription, but good for the writer in some supposed way. The experienced writer, on the other hand, looks upon the outline as his best friend and ally. As H. K. Glidden[3] reminds us, the outline is a *unique* writing tool which is:

Used to record thinking, but seldom expresses thoughts;
Used to organize the presentation of the results of the writer's thoughts;
Applicable to all stages of writing up to the final typing;
Assembled according to some accepted system or systems;
Built of topic words or sentences;
Flexible, and of best use if constantly improved;
Intended primarily for the writer's use, and seldom seen by another;
Useful in designating and evaluating the degree of respective topic importances;
Useful to detect errors, omissions, repetitions, superfluity, and illogic;
Useful in planning length of report by control of sub-headings;
A guide to writing topic sentences;
A guide in oral presentation.

Before starting an outline, the writer must agree that it is worth investing a considerable time to make a good one, then he must be willing to use the outline to the bitter end.

Using the outline effectively means that sub-headings and entries must be made (and revised) down to the level of individual paragraphs. After this preparation is completed, however, it pays off handsomely. The writer: 1) now has an idea with which to start each paragraph, and 2) he safely avoids the "tangency trap." The former reward is compensation enough, but the latter is a real time-saver because the writer is saved from crumpling up some "gems of prose" that turn out to be irretrievably far afield from the topic.

Writing the draft. Converting ideas into smooth-flowing words and phrases is a difficult task with any language, because of word alternatives, idioms, pseudo-meanings, and all the complex rules of English usage. In our communication context, the writer's problem is simply to minimize the semantic noise. But such minimization is best done by experts; just as experts in telegraphy or painting introduce less noise into their work than amateurs do.

How do we start writing then, since at the beginning we aren't all experts? The answer is disarmingly simple: Make a draft and then revise. This follows our design logic of analyze, synthesize, evaluate, and iterate.

The feedback, with iteration, quickly causes the output results to converge on a reasonably satisfactory solution.

There is no softening the blow of reaching for the pencil and starting to write. But the revision can be made more bearable by a series of suggestions[1] that clearly emerge from experience.

1. Don't review the draft to revise it until after a settling time of several hours or days.
2. Review the reader's needs, vocabulary and expectations before revising the draft.
3. Review first for logic and flow by reading through continuously, then later for the polishing of the mechanics; stop and go.
4. When logic errors bunch together, try rewriting the paragraph rather than patching it up.
5. Save format review until the last.

Idea development format. We have thought format to consider. The engineering design report is usually either *comparison* or *expansion: Comparison* of alternatives or *expansion* of the details and stages of a configuration. If the report is a comparison, the writer can compare whole-to-whole or part-to-part[3] as indicated below:

Whole-to-Whole

I. Portable power plant A:
   A. Initial cost
   B. Performance and economy
   C. Maintenance and life
   D. Prospect of sales
II. Portable power plant B:
   A. Initial cost
   B. Performance and economy
   C. Maintenance and life
   D. Prospect of sales
III. Summary

Part-to-Part

I. Initial cost:
   A. Plant A
   B. Plant B
II. Performance and economy:
   A. Plant A
   B. Plant B
III. Maintenance and life:
   A. Plant A
   B. Plant B
IV. Prospect of sales:
   A. Plant A
   B. Plant B
V. Summary

For expansion, the writer simply follows a logical unfolding from a beginning point through to a conclusion, stage by stage or part by part.

We suggest that the writer look back over his subject matter and over the alternatives we have just listed here, in order to find the most appropriate thought format for his report.

Paragraph structure. A graphic way to visualize the paragraph has been given by Glidden[3] showing how the separate idea of a paragraph is connected to the other parts of the text, before and after. The block in Figure 17.4 represents an important paragraph and the sentences describe what should go on at the beginning, middle, and end in that paragraph.

Graphics and illustrations. The physical nature of the engineering design project makes the report beg for illustrations. The cost of reproduction of sketches, photographs, curves, charts and plans is so reasonable* nowadays that the writer must have a good reason if he doesn't intersperse them freely. With the varieties of method available, he can achieve any level of professional graphic quality he chooses, incorporating simple free-hand sketches, professional renderings, production drawings, or photographs.

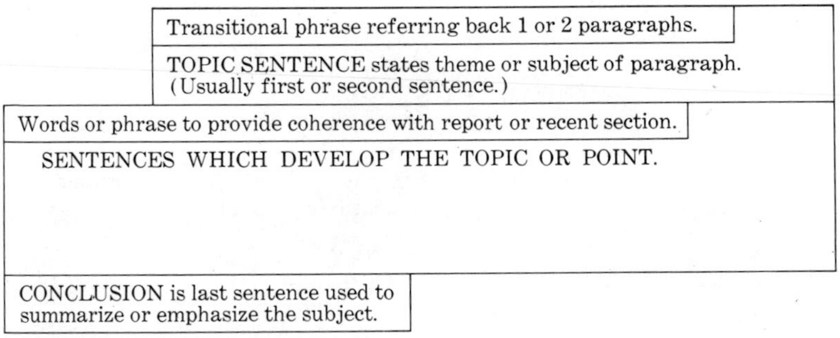

Fig. 17.4 Graphical display of paragraph structure, showing the usual uses of position.

The important questions are: "How much graphics do I include? Where do I put them and how do I use them?"

The inclusion of graphics, if it needs any defense, serves the purpose of multiple association, or in our communication sense, the creation of redundancy, the reduction of semantic and channel noise. If you give the message two ways, it is more likely to be understood. Describe the subject in words and relate those words to a picture. For example, the pinion (a) meshes with gear (b), where (a) and (b) are pointed out on a drawing.

All readers know the variety of graphic aids (see Chapter 11):

For all communication		*For oral presentations*
Tables	Curves	Flip charts
Charts	Sketches	Slides
Block diagrams	Drawings	Films
Schematics	Maps	Models, exhibits
Nomograms	Photographs	

* Blueprints or ozalids at $0.02/sq ft approximately; Xerox at $0.05/sq ft, approximately. Multilith offset at $1.00/$8\frac{1}{2} \times 11$ in. plate, plus $0.005/sheet thereafter. Hectograph (spirit duplicator) at $0.10/master, plus $0.002/sheet thereafter.

Each type of graphic aid has its own role to play, and that role depends upon the time to prepare, cost, available talent, audience, and urgency. Individually and collectively graphic aids are more than just pictures. They

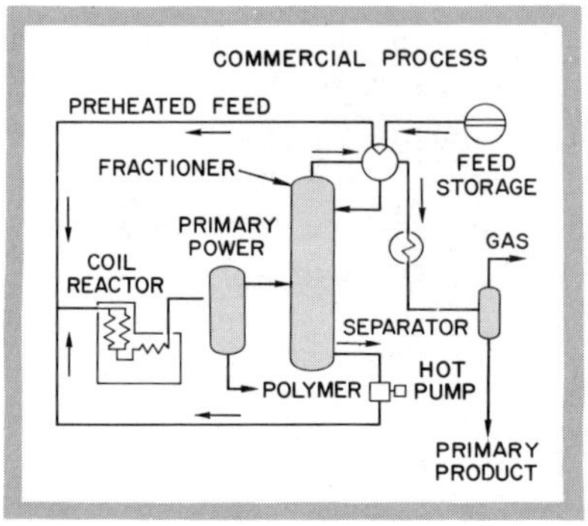

Fig. 17.5 Presentation standards for a typical diagram.[4]

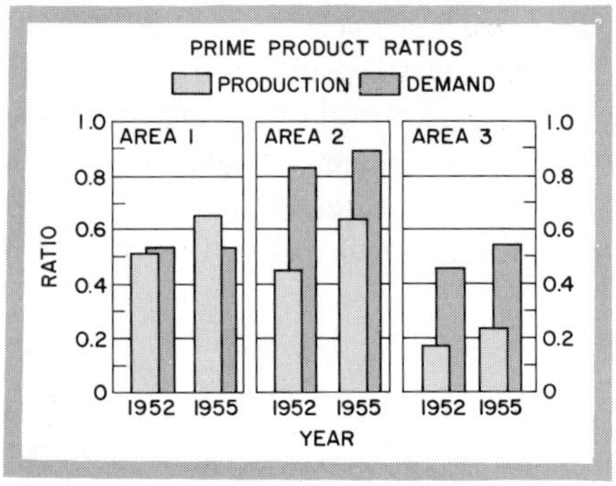

Fig. 17.6 Presentation standards for a typical bar chart.[4]

demonstrate relationships, function, location, constraints, concepts, and they provide visualization aids.

Much work has been done by many groups to improve the quality of published and spoken technical papers. One excellent pamphlet,[4] "Illustrations for Publication and Projection," has been produced for the American Standards Association and discusses recommended practices for graphics. Sections of this publication have been included on these pages.

COMPARISON OF CATALYST TYPES

FEED STOCK RANGE, 200–270°F		CATALYST		
CONDITIONS	YIELDS	A	B	C
200 PSIG 400°F	AROMATICS, VOL %	39	38	42
	CARBON, WT %	0.1	0.1	0.2
	LIQUID, VOL %	91	89	89
200 PSIG 600°F	AROMATICS, VOL %	53	—	52
	CARBON, WT %	0.9	—	3.5
	LIQUID, VOL %	81	—	66

Fig. 17.7 Typical table showing the maximum information which can be clearly shown on one projection slide.[4]

IMPORTANT RULES

1. EMPHASIS ON MAIN FINDINGS.
2. ONE MAIN IDEA PER SLIDE.
3. ONLY THE PERTINENT ITEMS.
4. SIMPLE TERMS, BRIEF EXPLANATION.
5. STANDARD DRAFTING METHODS.

Fig. 17.8 Rules for key word topic headings.[4]

Figure 17.5 and 17.6, Schematic Diagram and Bar Chart, illustrate the simplicity necessary in graphics for papers, slides, or charts. Only a few

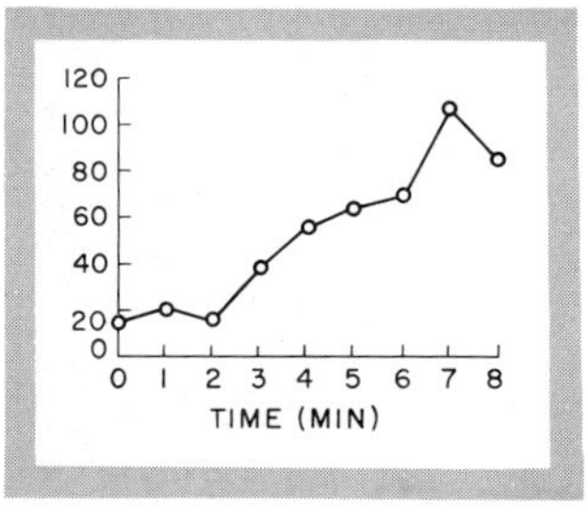

Fig. 17.9 Chart standards for observed relationships.[4]

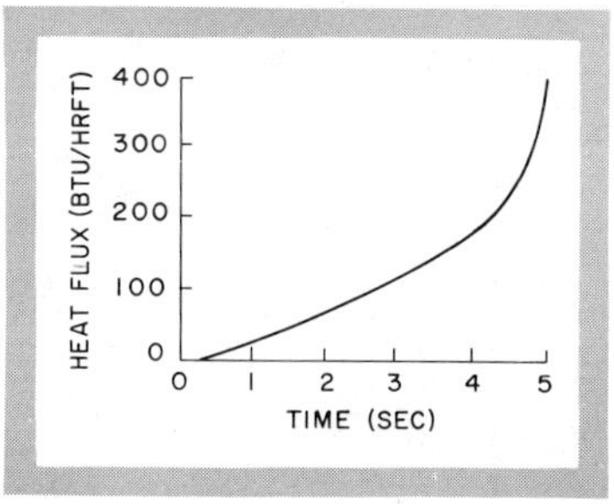

Fig. 17.10 Chart standards for a theoretical relationship.[4]

heavy lines and relatively little detail can be shown. Also Figures 17.7 and 17.8, a table, and a list of topic key words respectively, illustrate how very

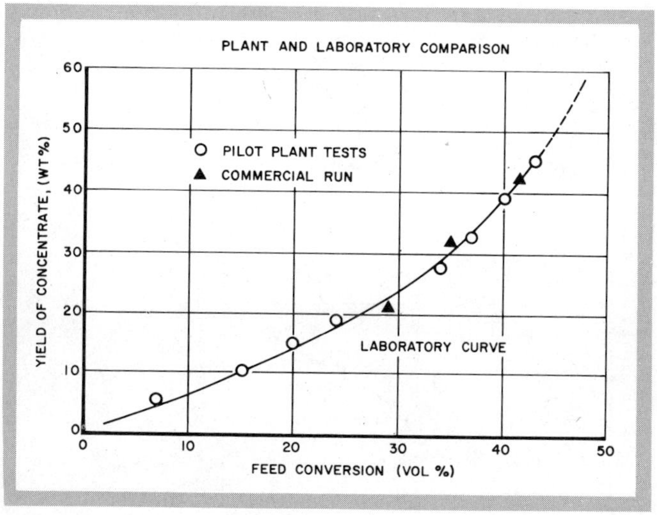

Fig. 17.11 Chart standards for an empirical relationship.[4]

simple charts must be. The lower half of the summary sheet, Table 17.2, page 344, tabulates the most important of these requirements.

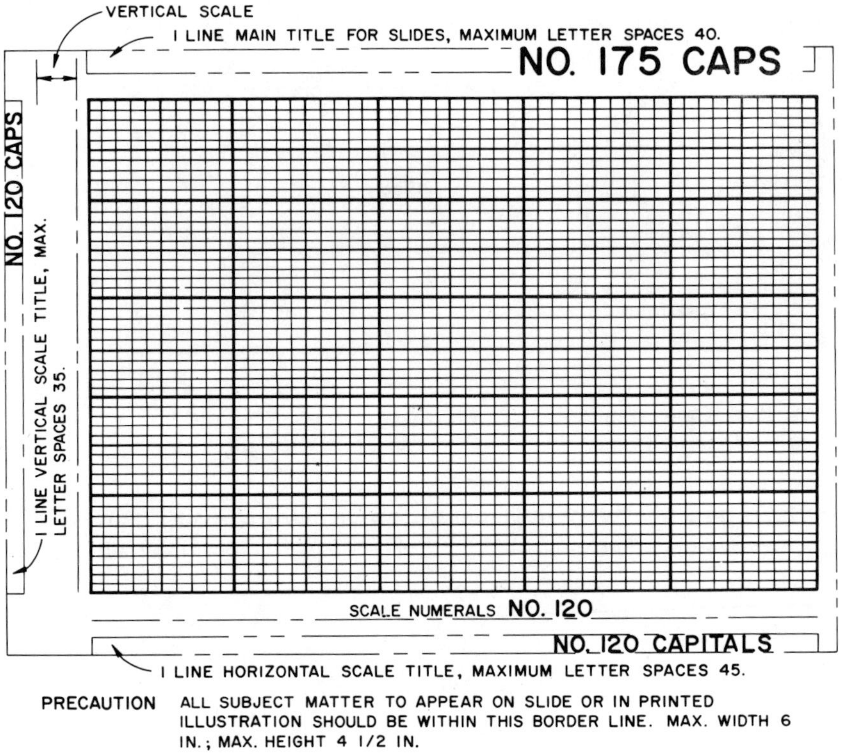

Fig. 17.12 A standard grid for curve plotting.[4]

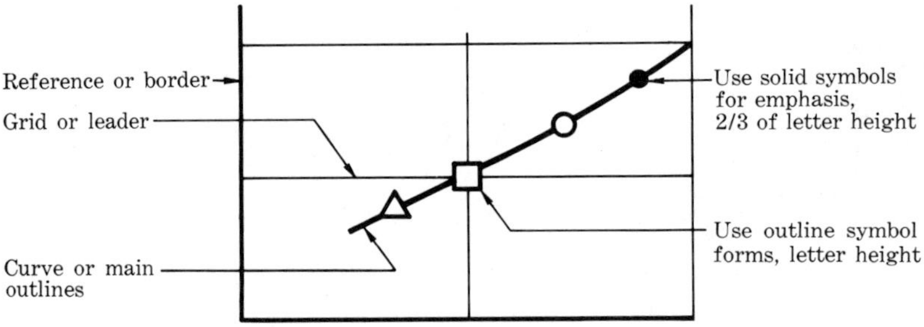

Fig. 17.13 Graphic point and line standards.[4]

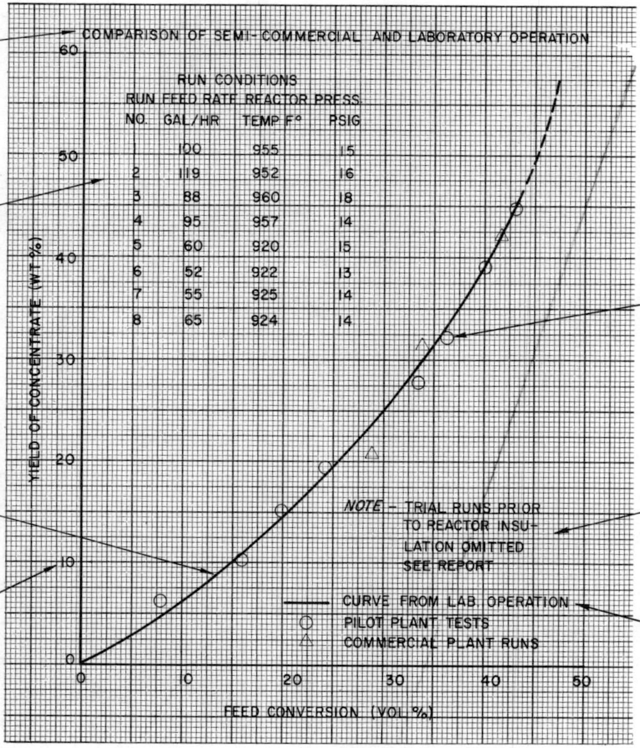

Fig. 17.14 Example of poor readability.[4]

Curves of functional relationships, $y = f(x)$, must be recognized to fall into at least three classes:

1. Observed, but little understood relationships (Figure 17.9): Simply connect the points.
2. Theoretical relationships (Figure 17.10): A smooth curve representing the plot of the mathematical model, (e.g., $y = kx^2$).
3. An empirical relationship, generally anticipated, (Figure 17.11): Adapt a smooth curve to the best fit of the measured points; take into account the expected theoretical shape.

Parenthetically, don't stretch a curve away from the points in item 3 above; you will probably find that hidden variables explain the behavior. Figures 17.12 and 17.13 show the grid pattern and the point symbols that give the best results on slides or charts. Figure 17.11 uses them.

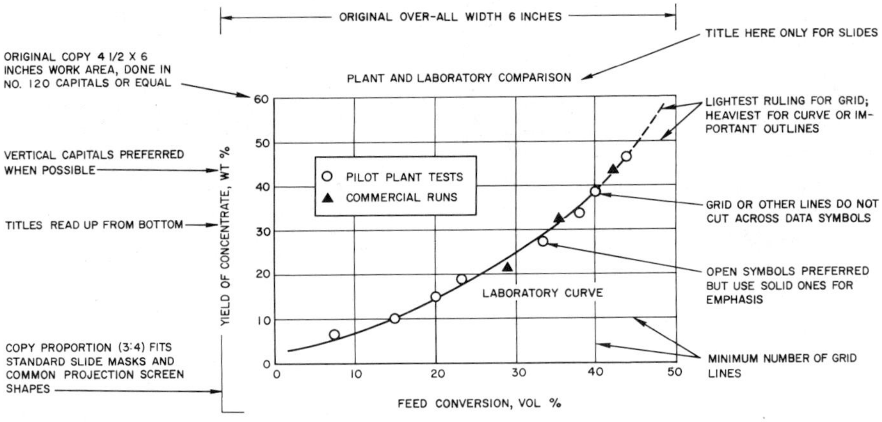

Fig. 17.15 Example of good readability.[4]

Figures 17.14 and 17.15 show by poor and good example how and why these standards are to be used.

The readability of a graphical illustration is very important to the reader, so it should be important to the engineer. What does a reader need in order to be able to use your chart or table? These are necessities:

Title	Units
Reference	Parameters, if Family of Curves
Identification of axes or column headings	(Writer's initials and Date, for a report)
Scales	

All the elements above should be complete enough, with the text, for the reader to know exactly what is being reported. The figure must be relatively self-sufficient; otherwise the reader may be drawing conclusions in the wrong subject area.

Even if the above content is complete, the chart may not be readable unless lettering, weight of lines, types of points, and distances between grid lines are given proper emphasis and care. Also, the independent variable belongs on the horizontal axis; the curve should be proportioned to fill the page allowed; the slope should allow interpolation or extrapolation to the desired accuracy; and the scale divisions should generally be decimal.

One more comment. The reader usually concludes that a steep slope means that the rapidity of change is highly significant. Be sure that you, as the writer, make the curves tell the story in conformity with the facts.

Presenting the Report Conclusions

Put your results and conclusions in a conspicuous place. They are the essence of the whole effort. Not only what you say but how you say it influences the reader. Thus, the conclusions he draws depend on a number of intangibles as well as upon the positive statements. For example, exaggeration, or florid, extreme descriptions even in minor areas repel the reader. Disorganized, cluttered, or messy presentations leave the reader unconvinced. Excessively formal or stiff reports, or unduly personal or relaxed reports make reading difficult and uncomfortable. Be natural, but find a middle ground in style. As mentioned above, the accuracy of language usage is very important for the best effect. While writing, be on guard against the common failings of careless antecedents, involved sentences, incomprehensible statements, poor spelling. Also ask these questions about the literary effect:

Are the emphases on proper parts of the presentation?
Is the writing style informal, direct, consistent, easy?
Is there logical unity and coherence:
  sufficient support?
  no unnecessary inclusions?
  all parts point in the same way?
Are the references to illustrations adequate and sequential?
Are the vocabulary terms defined or from common usage?
Are there color, variety, interest, appropriate analogies, figures of speech, change of pace, and paragraph headings?

In appropriate cases, the last pages of the report carry sample calculations, observed data, and a bibliography. Again, include just enough of these

Table 17.2 Illustrations for Publication and Projection.[4]

Planning:

1. Base your choice of illustration form on:
   a. The nature of the audience (interests, needs, size, technical background);
   b. The greater clarity of the graphics versus words for your particular message;
   c. Time to prepare and equipment available for slides, charts, etc.
2. Select the specific illustrations:
   a. To show one central idea each, in simple form;
   b. So each concentrates about the point it is to make;
   c. To carry a balance of illustrations for various portions of the whole presentation.
3. Organize each illustration by:
   a. Stripping out nonessentials (no more than 3 or 4 curves each);
   b. Using familiar forms rather than unfamiliar forms;
   c. Putting supplementary information in text, or by giving verbally;
   d. Showing curve in such a manner as to give proper impression, (i.e., steep for rapid change, adequate space for the important areas).

Executing:

1. Drawing size and proportion:
   a. Visual ratio is 3:4 (horizontal or vertical);
   b. Slides are $3\frac{1}{4} \times 4$ or 2×2 in. (Original drawings for slides must be $2\times$ or $3\times$ size and reduced photographically. Printed column ranges from 3 to $3\frac{1}{2}$ in wide in magazines.)
2. Lettering:
   a. Generally for lettering, use vertical caps of one size reading from left to right; or with the page turned to read from right-hand side if necessary;
   b. No more than 12 lines of words; four to six columns; 20 items in a table;
   c. Use lettering guide where possible, otherwise rough-in before finishing.
3. Lines and symbols:
   a. Use heavy, medium, light lines for appropriate emphasis;
   b. In curves, do not pass lines through lettering data points (circles, triangles, etc.);
   c. Use standard ASA math, mechanical, electrical, chemical, etc., symbols.
4. Titles, margin, scales:
   a. Put titles outside of working area. Lettering for title is $1\frac{1}{2}\times$ normal;
   b. Allow generally uniform margin all around not less than $\frac{1}{20}\times$ width;
   c. Use easily read scales, based on 1, 2, 5;
   d. Fit curve and scale to best use of whole area;
   e. Offset the zero only when carefully indicated. (i. e., 0, . . . , 80, 81, 82, 83, 84, etc.).

for the purpose. Keep the rest of your data and calculations in file for reference; don't pack it all in the report.

SUMMARY

Technical Report Writing

Break the writing job into pieces small enough for each to have an obvious starting point.

Make and follow an outline.

Secure and organize the necessary information.

Roughly sketch the curves and drawings.

Write a first draft (fast, not critically).

Revise, revise, revise.

Draw the illustrations and curves carefully.

Make sure the final form is neat, concise, interesting, complete.

Technical Report Content

Answer the specific questions asked.

Write directly to the expected reader.

Present it convincingly (if inconclusive, say so).

Anticipate the reader's questions.

Be objective, restrained, impersonal, but clear.

Technical Report Graphics

Plan the graphics while roughly outlining the report.

Where appropriate, use drawing or a summary curve sheet to present the final conclusions.

Make each illustration sheet relatively self-explanatory.

Refer adequately in the text to each illustration, calling the element out by symbols.

Take extra care in the craftsmanship of the curve or drawing.

Be sure the curves convey the correct impression and support the report.

REFERENCES

1. Rosenstein, A. B., R. R. Rathbone, W. F. Schneerer: "Engineering Communications," Prentice-Hall, Inc., Englewood Cliffs, N.J., 1964.
2. Tribus, Myron: *Engineering 4B Syllabus*, chap. 24, Department of Engineering, UCLA, 1961.
3. Glidden, H. K.: "Reports, Technical Writing and Specifications," McGraw-Hill Book Company, New York, 1964.
4. American Society of Mechanical Engineers, *American Standard Illustrations for Publications and Projection* (ASA Y15.1-1959), New York. Figures 17.5–17.15 have been used with permission of the publisher.
5. Shannon, Claude and Warren Weaver: "The Mathematical Theory of Communication," University of Illinois Press, Urbana, Ill., 1949.

Additional References

"Harper Handbook on Communication Skills," Harper & Row, Publishers, New York, 1957.

Piper, H. D. and F. E. Davie: "Guide to Technical Reports," Holt, Rinehart & Winston, Inc., New York, 1961.

Oral Reporting

Eighteen

"The rehearsal makes the show."
Theatrical axiom

Engineering Presentations

The engineer is frequently called upon to make oral presentations, and the more skilled he is and the more easily he speaks, the more frequently this duty is apt to arise. The occasion may be an informal status report to a supervisor, a briefing session; or it may be the presentation of a new product or system to executives, sales personnel, or clients. Of course, technical papers presented at professional meetings are presented orally also. Even if the engineer does not have facility in speaking, there will be times when it would be embarrassing for him to decline the invitation.

The oral presentation implies special importance to the subject and to the speaker. The listener or the audience is taking time for this one activity, rather than sandwiching the reading of a written equivalent in among the other demands of the day. Frequently the listener must travel a considerable distance to be present and this fact alone implies a special value for the occasion.

The intent of the speaker is to inform, advise, or instruct (command), but most frequently to inform.

The audience expects to be informed, advised, or instructed, as may be appropriate; but more importantly, the audience expects the benefits of oral communication. Otherwise it could follow the simpler, time-saving path of reading a digest of the material.

Advantages of Oral Presentation

The audience hopes to get a clearer, more thorough understanding of the subject, benefiting from the *dialogue* which is presumably possible during the question-and-answer period. A speech, in pure lecture form, is hardly an improvement over reading the report. The oral report is better than the written one largely because of the verbal exchange. In summary, the advantages of the oral report are:

1. Direct focusing on the special needs and characteristics of the persons present.
2. Heightened effects of the speaker's personality, his enthusiasm conveyed through intonation, gestures, and special emphases.
3. Thorough coverage through dialogue, questions, and answers.
4. Group participation to supplement a specialist's limitations, as in a panel or conference roundtable.
5. Better descriptive value of visual aids, motion pictures, graphic replicas, or actual working models and demonstrations.
6. Group interaction in persuasion and selling.
7. Economy of eliminating report preparation and distribution.

At the same time, there are some disadvantages which must be considered, when one has the choice between written and oral reports:

1. Less material can be covered in an oral report, and any numbers presented must be taken down as notes in order to be remembered by the audience.
2. The effect of the report is strongly influenced by delivery, by staging, props, interruptions, etc.
3. The listener cannot easily refer back to material presented earlier.
4. The speaker must be right the first time; corrections and changes are less effective due to first impressions.

Comparison with Written Communication

All the details of the communication diagram discussed in the last chapter still apply to oral presentations:

1. There must be a message; it must be encoded, transmitted, etc.
2. All the types of written-communication *noise* are present, plus a few more. (Now, the meeting-place environment, the other persons present, and the mechanics of delivery all may contribute noise.)
3. The speaker is the originator, encoder, and transmitter.
4. The need for logic, coherence, and unity of the oral report are the same as of its written counterpart; but the completeness, the organization and inclusion of special-interest material comprise noticeable differences.

The logic is particularly crucial, because the listener has no immediate opportunity to "turn back a page" and clarify the point. The speaker keeps on talking and the listener must keep up or become lost. Smooth development of the thought, from the audience viewpoint, is essential.

It is frequently true that a written report is or will become available, in which case the speaker concentrates on the conclusions, the novelty, the emphases, or the peripheral aspects of the subject. Special audience interest encourages particular inclusions or illustrations also.

Whereas the attitude of the reading audience is usually assumed to be neutral, a known favorable or hostile audience strongly modifies the speaker's approach.

The written report usually implies extensive preparation, but unfortunately, the oral report is frequently relegated to improvised, ad-libbed, or off-the-cuff presentation. This latter style, if assumed in giving an engineering report, doesn't properly convey due significance or seriousness. Verbal reports must be as well prepared as written ones, possibly better prepared because of their briefer summarizing and condensed nature.

Remember that it is the unavoidable revisions, which improve the written draft up to the point of publication quality. Such revisions must be matched by equivalent speaking rehearsals, under the delivery conditions, to provide a similar quality for oral presentations. It is *not* true that preparing a well written report qualifies the engineer to forthwith stand and make an equally good speech on the same subject.

Special Opportunities

With the audience in close personal contact, the skillful speaker can achieve much more than he can by the written word; cold, distant, easily-evaded informer or persuader that it is. We are speaking of valid and useful communication, not the high-pressure, confidence-artist selling usually associated with effective persuasion.

The well thought out information and arguments of a good talk, the favorable arrangements of the room displays, exhibits, and lighting, and the direct, convincing, sincere manner of a speaker all combine to maximize communication effectiveness. Nothing can match good personal contact.

Preparation

There are two almost equal portions in the preparation of an oral report: One is the text and the other is the illustration or visual aid. The latter contributes so much to a talk that, now, one seldom thinks of omitting charts in preparing for a speech. The charts summarize the text and by means of the visible topic-headings keep both the speaker and the audience on the subject. The communication value of diagrams and sketches needs no defending. Of course, projected slides may substitute for charts, but the slides require more equipment and facility preparation.

The written outline, either extracted from the source report or prepared especially for the purpose, naturally comprises the first step. This outline gathers the right amount of appropriately concentrated material into a first form, and is the framework for the initial suggestions of illustrating charts, diagrams, or sketches. The illustrations should start taking form early in the development of the talk, rather than late. The actual balance between words and graphics (whether listed topics, diagrams, or pictures) should be decided early, because considerable economy in talking and a noticeably deeper effect can come from well-chosen charts.

In case the reader is now visualizing the charts as professionally and expensively drawn on Bristol board, let us dispel his concern. Most engineers can produce a very acceptable graphics display with colored felt-point pens on plain or one-inch squared flip-chart pads, 27 × 35 in. With a little care in lettering and by use of a straightedge, very presentable charts can be made in a short time, particularly since only limited text material, bold-line curves and sketches will serve the purpose.

To return to the outline, any extra care devoted to its organization and to the supporting ideas and words is well worthwhile. The outline can be extracted for topic headings on the charts and for notes to be put on a 3 × 5 in. reminder card, if such added prompting becomes necessary. The thinking in the outline can follow any of the paragraphing alternatives suggested in the preceding chapters; but the logic should be carefully gauged to the audience's needs and expectations and to your purposes as the reporter-writer-speaker. One other consideration deserves mention. The outline should have built-in flexibility, allowing the speaker to delete or only lightly emphasize certain points, if time becomes pressing. The importance of time is one major difference with respect to written reports. Poor use of the closing moments should be avoided, with a safety factor put into the outline by underlining or some other personally understood coding.

Before proceeding to the visual illustrations, let us close with a reference to verbal illustrations. The eye responds to the visual stimuli, adding both understanding and memory. When visual illustrations are not convenient or possible, verbal illustrations should be substituted. Each general statement should be followed by a specific example, no matter how sophisticated the audience is. The design report always follows the design, just as the lab report follows the test or a financial statement follows an audit, or this detail-drawing follows that draftsman's layout. These latter items are more and more specific, although not particularly spectacular, verbal illustrations.

Visual Aids

The visual aids to communication run an extensive gamut, from inexpensive to very costly, from moderately effective to very effective. Consider the range presented by:

Flip-chart paper and felt-point pens.
35 mm (2 × 2 in.) slides.
35 mm (movie film) film-strip projector.
Poster-painted Bristol boards.
Colored *Scotch* tape on boards.
Pre-cut letters and symbols.
Felt stick-on boards.
Magnetic stick-on boards.
Enlarged photographs.
Fluorescent paints and ultra violet light.
Cutaway model sections.
Working models.
$3\frac{1}{4} \times 4$ in. glass slide projector.
Visual-cast reflection projector (any $8\frac{1}{2} \times 11$ in. page projected on a screen about 10 ft away with sizes up to 5 × 7 ft).
16 mm movie (with or without sound).
Magnetic tape recorder ($3\frac{3}{4}$ in./sec voice, $7\frac{1}{2}$ in./sec high fidelity).
Closed-circuit television.

As the speaker approaches the choice of the medium to help him in his oral report, he will naturally first be influenced by the time and cost needed to prepare, and then by the availability and the precise suitability. Many of the inexpensive forms above will perform interchangeably in most speaking situations. The number of times the talk will be given also is to be considered, as well as the importance of the audience and the subject.

Some of the choices can be made by calling upon the skills area dominated by the theater arts. Persons in that discipline are too little utilized by engineers, yet they have much to tell us about making impressions on the mind and moving people to action. One element of the good show to which we refer is *good staging*, and good staging should be sought in every engineering spoken presentation.

1. Make the visual material vivid and readable to all the audience.
2. Focus attention by easel or screen or draped exhibit table.
3. Stress good workmanship on charts and displays.
4. Integrate the audio-visual material smoothly into the story.
5. Keep charts and exhibits covered until ready to use.
6. Use assistants to expose or demonstrate the displays or exhibits, if necessary, to avoid confusion.
7. Make doubly sure that every article is in place and that all working parts function!
8. Remember the importance of good lighting, good sound, and good cues.

It is good practice to visualize how a dramatic presentation of your subject would be made by a theater or television show. If possible consult a staging specialist. You can adopt part of his counsel anyway, if you don't think all of it appropriate.

There are several limits on the size of charts and displays depending upon room size and audience. The first rule is: *Nothing to be read on a chart should be smaller than* $\frac{1}{250}$ *the distance to the farthest viewer.* Thus, for a distance of 25 ft, all letters should be $\frac{1}{10}$ ft or about 1 in. high; for 50 ft, about 2 in. high. These may seem large, but experience indicates you are wasting your time if you make them much smaller. Better advice is to make all of them 2 in. high. The lines per chart, margins and other specific recommendations are shown in Tables 17.2 and 18.1.

Generally, too little use is made of three-dimensional exhibits and working models. The audience is more subtly convinced by physical evidence than by mere words. This is why successful sales persons go to the trouble of carrying samples and working demonstrations.

As we mentioned above, one of the practical uses of visual aids is in *prompting*. This should remove some of the tension from a beginner (reduce his psychological noise input) and put the audience proportionally at ease.

In summary, concerning the audio-visual aids proper:

1. Be sure the aids fit into and support your theme and objective.
2. Make the aids vivid, easily read, and well executed.
3. Contact the audience through as many senses as possible (seeing, hearing, feeling, even smelling, and tasting if possible).
4. Show working models or components if at all possible.

The Room, Facilities, and Rehearsal

The writer of an engineering report can do nothing about the *noise* introduced by the reader's poor lighting or other discomfort. However, for the oral report the speaker usually has a considerable control over the noise injected into the listening situation. The room, the seating, the lighting, the ventilation, the visibility of the speaker, the clarity of the charts or slides, the adequacy of the hearing (sound system) all strongly influence the "channel noise" or "psychological noise" intruding into the performance. The ordinary speaker may feel that some of these factors are not under his control, but the careful speaker will go to considerable trouble to assure himself that all these environmental conditions are as agreeable as possible. Many times a good speech is utterly ruined because of failures in light or sound that could have been checked out earlier. Audience cynics observe: "Don't expect a good amplifier system if an electrical engineer is speaking . . . or a good air conditioner if it's a mechanical engineer."

The precautions that are in order must be checked by the speaker, if he is not sure the chairman has done so. It is a good idea to make a tactful casual check anyway, and get familiar with:

1. The lighting and light switches if they are to be used.
2. The speaker's location, chart easel, and table (if necessary).
3. Projection equipment, sound system, fusing adequacy, and window shades.
4. Visibility from all seating areas.
5. Noises from adjacent sources (expected at the time of speaking).
6. Preparation of charts, exhibits, slides, drapes, lectern, pointer, blackboard.

These may appear to be necessary only for formal talks or large audiences. Actually all of these precautions are important to consider, even for a talk with a few persons. Of course, closer contact with a few persons makes noise less intrusive, but the influences are just as real with an audience of any size. It may be unfamiliar necessity to the engineer, but he must learn the strategic importance of the physical arrangements in spoken communication.

As a concluding reminder, the following check list on potential sources of trouble stress the physical aspects of the presentation:

Voice inadequacies: too loud or too soft (the latter usually).

Obstruction of chart-view by the speaker; lectern-light glare on the projection screen.

Mechanical failure of equipment, or display; blown fuses; cordsets tripped over or pulled out; fallen or misplaced chart.

Poor entrances or exits. Misunderstood cues.

Not enough chairs for the latecomers, etc.

Reverberations or dead spots in the acoustics of the room.

The antidote for most of these potential troubles is the rehearsal, the dry run. If the talk has any importance at all to the speaker (and it should have, or *he* wouldn't be speaking), he should visit the room well ahead of time, to note the corrective or adaptive steps he must take. He should practice a dress rehearsal in the room, at least *once* with all "properties" there and operating. It is taking a chance all experienced performers avoid, to put on a presentation without a full dress rehearsal. In addition, of course, he should try out the material with an audience of associates or friends, and he should rehearse alone until satisfied. There is no substitute for rehearsal!

"The rehearsal *does* make the show!"

The Speaker

The key man in a successful presentation is the speaker. Whether he likes it or not, a great deal of the success stands or falls with him. The substance of the event is more, of course, than the content of the talk, the motions or pronunciations of the speaker, or the vividness of the displays. The essence of any presentation comes out through, and because of, the personality of the speaker. It is that person's individuality, imperfect, unpredictable creature that he is and we all are, that carries the interest and the appeal of even technical talks. The man who most tries to hide himself in the impersonality of speech and manner still portrays himself clearly for what he is or thinks he should be.

Thus we come to the central fact of person-to-person communication. A person's character, his conviction on the subject and his enthusiasms and animation most strongly draw out the response from the audience. The object of improvement in speaking skills is to let the real person show through, to remove the unnecessary covering of nervousness, ineptness, and unfamiliarity which mask the novice.

For the beginning public speaker, his uncertainty about audience response, about what constitutes good speaking habits and practices, about his own ability to carry off the assignment, these all tend to bring out undesirable platform behavior. No amount of our conversation can eliminate nervousness; but nervousness can be controlled by developing confidence in the project and by having visual aids standing by to guarantee that all the topics of the task will be covered. In other words, talking about an interesting project and having the talk outlined on flip-charts combine to provide the best antidote for stage fright. It is a simple matter to have one sheet cover the outline and to flip back to it when in need of prompting. If the rehearsals are done with this crutch, it will seem natural and easy to do during the performance.

The Presentation

When you speak, be natural (as natural as you can, that is, under the circumstances), because the conversational tone is the one you usually use and hence the one likely to be least strained. The audience does not expect an oration, and would be as uncomfortable as you would be if you suddenly launched into one. If there is a public address system, speak at a normal voice level six inches from the microphone usually. You can't hear yourself amplified, but the audience will signal if they can't hear you in the back of the room. If there is no microphone, speak up loud enough to cover the room; it usually seems too loud for the speaker when it is just right for the audience. Have your charts and props covered when you start and launch into your talk positively. Don't apologize; have your opening moves thought out; making strong opening statements, and remember to fall back on the flip-charts if your mind suddenly goes blank. Pay attention to audience response, so you can adjust your level, your coverage and emphasis as needed. Once you feel as if you are talking *with* them, you will lose some of your self-consciousness and the words will flow more easily. Expect your audience to meet you halfway; they usually do. Keep eye-to-eye contact all the time, moving your attention over the group as if you are talking to each individually. Don't look at the ceiling, or concentrate on the floor or walls; the people are your reason for being there.

Keep your props covered until the right time, then pick them up and demonstrate, passing them around if that won't distract the listeners. If preferable, invite inspection after the talk.

Inject a little humor if you can tell a story well, *and* if you can find an appropriate one. Leave the stories out unless they fit in well.

As to the words, you can memorize the talk word for word; it is horrible, however, if you forget anywhere. Or you can read the text; but that is almost as bad as memorized talk unless you have the speaking skill of a dramatic actor. Finally, you can make some notes on a card (or use the easel chart) and speak extemporaneously. This last method is the one urged by experienced persons. The best advice about gestures and hand positions is to forget them. Do what seems comfortable and natural. Hold a card or appropriate object in your hand if you feel safer.

Finally, finish on time; summarize or close with a punch-line or a restatement of your opening topic or subject. End positively, as you began. Have a good beginning and a strong ending if nothing much else in the speech.

In conclusion, note the special characteristics of spoken communication: Quick feedback by dialogue; importance of tone, emphasis, and gesture; added value of charts and models supplementing the words; impact of personal enthusiasm (or hostility); sensitivity to the presentation facilities; the time-nature of a talk vs. the space-nature of a written report.

Let us summarize the oral presentation with the cautions:

1. Prepare the outline from a written text.
2. Identify, measure, and adapt to the expected audience.
3. Check out the quarters and facilities thoroughly.
4. Use a reminder or prompter in the form of charts.
5. Add enough visual aids to get the idea across easily.
6. Go through several rehearsals with a critic.
7. Speak as if talking to the group one by one.

Table 18.1 Illustrations for Display: Visual Aids.

Follow the recommendations of ASA Y15.1-1959,* and in addition:

Planning:

1. Prepare for proper equipment. Depending on size of audience, you can use:
   a. "Visual cast," if no more than 30 persons (15-ft visibility);
   b. Charts, if no more than 60 persons (25-ft visibility);
   c. Slides on 6 × 8 ft screen, for 150 persons (50-ft visibility). Use proportionally larger screens for larger audiences.
2. Organize material to take advantage of color.

Executing:

1. Charts are usually 27 × 35 in. They can be horizontal or vertical.
2. Smallest letter should be $\frac{1}{36}$ to $\frac{1}{40}$ the vertical height of chart or $\frac{1}{250}$ viewing distance from farthest spectator.
3. Lightly pencil in and arrange the material on the chart before using the colored ink felt-point pen.
4. Cover each sheet, if it otherwise would be exposed prematurely before presentation.
5. Mount charts securely on an easel in advance of presentation. Be sure easel is high enough for visibility.
6. If using "visual-cast" be sure machine does not obstruct the view of part of audience.
7. Leave some lights on in the room (rear at least) while projecting slides. Post an associate to take care of lights.
8. Make a dry-run rehearsal to check all equipment, well before the actual time.

* See Figure 17.15.

Open Challenges to Engineering

Nineteen

"Make no small plans, set no timid goals.
They have not the power to put
a gleam in the eye, to quicken the step,
and to fire the spirit."

Appearances

Immersed in the material abundance of western civilization, it is easy to become lulled into a technical complacency. Progress is visible. Billions of dollars are spent on research and development in the United States alone[1] (Figure 19.1). Gross National Products are climbing steadily. The whole world seems to share the forward motion. How can gloom-mongers be serious? Why stir oneself with talk of challenges?

Yet one only has to look under the thin skin of visible accomplishment to discover unsolved most of the pressing problems of mankind's long and tangled past. Hunger, disease, underemployment, ignorance, and inequity are still the rule. The large portion of the world that goes to bed hungry each night is a swelling number, as the underdeveloped populations press against their restraints of food supply and borderline health. Unthinking power-hunger still generates its present hates and future rebellions. The industrialized nations face the new problems of air, water, and soil pollution, coupled with the inexorable depletion of natural resources. The daily communication of our jet age stimulates the underdeveloped nations to strain to catch up, and their people now have frequent reminders of how relatively deprived they actually are. Figure 19.2 dramatizes the contrasts.[2] All nations wrestle with the enigma of their great cities, which at least perpetuate if not actually promote ill-health, accidents, social violence, and ugliness—ironic contrasts to the "good life."

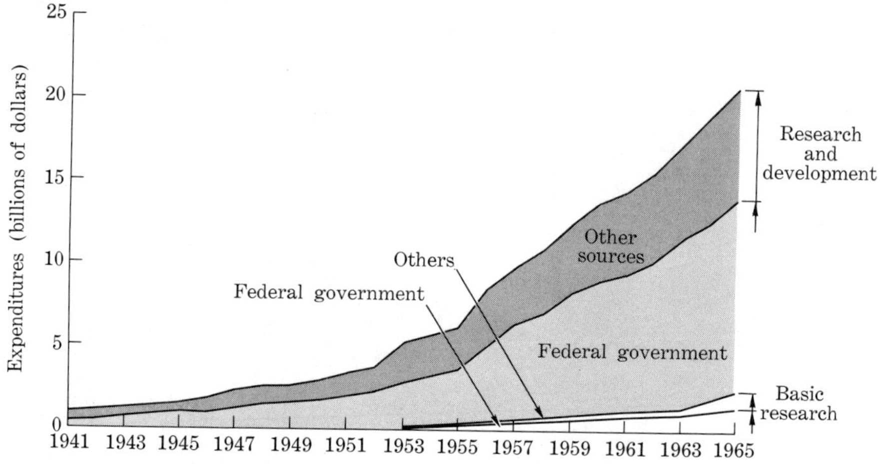

Fig. 19.1 United States spending for science. Figures for 1953–1963 are from the National Science Foundation; other figures are estimates. (From *Scientific American*, July 1965.)

Such problems face all citizens, not just engineers. But because engineers have indifferently aggravated certain problems, and because many problems need sophisticated technology to achieve any reasonable solution, engineers have unique roles to play in any concerted attack.

Realities

The major research problems facing our profession in the United States have been summarized by the Engineer's Joint Council,[3] and I paraphrase them here as challenges to engineers.

1. *Energy:*
   Limited solid, liquid, gaseous fossil fuels; uneconomical solar energy; hazardous nuclear energy.
2. *Raw Materials:*
   Depletion; inefficient extraction, and salvage; increased cost; unexplored oceans.

3. *Urban Environment:*
   Central-city blight; noise; crowding; time loss; pollution; ugliness; disregard of human values.
4. *Transportation:*
   Irrational local, regional, national systems; hazards; costs.
5. *Information Handling:*
   Inaccessibility; limited retrieval and use; duplication.
6. *Biotechnology:*
   Untapped chemical processes; undeveloped interdisciplinary exchange.
7. *Health, Medicine:*
   Relative isolation from the resources of modern engineering methods and skills.
8. *Education:*
   Urgency to involve engineering in improving and accelerating education at all levels.
9. *Developing Nations:*
   Population explosion; diminishing food calories per capita; low skills and income; inadequate economic infrastructure; inflation.

This array of topics has been thought out by engineers for engineers. It is a challenge to the profession and eventually to those individual engineers who sense a personal opportunity, to select a point and plunge in. Most of these subjects require both the massed and concentrated talents of teams to make significant inroads. Here and there, however, are the possibilities for single individuals to make meaningful impact.

Most persons don't fully appreciate that basic improvements always start with an idea, a cogent, daring, well-thought-out idea. The idea and the persuasiveness of its proponent must be strong enough to convince those who are able to do something about it. Putting such ideas to work always requires men, money, and time; and these in turn hinge upon knowledge and approval. Only individuals who are sufficiently wealthy may act independently in following their personal preferences; all others must persuade and cooperate.

Let us bring out some of the salient facts about the needs listed above.

Energy

We need energy to heat our living and working spaces, to operate our private vehicles and public transportation systems, to drive our mining, purification, and manufacturing processes, and to power all the communication, water, and waste systems, etc. Figure 19.3 shows the amount of

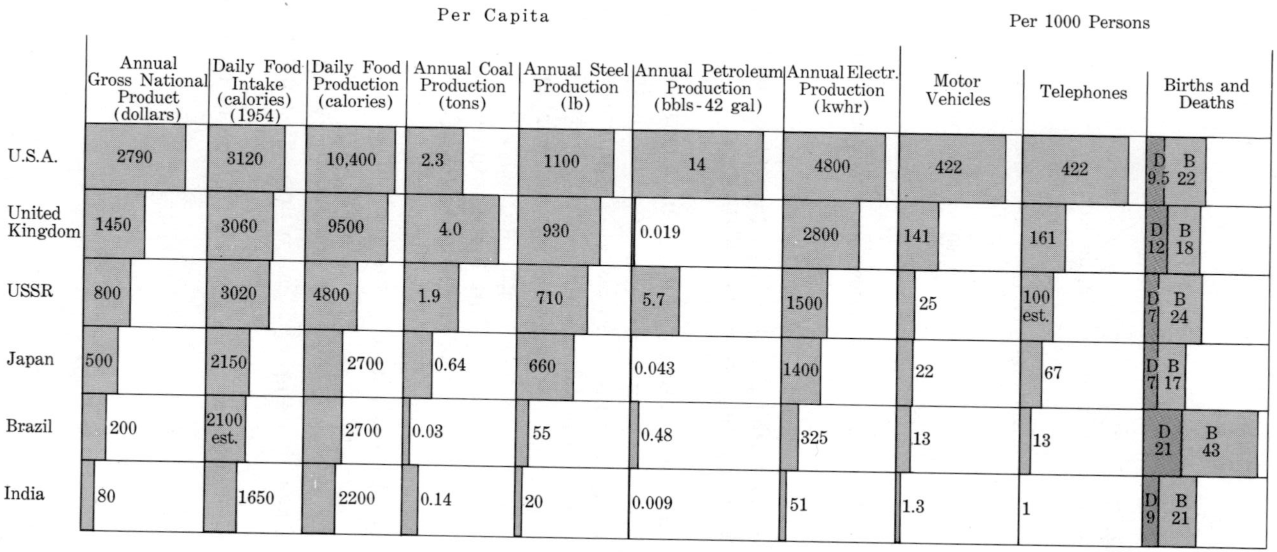

Fig. 19.2 Summary and comparison of significant indices of national industrial development. Data for 1961 except as noted. (From "Statistical Abstract of the United States," 1963, U.S. Government Printing Office, and Harrison Brown, James Bonner, John Weir: "The Next Hundred Years," The Viking Press, Inc. New York, 1961.)

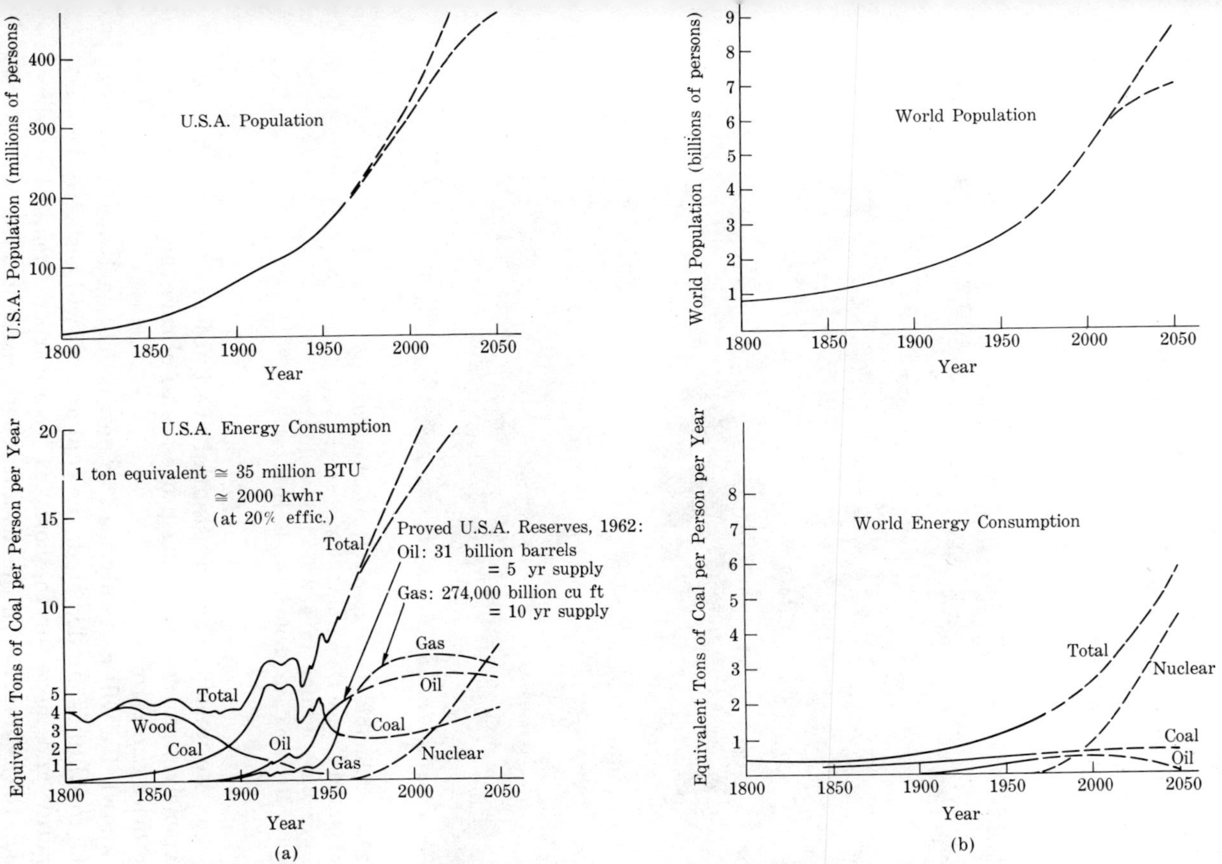

Fig. 19.3 Population and energy consumption, United States vs. the world; present and possible patterns. (From "The Next Hundred Years," Ref. 2; "Statistical Abstract of the United States," Ref. 11; "Preliminary Design of a Small Nuclear Power Plant," Ref. 12.)

energy derived from the several sources for the United States, together with some U.S.A. and world projections for the next hundred years.[4]

More energy is needed, not only for more labor-saving devices, but also for the increasing effort in reaching, extracting, and separating our minerals. As the richer and more accessible deposits of coal, iron, aluminum or copper are used up, we must go deeper into the crust; we must mine thinner veins and lower percentage ores. Even now, relatively few years into the

Fig. 19.4 Columbia basin project, Washington. View of Grand Coulee Dam from the west side showing power lines leading to a 230-kv switchyard. (Bureau of Reclamation photo by E. E. Hertzog)

industrial age, we must remove substantial overburden, crush much rock, and use expensive, power-consuming processes to obtain the raw materials for our factories and farms.

New knowledge will certainly lead to new economical energy sources. Also, as the fossil fuels are depleted, the coal and oil will certainly be used less for energy and more for the complex hydrocarbon chemical building blocks which they comprise. In the distant future, man's energy must eventually be supplied by the sun because even uranium and thorium deposits are finite; and hydro-electric power (Figure 19.4) will provide only about 6 per cent of the world's future needs.

Solar energy,* received at a rate of 1 cal/cm²/min at the earth's surface, is equivalent to 2300 kw/acre, or enough incident on 100 acres to supply a present U.S. city of 100,000 persons (assuming daylight operation and energy storage). Thus the challenge is to conserve the chemical resources of our fossil fuels for use and re-use in our material's industries, and to develop nuclear and eventually solar energy for our power sources.

Raw Materials

An industrialized society consumes vast quantities of materials each year, and that quantity grows annually. In the United States, the average annual production of steel has doubled in the last 30 years. The present consumption of materials in the United States is shown approximately in Table 19.1.[4]

Table 19.1 Per Capita U.S. Consumption of Materials (mid 1950's).

Metal	*Pounds/ person/year*	*Non-metal*	*Pounds/ person/year*
Steel	1290.	Stone, sand, gravel	9500
Aluminum	22.	Cement	640
Manganese	21.	Common salt	260
Copper	15.	Phosphate rocks	175
Lead	14.	Lime (other than for cement)	120
Zinc	11.	Gypsum	105
Nickel	1.2	Sulfur	72
Tin	1.0	Potassium salts	22
Antimony	0.15	Pyrites	16

When man first started to use metal tools, several thousand years B.C., he was able to find relatively pure copper ores on the surface of the ground. As these supplies were used up, he had to go underground for the metal. But even as late as 1900, the concentration of copper was still as high as about 5 per cent. By 1950 the demand had pressed the suppliers to work ores as low as 0.8 per cent and even 0.6 per cent. This downgrading meant more overburden-rock had to be removed (or deeper mineshafts dug) and much greater gross tonnage handled per ton of ingot produced, leading to as much as a 50 times increase in operation's activity.

Water was ample and cheap 50 years ago. Now the increased agricultural and industrial requirements, plus pollution of sources, requires extensive dams, pipe lines, pumps, and treatment plants. Irrigation water currently costs $10 per acre foot,† and 2–5 feet are applied each year. Drinking

* In space, at the earth's mean distance from the sun, the incident solar energy is 2.00 cal/cm²/min.

† See Fig. B.14, page 396, for costs in a specific project.

water costs $50–100 per acre foot from natural lake or river sources. If drinking water is made by desalinizing sea water (35,000 parts salt per million), the cost is doubled or tripled.[6] Underground brackish water (5,000 parts salt per million) is found in large water-hungry areas of the United States, and the world, and its purification costs only about 50 per cent more than the treatment of natural fresh water. Thus, suitable plants and tests are in process now to hasten the successful conversion of brackish

Fig. 19.5 Salt water conversion plant at Roswell, New Mexico. With an output of one million gallons per day, this plant reduces the salt concentration from 18,000 ppm to less than 10 ppm. (Photo by J. W. McCutchan)

water. Present cities require from 150–250 gal. of "drinking" water per person per day. Hence, very progressive long-term thinking, large systems and large investments are the challenges of the water problem. See Figure 19.5.

Mineral and vegetable raw materials from virgin sources have been casually treated and opportunistically exploited until recent years. We must soon come to realize that when the oil, gas, and coal are gone, there is no more. Even the vegetable-growing land resources can be destroyed. Timber can be regrown, but the logging operations must not kill the ground cover and cause soil erosion. The clear challenge to engineering is the reduction of extraction and use-waste and the conservation of food and fiber growth-environments.

Urban Environment

One sixth of all the people in the world now live in cities of over 100,000 population. There are 36 cities having populations of 2,000,000 or more! These large agglomerations of people are more and more faced with the

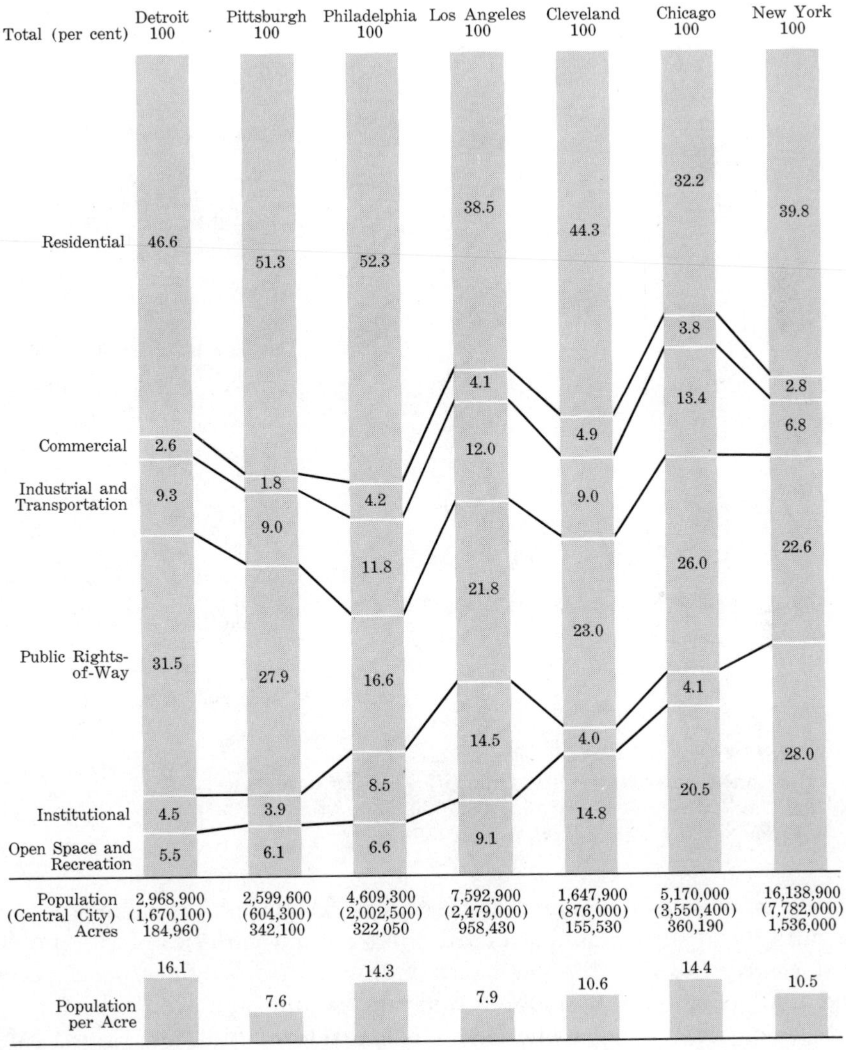

Fig. 19.6 Land use in metropolitan regions, arranged so that percentage of open space increases from left to right. (From *Scientific American,* Sept. 1965.)

physical problems of movement and access, with the economic problems of values and markets, and with the human problems of pollution, noise, confusion, depersonalization, frustration, and alienation.

The planning boards and commissions of cities are broadly responsible for planning the land use, the metabolic processes of a city's life, and the

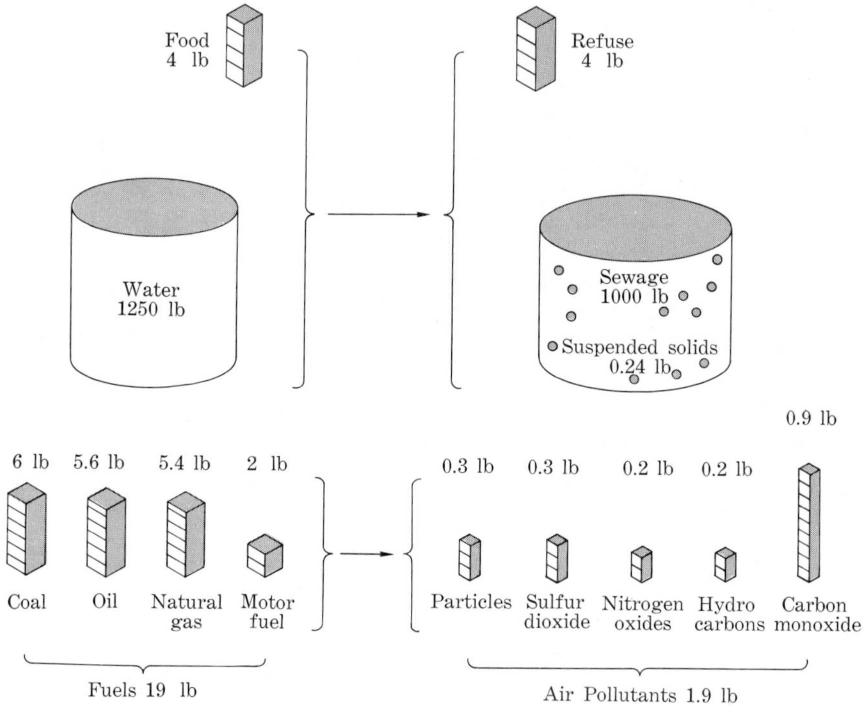

Fig. 19.7 Partial balance of materials transformation in a United States city. Quantities are per person, in an assumed city of 1,000,000 population. (From *Scientific American*, Sept. 1965.)

renewal of obsolescent and decaying areas. Engineers are members of and consultants to such planning boards; but certain varieties of the problems most clearly fall into the engineer's bailiwick. Non-engineering aspects of land use,[7] (Figure 19.6) involve taxation, zoning, rent controls, financial, and sociological issues; whereas, transportation, building codes, safety, water, waste, and atmosphere depend heavily upon technical and scientific work.[13]

The average daily inputs of water, food, and fuel are shown in Figure 19.7 at the rates applicable to a city of about a million. The inputs are transformed into outputs of sewage, solid refuse, and air pollutants. Of the fresh water withdrawn from the available supply, from 80–95 per cent can be treated and returned to the water table for re-use. As much as 20 per cent is diverted to lawns or otherwise rendered non-recoverable. Sewage is treated to different degrees by different cities, depending upon the economics and the geography. Such treatment ranges from direct discharge of raw sewage into the ocean, to full treatment yielding organic fertilizer.

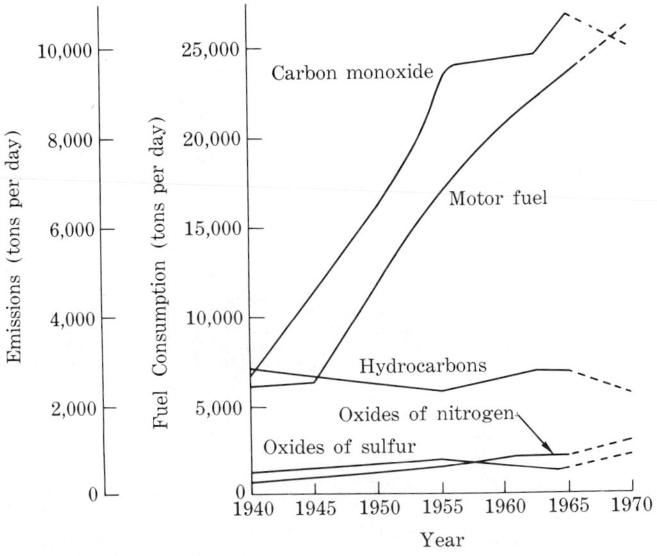

Fig. 19.8 Atmosphere pollutants measured in Los Angeles. Notice the parallelism between carbon monoxide emissions and motor fuel consumption. (From *Scientific American*, Sept. 1965.)

Air pollution is caused almost entirely by combustion end-products; and the constituents which in turn cause eye irritation, health hazard, or plant damage are those shown in Figure 19.8, which presents the conditions in Los Angeles. In that figure, note the parallel rise of carbon monoxide with that of motor fuel consumption.

There seem to be only three ways to control the pollution entering the atmosphere. First, change from a fuel high in undesirable end-products to one with fewer such constituents. For instance, gas produces less sulfur dioxide than fuel oil. Second, change technologies. Nuclear power produces none of the atmospheric contaminants that fossil fuels do; but of course the long-term environmental effects of radioactive wastes are not known. Third, remove the undesired components from the emissions. Much can be done and has been done in large plant installations, particularly relative to fly-ash. But, how do we deal with the millions of small offenders—the automobiles pouring out hydrocarbons and carbon monoxide, and the homes with poor combustion in their central heating systems?

The challenges of the city are the rational planning of land use, conservation against waste, the sensible reduction of pollution, and the relaxing of rigid, dehumanizing influences.

Transportation

Standards of comfort and speed have steadily risen since World War II in the movement of people from home to job or from city to city. Freight

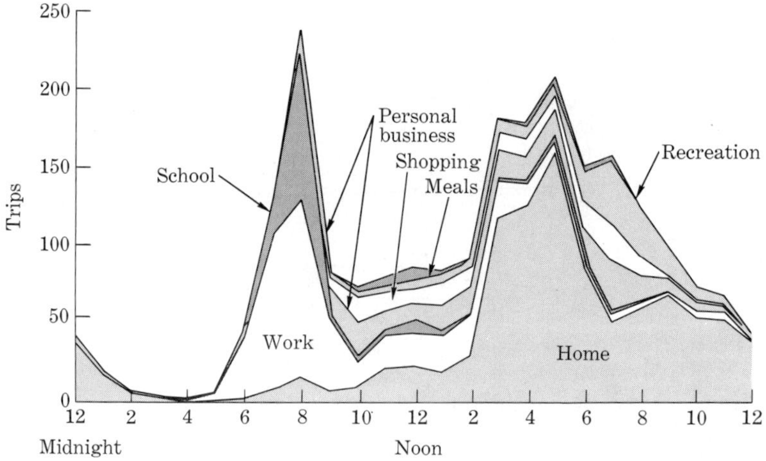

Fig. 19.9 Commuter transportation problem, differentiating purposes of the trip by destination. (From the Pittsburgh Area Transportation Study quoted in *Scientific American*, 1965.)

movement has also speeded up, but with nothing like the same degree of change. Passengers travel across the country now in five hours, adding, with luck, only an hour more at each end to go to and from the airport. Yet, we have substituted the freedom and status of the individual car for the simplicity, safety, and relaxation of the public rapid-transit system. The automobile driver faces congestion, fumes, delays, and danger in exchange for his freedom of schedule and destination. The problem is not simple, because suburban living, with its decentralized factories, offices, and shopping centers, and increased expectations all reduce the serviceability and appeal of mass rapid-transit systems of any kind. The commuter epitomizes the dilemmas of all passenger transportation systems. His journey is a mixture of speed-up, slow-down, luxurious thoroughfare, back-alley bottlenecks, crowd-in, spread-out. Figure 19.9 shows the peaking problem published in the Pittsburgh Area Transportation Study.[8]

Intercity transportation has been given a technical impulse by the successful operation of the new 125 mph Tokaido railway line out of Tokyo.[9] Other high-speed systems are under consideration, particularly for such high-traffic-density areas as the "Northeast Corridor" (Boston-New York-Washington). In high-speed air travel, however, the supersonic transport plane (SST) will be a disappointment, unless its 2000-mph speed is matched by appropriate air-terminal service.

The challenges in transportation are 1) pollutant-free individual local vehicles, 2) intercity high-density mass transport, 3) aircraft safety and convenient terminal service, and 4) the unification of all these into an integrated transportation concept.

Information Handling

It is estimated that probably two-thirds of the 600,000 engineers in industry are *not* performing creative nonroutine work.[3] Much of the engineer's freedom to create may be restored by putting repetitive information processing onto the computer, where it now belongs. Such work as design from formulae, compliance to legal codes, tabulating bills of materials, and preparing manufacturing routines can, in most cases, be done better and faster by machine. Motors, transformers, and computer circuit-boards are a few of the products now designed and specified for manufacture by computer.

The needs lie in storage, retrieval, and scanning of vast libraries of information for the engineer and his associates. Also applying machine routines for operation, maintenance, and surveillance tasks will do much to restore the independence and the exercise of ingenuity to our technical brainpower. System simulation, pattern recognition, and graphic manipulation also will help free the engineer. New machine-machine linkages and common computer languages will make possible amplification of memories and ready access by the questing engineer. These all combined will offset the currently static engineering school enrollments, yet increase the quality and effectiveness of already-trained engineers.

Biotechnology

Biological systems and processes are promising fields of new materials and techniques for engineering. Not only the application of cultures of organic matter (bacteria, molds, or enzymes), but the fields of animal sensing, of artificial intelligence, and interspecies communication can stimulate, contribute to, and be harnessed for engineering purposes.

The vista is extensive: Neurological control systems, energy conversion of muscles, central nervous system filtering and pattern recognition, learning and decision theory, memory, etc.

The challenge is to acquire and use interdisciplinary knowledge in both biological and engineering fields.

Health, Medicine

An extension of our previous subject is the effective use of engineering in the institutions of health: The hospitals, research centers, and clinics.[3]

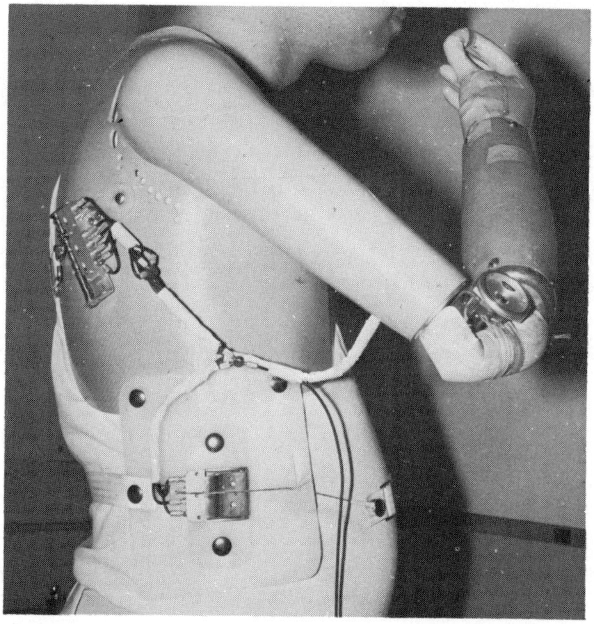

Fig. 19.10 Pneumatically-operated Heidelberg artificial arm. (From "Bulletin of Prosthetics Research," Spring 1966.)

The need is for further application of inventive design engineering to instrumentation, prosthetics (aids for the handicapped), function machine (heart-lung, kidney substitutes, etc.) and information acquisition, processing and correlation. Figure 19.10 shows an artificial arm operated in response to purposefully-trained back muscles. The total needs extend from electrical, mechanical, chemical, acoustic, and nuclear hardware to facilities layout, operations improvement, and industrial engineering in general.

The need is to reduce the cost and improve quality of a family's health services, an item which is over 5 per cent of the average budget now, and the most rapidly growing.

Education and the Developing Nations

Engineering can contribute to the vast educational needs of the world in at least two distinct ways.[3] The first and simplest is by the application of programmed learning through simple teaching machines or through computerized adaptive teaching systems. These may somewhat offset the teacher shortage in our own country as well as overseas.

The other contribution is as vital, and it consists of working closely with willing individuals and groups in developing countries, teaching them how to improve their local physical and educational facilities or helping them to start small industry. Engineers may work with larger overseas development agencies in regional or national planning, as employees or as consultants.

In either of these above situations, the engineer's duty is primarily educational, even though he may act as a technical consultant. He is an educator, because he eventually will leave, and the local nationals must carry on and further accelerate the development without his help.

The enormity of the task facing the developing nations is illustrated by the following figures. Assume one billion persons in emerging countries, and a wage-earner is to be provided for one out of five persons. The average investment needed per worker is about $2000 even in low technologies. As an estimate, we have about 20 years to make major economic inroads, before the common man resorts to violence for relief from his hunger and misery. (Maybe we have only ten years.) The conclusions are that $20 billion must be invested per year in some kinds of production facilities, and 10 million wage earners must be put to work each year. This includes 1 million foremen, 100,000 factory or business managers, and large numbers of assemblers, machine operators, printers, weavers, painters, cooks, tool makers, welders, electricians, mechanics, and other craftsmen. All countries now have large numbers of indigenous craftsmen, but not oriented toward improved or labor-saving methods.

Engineering has already contributed[10] to the small industry development in several areas of the world. See Figure 19.11. Further opportunities are wide open through the U.S. Agency of International Development,

(a)

(b)

Fig. 19.11 (a) Technical assistance for developing nations. Joint United States and Brazil team presentation of feasibility study to potential small industry entrepreneur group in northeast Brazil; (b) Technical assistance implementation. Construction phase of small radio and motor plant in northeast Brazil, organized and locally financed as an outcome of feasibility studies and plant design work illustrated in (a) above. (UCLA photos) through private corporations, private foundations (Ford Foundation, Rockefeller Foundation) or directly through the national ministries.

SUMMARY

The first engineers were ingenious craftsmen who responded to and met the technological needs of early man. Today, professions exist and are recognized and honored because their members responsibly use their special skills for the benefit of man. The engineer is such a professional when he places the interest of his client and of society on a par with his own, when he considers his special knowledge as a trust from the masters of the past, accepts his obligation to his art, and relates ethically to his peers and to the society in which he lives.

REFERENCES

1. Wolfle, Dael: The Support of Science in the U.S., *Scientific American*, July 1965.
2. Brown, Harrison, James Bonner, John Weir: "The Next Hundred Years," p. 12, The Viking Press, New York, 1961.
3. "The Nation's Engineering Research Needs 1965–1985" Summary Report of the Engineering Research Committee, Engineers Joint Council, 345 East 47th Street, New York City, 1962.
4. Reference 2, p. 110.
5. Reference 2, p. 157.
6. Friedlander, Gordon D.: Science and the Salty Sea, *IEEE Spectrum*, August 1965.
7. Abrams, Charles: The Use of Land in Cities, *Scientific American*, p. 154, September 1965.
8. Dyckman, John W.: Transportation in Cities, *Scientific American*, p. 166, September 1965.
9. Kawakami, Toshikaza: Electrical Features of the New Tokaido Line, p. 57, *IEEE Spectrum*, January 1966.
10. Asimow, Morris, John S. McKnown: Engineering Education and International Development, p. 65, *Journal of Engineering Education*, November 1965.
11. "Statistical Abstract of the United States," 1963, U.S. Department of Commerce; U.S. Government Printing Office, Washington, D.C.
12. Smith, Craig B.: "Preliminary Design of a Small Nuclear Power Plant for Latin American and other Developing Areas," Ph.D. thesis, 1965, UCLA, Los Angeles, California.
13. Branch, Melville C.: "Planning, Aspects and Applications," John Wiley & Sons, Inc., New York, 1966.

Common Engineering Reference Data

Appendix A

Work, Power, and Heat Conversions

The following data are primarily for order-of-magnitude approximations. Further information is obtainable in recent editions of engineering handbooks; also, certain engineering periodicals carry current cost information (e.g., *Iron Age*, *Materials in Design Engineering*, *Construction Engineering*). Costs apply in the United States during the mid 1960's.

1. *Work, Energy* = *Force* × *Distance acted in direction of the force*

   Units:

   Foot × Pound: ft-lb

   British Thermal Unit: 1 Btu = 778 ft-lb
   (1 Btu raises 1 lb water 1 °F)

   Calorie, gram: 1 gm cal = 3.1 ft-lb = 0.004 Btu
   (1 gm cal raises 1 gm water 1 °C)

   Kilowatt-hour: 1 kwhr = 2,650,000 ft-lb = 3414 Btu = 860,000 gm cal

2. *Power = Work per unit time*

Units:

Foot-pound per second: ft-lb/sec

Horsepower: 1 hp = 550 ft-lb/sec = 0.746 kw = 2545 Btu/hr

Kilowatt: 1 kw = 1.34 hp

3. *Fuels*

Fuels	*Heating Value*	*Cost*
Gasoline (kerosene):	120,000 Btu/gal	$0.15/gal (without highway tax)
Diesel oil:	160,000 Btu/gal	0.12/gal
Crude oil:	145,000 Btu/gal	0.05–0.07/gal (2.90/42 gal bbl)
Natural gas:	1,000 Btu/cu ft	0.08/100 cu ft
"Bottle gas:"	94,000 Btu/gal	0.30/gal
Coal (carbon):	14,000 Btu/lb	10.00–20.00/ton
Ethyl alcohol:	13,000 Btu/lb	1.00/gal

4. *Power Cost*

Electricity:

Home	$0.04/kwhr approx.
Commercial	0.02/kwhr approx.
Industrial	0.01/kwhr approx.

5. *Efficiency*

Electric motors ($\frac{1}{8}$–1 hp): 65% approx.; (5–50 hp): 80% approx.

Diesel or gasoline engines (5–25 hp): 20% approx.

Centrifugal water pumps (75–250 gpm): 60%; (250–3000 gpm): 70%

6. *Prime-Mover cost*

Electric motors ($\frac{1}{10}$–$\frac{1}{6}$ hp): $120 to $60/hp (single-phase)

Electric motors ($\frac{1}{4}$–$\frac{1}{2}$ hp): $45/hp (single-phase) (max. production volume)

Electric motors (1–10 hp): $80 to $20/hp (single-phase)

Electric motors (10–100 hp): $15/hp (three-phase)

Gasoline engines (1–10 hp): $15/hp

7. *Building Cost*

Factory (without equipment): $10–20/sq ft (High bay: 1$\frac{1}{2}$–2 stories)

Residence: $8–15/sq ft

Physical Properties and Costs of Materials

Metals	*Cost** ($/lb)	*Density* (lb/cu in.)	*Tensile Strength* (K-psi)	*Linear Coeff. Expans.* (in./in./°F × 10^{-6})	*Specific heat* (Btu/lb/°F or cal/gm/°C)
Platinum	1200.00	0.78	30	5	.03
Gold (annealed)	500.00	.70	19	8	.03
Silver (annealed)	15.00	.38	22	11	.06
Titanium (HR alloy)	5–10.00	.17	75–150	5	.14
Mercury (liquid)	3.50	.49		34	.03
Tool Steel	.75–1.50	.28	150–250	7	.11
Magnesium (sheet)	0.80	.06	30–40	16	.25
Phos Bronze (94Cu 5Sn)	.75	.30	50–80	10	.09
Copper (sheet)	.56	.32	30–50	10	.09
Brass (70 Cu, 30 Zn)	.52	.31	50–60(1)	11	.09
Stainless Steel (302 sheet)	.50	.28	75–200	8	.11
Aluminum (24S sheet)	.45	.10	70(2)	13	.21
Zinc Alloy (die cast)	.15	.25	35–45	15	.09
Lead (antimony)	.13	.41	7	16	.03
Cast Iron	0.05–.20	.26	20–50	6	.11
Spring Steel	.10–.15	.28	100–300	6	.11
Cold Rolled Steel	.07–.08	.28	60–90	7	.11
Hot Rolled Steel	.05–.07	.28	45–75	6	.11
Non-metal Solids					
Teflon TFE	5.00	0.077	3.–5.0	55	.25
Nylon	1.40	.041	10.0	60	.4
Neoprene	.40	.05		340	.3–.5
Acrylics	.50	.043	8.0	50	.35
Polyethylene	.30	.033	2.0	150	.53
Polyvinyl chloride	.30	.048	8.0	33	.3
Bakelite	.20	.047	6.–9.0	25	.3–.4
Rubber, natural	.30	.054	3.0	370	.3–.5
Rubber, synthetic	.25	.05	3.0	310	.3–.5

Cork (sheet)	1.00	.008			.49
Graphite	.15–.30	.07	.4–1.0	1–9	.2
Glass (common)	.10–.15	.09	10.0	4	.2
Porcelain	.10	.07	1.8	2	.22
Granite (block)		.095		4–7	.2
Limestone		.09		2–5	.22
Sand	.001	.058			.2
Wood, oak	.08	.024	10. (3)	3	.57
Pine	.04	.016	5.	3	.67
Concrete (mix)	.003	.083		8	.16
CO_2 (solid)	.05–.10	.055			246 Btu/lb at −110 °F and atm pressure
Liquids	($/gal)				
Alcohol	1.00	0.028		.61	.6
Gasoline (cost without tax)	.15	.027		.50	.5
Motor oil	.75	.033		.40	.4
Vegetable oil	1.00	.033		.41	.4
Water	.0004	.036		.11	1.0
Gases	($/cu ft)	(lb/cu ft)			(C_p above 32 °F S.T.P)
Hydrogen	.013	.005			7 Btu/mol
Oxygen	.01	.083			7 Btu/mol
Air		.075			7 Btu/mol
Acetylene	.026	.067			7 Btu/mol
Helium	.13	.010			

* Costs are 1964 base prices (from a variety of sources) for commercial purchase; relatively large quantities; F.O.B. factory. For further detail see Joseph E. Shigley, "Mechanical Engineering Design," McGraw-Hill Book Company, 1963, p. 220; or *Materials in Design Engineering*, mid-October 1965, p. 45.

(1) $\frac{1}{4}$ H: quarter hard.
(2) T3: heat treatment.
(3) EL: Elastic limit.

Common Engineering Reference Sources

Below is listed a group of general-purpose handbooks and similar useful reference sources, available in many libraries.

American Institute of Physics	American Institute of Physics Handbook (1)
American Society of Mechanical Engineers	ASME Handbooks: Engineering Tables (1) Metals Engineering—Design (1) Metals Engineering—Processes (1) Metals Properties (1)
American Society of Tool Engineers	Die Design Handbook (1) Manufacturing Planning and Estimating Handbook (1)
Berry, Bollay and Beers	Handbook of Meteorology (1)
Brady	Materials Handbook (1)
Chemical Rubber Publishing Co.	Handbook of Chemistry and Physics (2)
Eshbach	Handbook of Engineering Fundamentals (3)
Etherington	Nuclear Engineering Handbook (1)
Greenwood	Mechanical Details for Product Design (1)
Henney	Radio Engineering Handbook (1)
Herkimer	Thesaurus (8000 devices) (2)
Ireson, Grant	Handbook of Industrial Engineering and Management (4)
Kidder, Parker	Architects' and Builders' Handbook (3)
Knowlton	Standard Handbook for Electrical Engineers (1)
Koelle	Handbook of Astronautical Engineering (1)
Lalonde, Janes	Concrete Engineering Handbook (1)
Laughner and Hargan	Handbook of Fastening and Joining of Metal Parts (1)
LeGrand	The New American Machinist's Handbook (1)
Marks, Baumeister	Mechanical Engineers' Handbook (1)
Materials Selector Issue (each October)	Materials in Design Engineering
Merritt	Building Construction Handbook (1)
Miner, Seastone	Handbook of Engineering Materials (3)
Pender	Electrical Engineers' Handbook (3)
Society of the Plastics Industry, Inc.	Plastics Engineering Handbook (5)
Stanley	The Consulting Engineer (3)
Truxal	Control Engineers' Handbook (1)
Urquhart	Civil Engineering Handbook (1)

(1) McGraw-Hill Book Company, 330 West 42nd Street, New York.
(2) Chemical Rubber Publishing Company, Cleveland, Ohio.
(3) John Wiley & Sons, Inc., 605 Third Avenue, New York.
(4) Prentice-Hall, Inc., Englewood Cliffs, N.J.
(5) Reinhold Book Division, 430 Park Avenue, New York.

Items in Construction Job Costs

Direct

1. Materials
2. Labor

Indirect

3. Costs of drawings and estimation
4. Permits and inspection
5. Taxes and licenses
6. Rental of equipment
7. Freight and storage
8. Bonds and insurance
9. Supervision
10. Utilities
11. Supplies
12. Maintenance and repair
13. Security
14. Traveling and communications
15. Advertising
16. Testing and consulting
17. Patents and legal fees
18. Rents
19. Cleanup and miscellaneous
20. Contingency

Elements of Cost in Factory Mass Production

The following items are used in the analysis of cost in typical cost accounting practice. Approximate costs ($\pm 5\%$) are projected at the end of a product development and during the tooling period. After the product has been in production for a short time (long enough to stabilize reasonably), accurate cost figures are compiled and a "standard" set of costs are agreed upon and released. These standard costs are shown below as No. 8: Shop cost, and No. 16: Standard cost of goods sold. Such standards are the bases to which current operations are compared. As cost reductions or increases occur they are reported as "variances," favorable or unfavorable. The standards are changed only once a year or on occasion of major model changes. The cost breakdowns shown below are obtained from actual invoices or charges when economical to do so; otherwise they are computed by accepted percentages.

The figures presented on the next two pages are typical for an engineered, mass-produced consumer durable product retailing for $200.00.

Direct Materials	*Typical Figures*	
1. Material (actually consumed in product)	$61.00	
2. Incoming freight (3%)	2.00	
3. Losses (1%)	1.00	
4. Total material		$64.00
Direct Labor (expended in proportion to production)		
5. (Machine operators, assemblers, inspectors, etc.)	6.00	
Indirect Manufacturing (Burden) (250% of direct labor)	15.00	
		21.00

6. *Indirect labor*

   Supervision, foremen, etc.
   Tool set-up
   Rework processing
   Storekeeping, shipping
   Maintenance, custodial

   Indirect Materials

   Process expendables (cutting fluids, solvents, etc.)
   Maintenance supplies (lighting, cleaning, etc.)

   Overhead (all costs not in above classes)
   Factory management
   Purchasing, accounting, payroll, planning
   Tool-room, purchased services, etc.
   Utilities, property taxes, insurance, rents

4. Total material						$64.00
5. (Machine operators, assemblers, inspectors, etc.)					6.00	
6. *Indirect Manufacturing* (Burden) (250% of direct labor)					15.00	
7. Total labor and burden						21.00
8. Shop cost		85.00				
Engineering Costs and Expenses						
9. Engineering (3 to 7% standard cost)	4.00					
10. Special tools amortization	2.50					
11. Patents and licenses	0.50					
12. Service costs and warranty	4.00					
13. Total engineering costs		11.00				
Standard Costs						
14. Standard manufacturing cost			96.00			
15. Shipping cost (national average)			4.00			
16. Standard costs of goods sold				100.00		
Administrative Costs and Margins						
17. Federal excise tax (5% where applicable)				5.00		
18. General and administrative; advertising and sales promotion				12.00		
19. Cost of goods sold plus G and A					117.00	
20. Gross margin (net income before taxes)					13.00	
21. Selling price to distribution						130.00
22. Federal income tax (52% of gross margin)					6.75	
23. Net margin for manufacturer					6.25	

Recommended Retail Selling Price

24. Price to distributor + 10% (for distributor) + 40% (for retailer)
$130 + ... $13 + ... $57 .... = $200.00

Student Design Project

Appendix B

Criteria for Choosing a Project

1. Choose a project in which you are interested (within the rules set up by the instructor).
2. Stay within your general area of knowledge, experience, and motivation.
3. Find out whether the project is individual or "group" and whether others are competing with you in its solution.
4. Start with a relatively nebulous need, rather than a well-defined problem. Problem definition comes later.
5. Individual work and reporting are to be identified, so partition a group project into relatively equal logical parts (i.e., active mechanism, control, structure, etc.).
6. Choose a project that has opportunity for some experimentation during the development, or a test at the end with a partial or whole prototype.
7. Consider the availability of information. You must be able to get sufficient quantities conveniently.
8. Choose in the same technical or product area as the rest of the class for uniformity of communication.
9. Be prepared to present regular progress reports in writing and a terminal oral report on your work.

Suggested Project Areas

1. LABOR-SAVING device for *one area* in the home or farm: Kitchen, yard, garage, barn, tool shop, family room, automobile, beach.
2. Instead of labor-saving, the project might be directed toward COST REDUCTION or QUALITY-IMPROVEMENT or SAFETY.
3. Specific classes within the above categories include:

Furniture	Food preparation	Food preservation	Tools
Storage devices	Cooking	Materials handling	Heaters
Cleaning devices	Utensils	Measurement	Timers
Openers	Cutters	Stirring and mixing	Sharpeners
Ladders	Racks	Drying	Coolers
Reminders	Communication	Inventorying	Ventilators
Mowers	Trimmers	Rakes	Saws
Barbecues	Tillers	Fertilizers	Sprinklers

4. Hobbies:

   Collections, general science, meteorology, photography, astronomy, radio, sports, autos.

5. Specific project titles used in the past:

Educational toy (math, english, physics, etc.)	Shoe shiner
Egg incubator	Window washer
Portable cement mixer	Burglar alarm
Fish bait bucket	Chemical still design
Hubcap theft prevention	Swimming pool heater
Physiologic function instrumentation	Beach cleaner
Coin sorter	Fireplace log splitter
Cryogenic storage container (N_2)	Isotope shipping container
Student lamp	Motorcycle book-rack
Jar top opener	Pick-up and reaching device

Supplemental Information on Projects at End of Chapter One*

Project A

Figures B.1–B.5 show various views of the Gemini spacecraft, auxiliary modules, cabin clearances, and cabin layout.

Project B

The Atomic Energy Commission requires shielding such that the radiation is a maximum of 200 milliroentgen (mr)/hr from any accessible surface, and 10 mr/hr at any point one meter distant from any surface. (See *Nucleonics*, November 1961, and APEX-176, Atomic Energy Commission, Oak Ridge, Tennessee, for information on lead shielding.)

* Supplemental information on projects is continued through page 394.

Fig. B.1 Gemini spacecraft and auxiliary modules. (Courtesy National Aeronautics and Space Administration, Manned Spacecraft Center, Houston, Texas)

Fig. B.2 Astronaut positions prior to launch; Gemini spacecraft. (Courtesy National Aeronautics and Space Administration, Manned Spacecraft Center, Houston, Texas)

Fig. B.3 Lead astronaut (left-hand) in cruise position; Gemini spacecraft. (Courtesy National Aeronautics and Space Administration, Manned Spacecraft Center, Houston, Texas)

Fig. B.4 Astronaut sighting, showing cabin clearances; Gemini spacecraft. (Courtesy National Aeronautics and Space Administration, Manned Spacecraft Center, Houston, Texas)

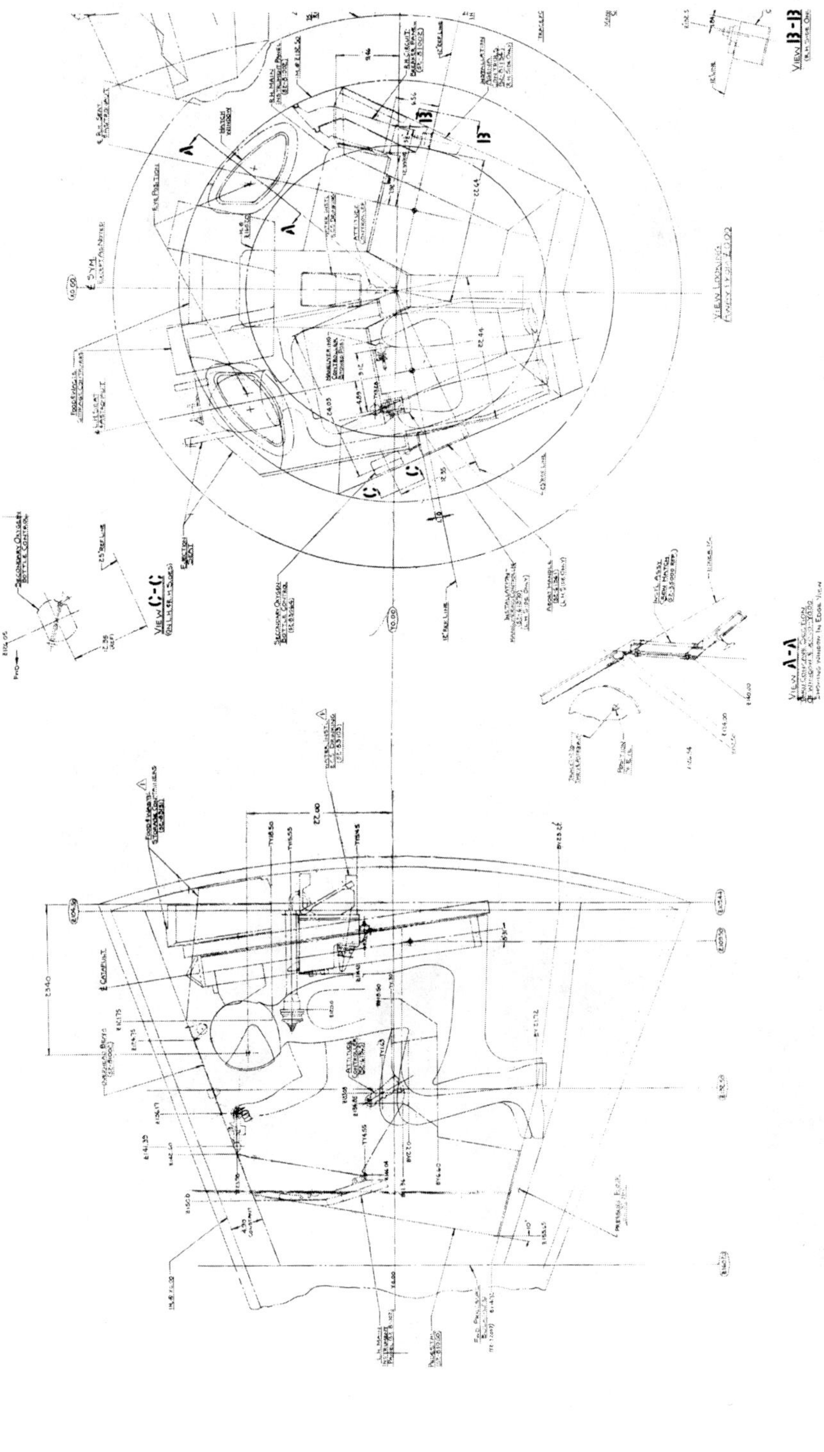

Fig. B.5 Longitudinal and Perpendicular cross section layouts; Gemini spacecraft. (Courtesy National Aeronautics and Space Administration, Manned Spacecraft Center, Houston, Texas)

ANTHROPOMETRIC DATA — ADULT MALE STANDING AT CONTROL BOARD

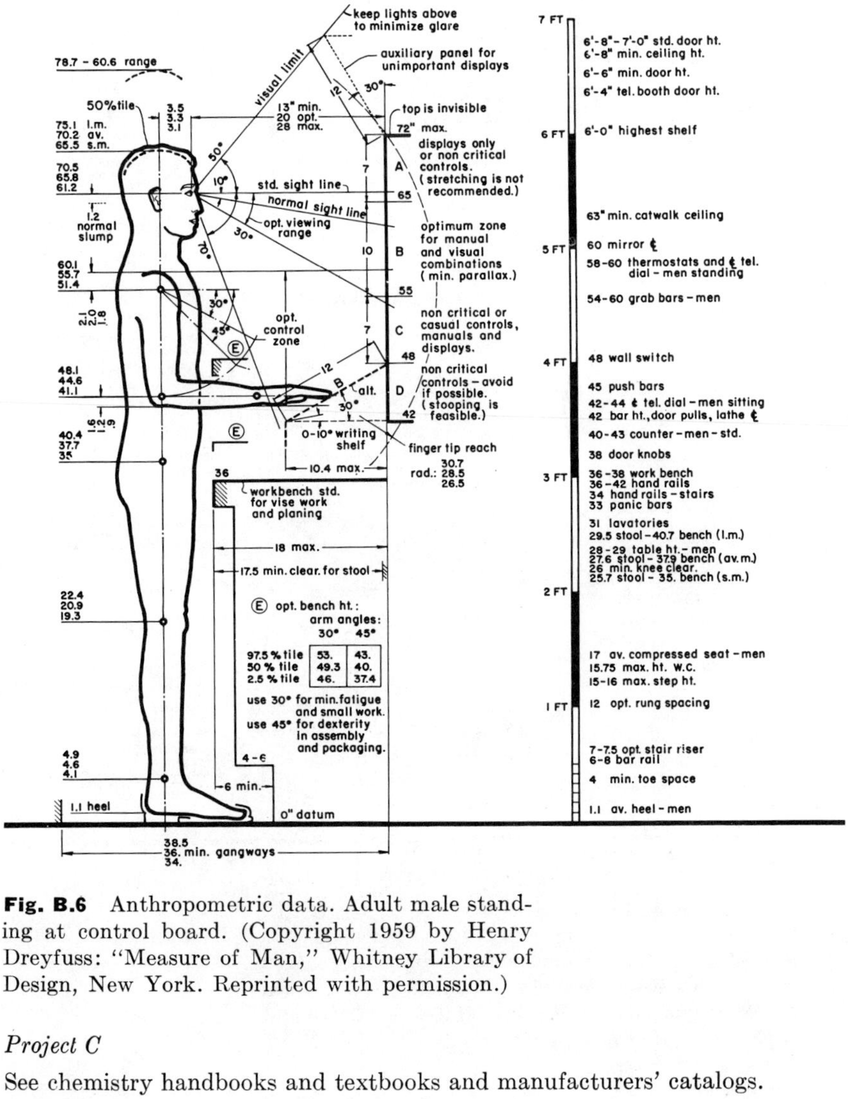

Fig. B.6 Anthropometric data. Adult male standing at control board. (Copyright 1959 by Henry Dreyfuss: "Measure of Man," Whitney Library of Design, New York. Reprinted with permission.)

Project C

See chemistry handbooks and textbooks and manufacturers' catalogs.

Project D

As aids in determining dimensions for adapting to human use, Figures B.6 and B.7 are supplied showing the size ranges for the standing and sitting adult male.

Project E

See chemistry textbook for Ferric Oxalate process and machine element encyclopedia for electro-mechanical devices.

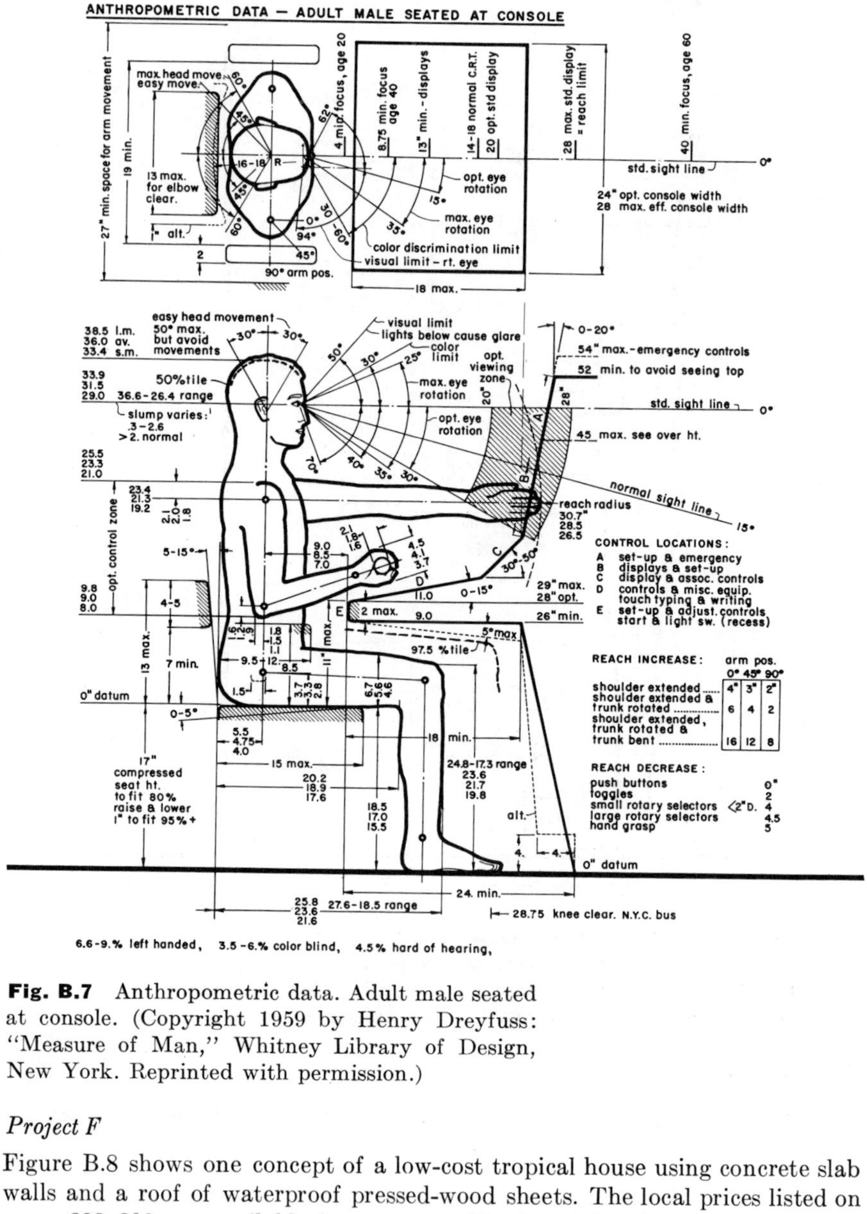

Fig. B.7 Anthropometric data. Adult male seated at console. (Copyright 1959 by Henry Dreyfuss: "Measure of Man," Whitney Library of Design, New York. Reprinted with permission.)

Project F

Figure B.8 shows one concept of a low-cost tropical house using concrete slab walls and a roof of waterproof pressed-wood sheets. The local prices listed on pages 388–390 are available for one area, Northeast Brazil.*

Cement: \$1.25/bag (1 cu ft)

Brick: \$.01/brick, 7 × 15 × 30 cm (local adobe quality)

* Source: UCLA Brazil Project, Preliminary Designs, 1962 (1965 exchange rate: 1850 cruzeiros/\$U.S.). Tropical House Design by Leonard Horn.

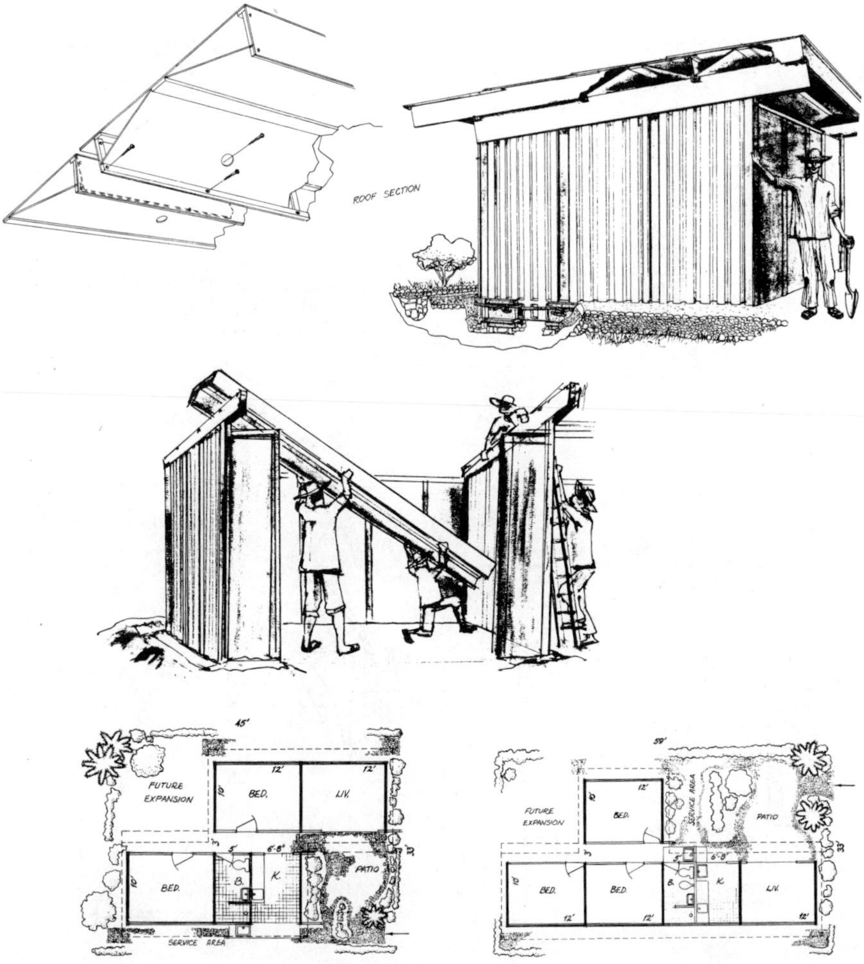

Fig. B.8 Low-cost tropical house. Several suggested details. Two- and three-bedroom layouts at lower left and right, respectively. (From Leonard Horn.)

Roof tile: \$.01 tile, 15 × 50 cm long, curved

Sewer tile: avg. \$.06/tile; 5, 8, 10, 15 cm diameter

Pressed wood: \$.75/sheet, 1.22 × 2.5 m × 2.5 mm thick
\$1.50/sheet, 1.22 × 2.5 m × 5.0 mm thick

Labor: \$2.00/day; general skilled: carpenter, mason, etc.

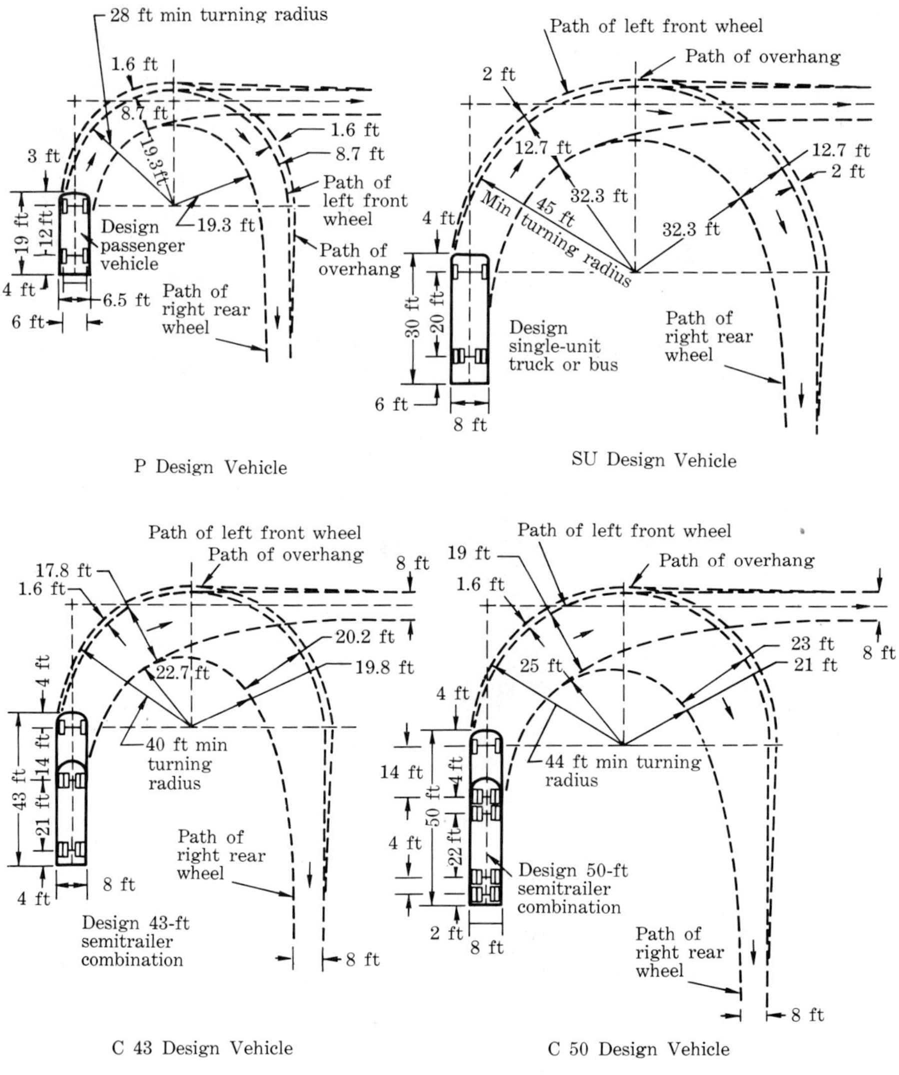

Fig. B.9 Minimum turning paths for design vehicles. (From Perry, R. H.: "Engineering Manual," McGraw-Hill Book Company, New York, 1959. Used with permission.)

Average building costs: (general industry building: \$20.00/sq m floor area)

Brick wall: \$1.00/sq m for 15 cm thick

Concrete wall: \$40.00/cu m (or foundation or floor)

Roof covering: \$2.50/sq m (asbestos-cement sheets)

Windows and doors: \$5.00/sq m

Floor tile: \$5.00/sq m

$$m = \frac{5730}{D} \text{ vers } \frac{SD}{200}$$

Also $m = R \left(\text{vers } \frac{28.65S}{R}\right)$ and $S = \frac{R}{28.65} \cos^{-1} \frac{R-m}{R}$

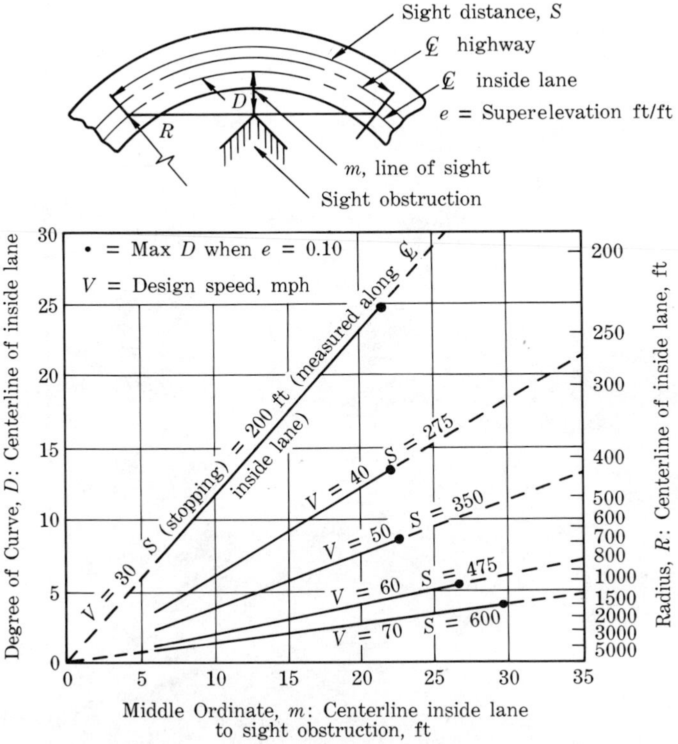

Fig. B.10 Stopping distances on horizontal highway curves. (From Perry, R. H.: "Engineering Manual," McGraw-Hill Book Company, New York, 1959. Used with permission.)

Projects G, H

Electrical and electronic textbooks, handbooks and manufacturers' catalogs carry the principles, circuits and suggested circuit constants for various signal systems.

Project I

Refer to textbooks and educators on child psychology, as well as to parents of young children. Visit a plant manufacturing and stores selling modern educational toys.

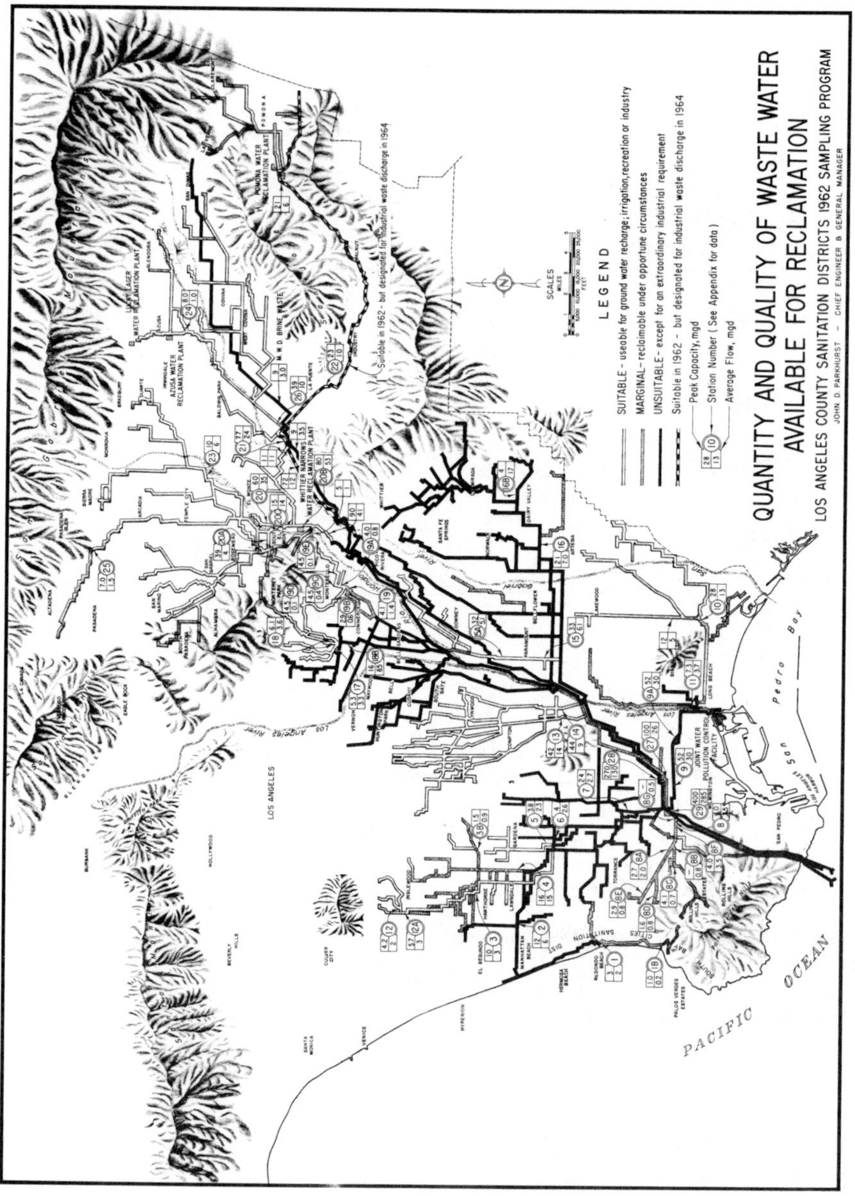

Fig. B.11 Quantity and quality of waste water available for reclamation. (From Parkhurst, John D.: "Plan for Water Re-use in Los Angeles County," 1962)

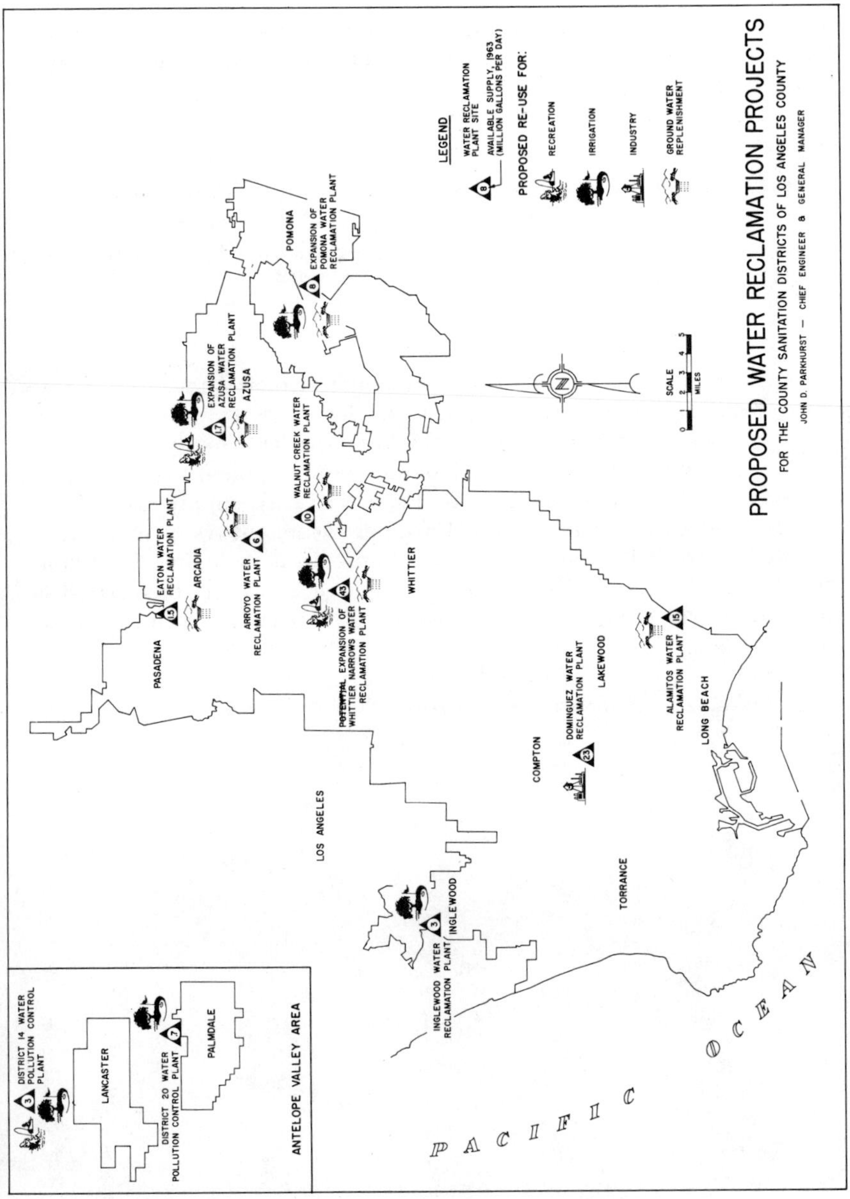

Fig. B.12 Proposed water reclamation projects. (From Parkhurst, John D.: "Plan for Water Re-use in Los Angeles County," 1962)

Project J

Measure the unscrewing torques and the sizes of various jars and bottles. (See Hill, Percy H., "Engineering Graphics and Design," The Macmillan Company, New York, 1964, p. 5.)

Project K

Figure 1.4 is an aerial view of a district in which preparatory work is being done to construct a freeway along the dashed line. Figures B.9 and B.10 show dimensional standards for vehicle turning and stopping adopted by the American Association of State Highway officials.

Project L

Water may be recovered from sewage treatment plants and returned to the water table by irrigating turfed recreation areas. Such a plan* was proposed by John D. Parkhurst, Chief Engineer of the Los Angeles County Sanitation Districts in July 1963 Figures B.11–B.13 present information of water sources, uses and costs for this project. Figure B.11 shows in gray, supply lines where the water may be treated and re-used. Dark lines represent water unsuitable for such use. Figure B.12 shows prospective re-use areas including the Azusa area where 1.7 million gal/day is available. Water generally can be recycled 4 or 5 times, after which its salt content reaches 1000 parts per million and it must be discarded.

* John D. Parkhurst, "Plan for Water Re-use," Los Angeles County Sanitation Districts, July 1963.

Existing Facilities

The Azusa Water Reclamation Plant is located at the westerly end of 1st Street in the City of Azusa. On the 7.5-acre site are a 0.75-mgd capacity standard rate trickling filter treatment plant and four spreading basins having a total surface area of almost 2 acres.

Source of Reclaimable Water

About 1 mgd of good quality flow from the Glendora area is now diverted from the Base Line trunk into the Dalton trunk, by-passing the existing treatment plant. The waste water from this area is of good quality and suitable for reclamation. A summary of results for Station 24 in the 1962 sampling survey shows the key quality characteristics to be:

A. Dissolved solids	613	ppm
B. Sodium-adsorption-ratio (SAR)	3.2	
C. Boron	0.8	ppm
D. Chlorides	80	ppm
E. Chlorides + sulfates	138	ppm

Proposed Use

A proposal has recently been made that a portion of the reservoir area behind the Santa Fe Dam be developed for recreational uses by the County Department of Parks and Recreation. This proposal is currently under serious study, but no final decision has been reached as to whether the project will be implemented. The area consists of about 1000 acres, and it is anticipated that about 600 acres will be turfed and will have an irrigation requirement of about 2 mgd.

By constructing a relatively short section of relief sewer, the existing 1 mgd of good quality flow in the Base Line trunk sewer could be conveyed to the Azusa Treatment Plant site for treatment and discharge to the park area.

If the entire 1.7 mgd that could then be reclaimed at the site were used to irrigate the recreational facilities, most of the water needs of the development could be met.

An alternate use for the water is ground water recharge either at the plant site or in nearby river channels.

Required Facilities

In addition to the facilities required by the Sanitation Districts to treat the water, a pumping plant and pipeline will be required to deliver the water from the plant site to the recreational area. Approximate costs for facilities to be furnished by the Department of Parks and Recreation are based on delivery of the water over a 12-hr period at a pressure of 100 psi to an assumed water distribution system within the park about a mile distant from the plant. It should be noted that if reclaimed water were not used, wells and pumping facilities would have to be provided for irrigation of the recreational area or imported supplies purchased, if available.

The cost estimate in Figure B.14 is based on construction in 1968 at an ENR Construction Cost Index of 1000. Plant facilities and trunk sewers have been amortized over a 50-yr period at 4 per cent interest. The unit water cost is based on the construction of an additional 1 mgd reclamation unit, with the production of 1.7 mgd of salable water.

Fig. B.13 Azusa water reclamation plant expansion conditions.

Sanitation District Facilities	*First Cost*	*Annual Cost*	*Water Cost Per mg*	*Water Cost Per ac-ft*
Capital Items:				
Activated Sludge Reclamation Plant	\$300,000	\$13,970	\$23.70	\$ 7.72
Foam Fractionation	6,000	280	0.48	0.16
Influent Sewer	20,000	930	1.57	0.52
Sub-total	\$326,000	\$15,180	\$25.75	\$ 8.40
Operation & Maintenance:				
Basic Plant Operation	—	15,790	26.76	8.70
Foam Fractionation	—	4,720	8.00	2.60
Maintenance of Equipment	—	3,100	5.24	1.70
Sub-Total	—	\$23,610	\$40.00	\$13.00
Total Cost for Capital and O & M	\$326,000	\$38,790	\$65.75	\$21.40
Less				
Stipulated Savings in O & M at Joint W.P.C. Facility	—	1,120	3.07	1.00
Sub-Total	\$326,000	\$37,670	\$62.68	\$20.40
Add				
Emergency Preferred Service Power Connection	—	1,430	2.46	0.80
	\$326,000	\$39,100	\$65.14	\$21.20
Less				
Maximum Allowable Credit for Capacity Savings in Downstream Facilities	—	20,100	34.32	11.20
Net Total Cost (Sanitation Districts)	\$326,000	\$19,000	\$30.82	\$10.00
Add Facilities Furnished by Dept. of Parks and Recreation				
Capital Items:				
Pumping Plant, 4 mgd @ 280 ft. head	\$113,000	\$ 5,260	\$ 8.90	\$ 2.90
14″ pipeline to Recreation area, 6000 ft.	105,000	4,890	8.30	2.70
Sub-Total	\$218,000	\$10,150	\$17.20	\$ 5.60
Operation & Maintenance:	—	9,060	15.35	5.00
Net Total Cost (Dept. of Parks & Recreation)	\$218,000	\$19,210	\$32.55	\$10.60
GRAND TOTAL (Delivered Water Cost)	\$544,000	\$38,210	\$63.37	\$20.60

Fig. B.14 Azusa water reclamation plant expansion costs (1 mgd expansion for 1.7 mgd total capacity).

Information Sources

Appendix C

Library of Congress Classification System

The Library of Congress uses letters for the major classifications, followed by numbers and decimals for the sub-classifications:

A	General works (Encyclopedia)	N	Fine Arts
B	Philosophy and Religion	P	Language and Literature
C–D	History, Biography	Q	Science
E–F	America	R	Medicine
G	Geography and Anthropology	S	Agriculture and Allied Industry
H	Social Sciences (Economics)	T	Technology (Engineering)
J	Political Science	U	Military Science
K	Law	V	Naval Science
L	Education	Z	Bibliography and Library Science
M	Music		

The T (Technology) classification above is broken down into the sub-classes:

TA	General Engineering, Civil Engineering
TC	Hydraulic Engineering
TD	Sanitary and Municipal Engineering
TE	Roads and Pavements
TF	Railroad Engineering and Operation
TG	Bridges and Roofs
TH	Building Construction
TJ	Mechanical Engineering and Machinery
TK	Electrical Engineering and Industries
TL	Motor Vehicles (Aeronautics)
TN	Mineral Industries (Mining and Metallurgical)
TP	Chemical Technology
TR	Photography
TS	Manufactures
TT	Trades
TX	Domestic Science

See also:

Engineering Index
Science Abstracts
Applied Science and Technology Index
Sweets File
Thomas Register of Manufacturers

A Few Useful Sources

1. *United Engineering Center, 345 East 47th Street, New York, N.Y. 10017.
2. American Standards Association, 10 East 40th Street, New York, N.Y. 10016.
3. World Almanac, New York World Telegram and Sun, 125 Barclay Street, New York, N.Y. 10015.
4. Superintendent of Documents, U.S. Government Printing Office, N. Capitol and H Streets, Washington, D.C. 20046 (for all Government Documents, Patents, etc.).
5. U.S. Chamber of Commerce, 1615 H Street, N.W., Washington, D.C., 20006.

* The United Engineering Center at this address is headquarters for a number of major engineering societies: AIMME, ASCE, ASME, IEEE (AIEE and IRE); see the following table.

6. Library of Congress, Washington, D.C.
7. National Association of Manufacturers, 2 East 48th Street, New York, N.Y. 10017.
8. Atomic Energy Commission, Washington, D.C.
9. Interstate Commerce Commission, 12th Street and Constitution Ave., Washington, D.C.
10. Federal Communications Commission, 12th Street and Pennsylvania Ave., Washington, D.C.
11. National Aeronautics and Space Administration, 400 Maryland Ave., S.W., Washington, D.C.
12. U.S. Department of Agriculture, 14th Street and Independence Ave., S.W., Washington, D.C.
13. U.S. Department of Commerce, 14th Street and E Street, N.W., Washington, D.C.
14. U.S. Department of Health, Education and Welfare, 330 Independence Ave., S.W., Washington, D.C.

National and International Engineering and Technical Societies*

AAAS	American Association for the Advancement of Science
ACI	American Concrete Institute
ACM	Association for Computing Machinery
ACS	American Chemical Society
AES	Aerospace Electrical Society
AGI	American Geological Institute
AIAA	American Institute of Aeronautics and Astronautics, Inc.
AIA	The American Institute of Architects
AICE	American Institute of Consulting Engineers
AIChE	American Institute of Chemical Engineers
AIIE	American Institute of Industrial Engineers, Inc.
AIMME	American Institute of Mining, Metallurgical, and Petroleum Engineers, Inc.
AIP	American Institute of Physics
AIPE	American Institute of Plant Engineers
AISI	American Iron and Steel Institute
ANS	American Nuclear Society, Inc.
APCA	Air Pollution Control Association
API	American Petroleum Institute
APWA	American Public Works Association

* Courtesy of Engineers Joint Council, 345 East 47th Street, New York, N.Y. 10017.

AREA	American Railway Engineering Association
ASA	American Standards Association
ASAE	American Society of Agricultural Engineers
ASCE	American Society of Civil Engineers
ASEE	American Society for Engineering Education
ASHRAE	American Society of Heating, Refrigerating and Air-Conditioning Engineers, Inc.
ASM	American Society for Metals
ASME	The American Society of Mechanical Engineers
ASQC	American Society for Quality Control, Inc.
ASSE	American Society of Sanitary Engineers
ASTM	American Society for Testing and Materials
ASTME	American Society of Tool and Manufacturing Engineers
AWS	American Welding Society
AWWA	American Water Works Association, Inc.
CCPE	Canadian Council of Professional Engineers
ECPD	Engineers' Council for Professional Development
ECRC	Engineering College Research Council
ECS	The Electrochemical Society, Inc.
EI	Engineering Index, Inc.
EJC	Engineers Joint Council
EMC	Engineering Manpower Commission
ESL	Engineering Societies Library
GSA	The Geological Society of America
IAESTE/U.S.	International Association for the Exchange of Students for Technical Experience/United States, Inc.
IEC	International Electrotechnical Commission
IEEE	The Institute of Electrical & Electronics Engineers, Inc.
IES	Illuminating Engineering Society
ISA	Instrument Society of America
ITE	Institute of Traffic Engineers
JETS	Junior Engineering Technical Society, Inc.
NACE	National Association of Corrosion Engineers
NAE	National Academy of Engineering
NAS/NRC	National Academy of Sciences—National Research Council
NEC	National Electronics Conference
NICE	National Institute of Ceramic Engineers
NSPE	National Society of Professional Engineers

ORSA	Operations Research Society of America
OSA	Optical Society of America
SAE	Society of Automotive Engineers
SAM	Society for the Advancement of Management
SAME	Society of American Military Engineers
SAVE	Society of American Value Engineers
SESA	Society for Experimental Stress Analysis
SFPE	Society of Fire Protection Engineers
SFSA	Steel Founders Society of America
SIAM	Society for Industrial and Applied Mathematics
SMPTE	Society of Motion Picture and Television Engineers
SNAME	Society of Naval Architects and Marine Engineers
SNT	Society for Nondestructive Testing
SPE	Society of Plastics Engineers, Inc.
SWE	Society of Women Engineers
TAPPI	Technical Association of the Pulp and Paper Industry
UET	United Engineering Trustees, Inc.
US/WPC	U.S. National Committee/World Power Conference
VITA	Volunteers for International Technical Assistance, Inc.
WPC	World Power Conference
WSE	The Western Society of Engineers

Units and Dimensions of Quantities

Appendix D

The dimensions of the terms in a fundamental equation must be the same on both sides.
That is,

$$F = m \times a$$

$$F = \frac{W}{g} \times a$$

or

$$\text{lb} = \frac{\text{lb}}{\text{ft/sec/sec}} \times \text{ft/sec/sec}$$

thus

$$\text{lb} = \text{lb}$$

The units of each quantity F, m, or a can be derived from three fundamental dimensions L, M, T (length, mass, time), in the case of mechanical equations. Thermal and electrical expressions also use L, M, T but have slight additions, as will be shown. (see page 246, Chapter 14.)

Dimensions of Mechanical Quantities (Fundamentals: L, M, T)

Area	L^2	Momentum	MLT^{-1}
Volume	L^3	Angular momentum	ML^2T^{-1}
Frequency	T^{-1}	Density	ML^{-3}
Wavelength	L	Work	ML^2T^{-2}
Velocity	LT^{-1}	Power	ML^2T^{-3}
Angular velocity	T^{-1}	Angle	number
Acceleration	LT^{-2}	Unit deflection (strain)	number
Angular acceleration	T^{-2}	Moment of inertia	ML^2
Force	MLT^{-2}	Elastic modulus	$ML^{-1}T^{-2}$
Pressure (stress)	$ML^{-1}T^{-2}$	Compressibility	$M^{-1}LT^2$
Torque	ML^2T^{-2}		

Dimensions of Thermal Quantities (Fundamentals: L, M, T, plus t, temperature)

Temperature	t	Temperature gradient	tL^{-1}
Quantity of Heat	ML^2T^{-2}	Thermal conductivity	$MLT^{-3}t^{-1}$
Heat capacity (per unit vol.)	$ML^{-1}T^{-2}t^{-1}$	Heat transfer rate	ML^2T^{-3}
Heat capacity (per unit mass)	$L^2T^{-2}t^{-1}$	Specific heat	number

Dimensions of Electrical Quantities (Fundamentals: L, M, T, plus K, electric permeability, and μ, magnetic permeability)

Voltage	$M^{1/2}L^{1/2}T^{-1}K^{-1/2}$	Inductance	$L^{-1}T^2K^{-1}$
Current	$M^{1/2}L^{3/2}T^{-2}K^{1/2}$	Capacitance	LK
Charge	$M^{1/2}L^{3/2}T^{-1}K^{1/2}$	Permeability	$\mu = L^{-2}T^2K^{-1}$
Resistance	$L^{-1}TK^{-1}$	Magnetic flux	$M^{1/2}L^{1/2}K^{-1/2}$

Summary of Graphics Essentials

Appendix E

The graphics summary sections which follow are primarily for reminding rather than instructing. They are abbreviated and compressed; but they cover the essential elements of free-hand sketching, orthographic projection, descriptive geometry, and an introduction to vectors. Most of the engineer's direct space-measurement problems are shown and solved here. This includes true-length, true-shape, true-angle, distance between lines and intersections.

The conventions followed for identifying points in successive orthographic views are as follows:

a^T = Top view of point "a"

a^F, a^{A_1}, a^{A_2} similarly represent Front, Auxiliary 1 and Auxiliary 2 views respectively

${a^F}_{R_1}$, ${a^F}_{R_2}$ = front view rotated once and twice respectively.

Extension-line arrows leading from a point in one viewing plane toward a point in another viewing plane indicate the originating and dependent or succeeding points respectively.

Much of the material in this Appendix E is from P. H. Hill's "Engineering Graphics and Design," The Macmillan Company, New York, and is used with permission of the publishers.

1. Width and Character of Lines

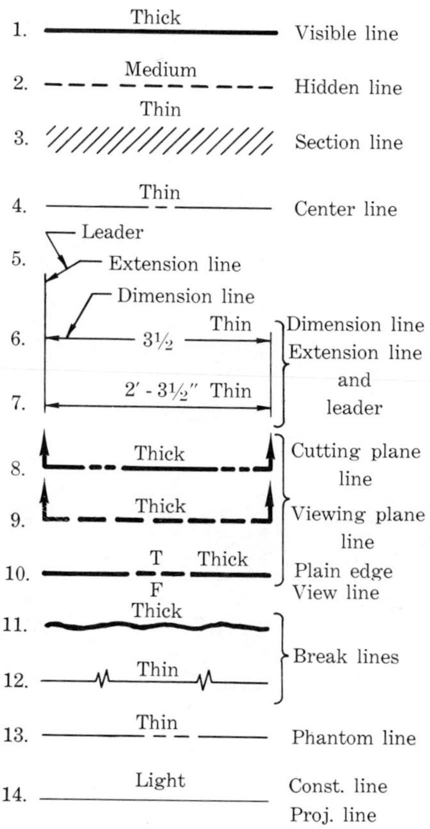

Hidden Line Junctures:

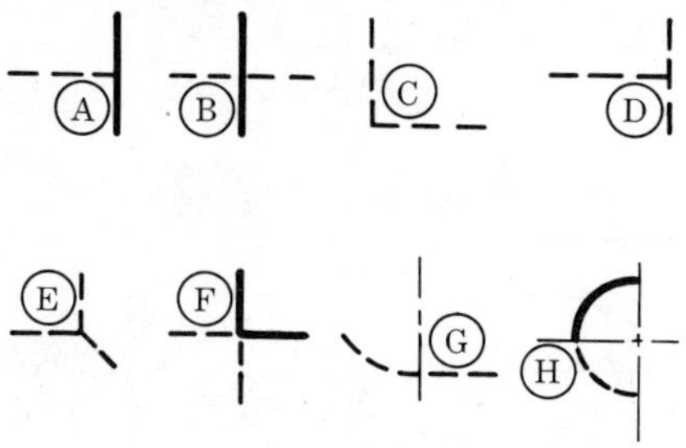

2. Free-hand Sketching

Straight-line: *A* to *B*

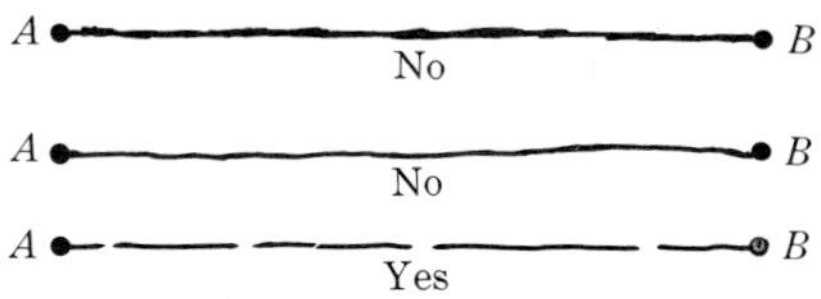

(i.e., stop and restart, by sections)

Circle:

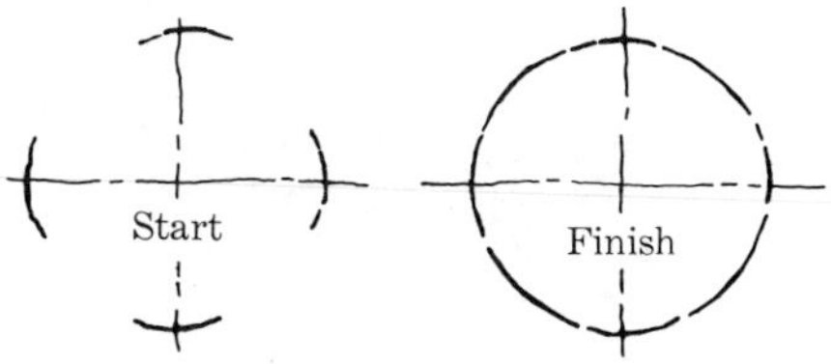

Block Outline:

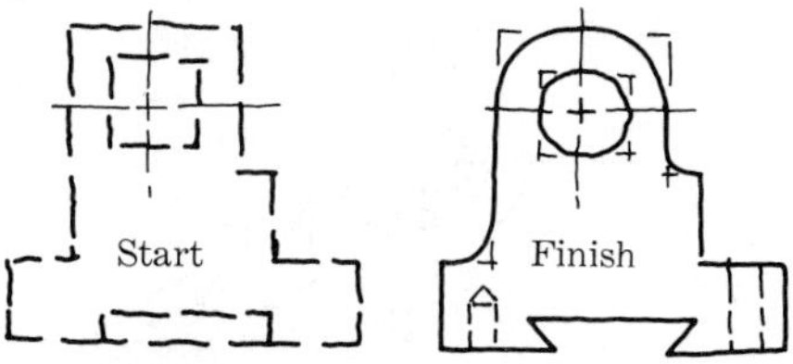

Oblique: The axis of the ellipse is along the diagonal of the rhombus:

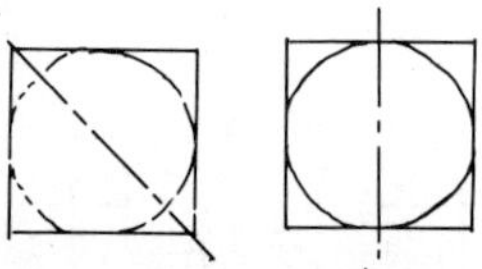

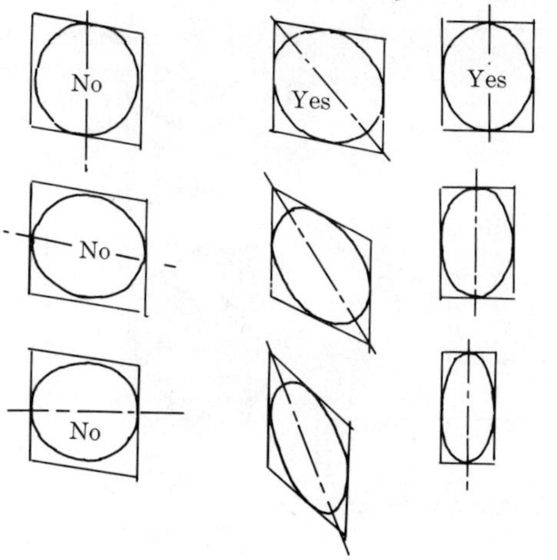

3. Isometric Sketching

Isometric layout is probably the most rapid and effective medium of communicating design ideas to one's self as well as to others.

Layout:

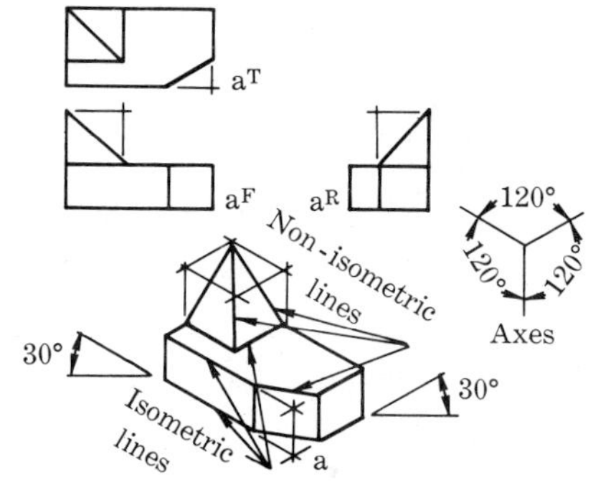

Sections:

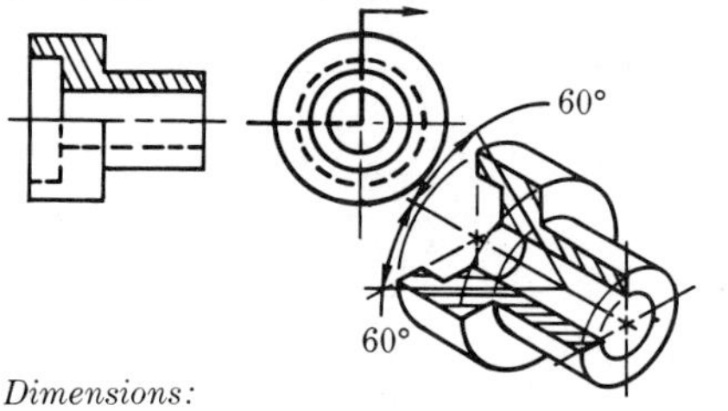

Dimensions:

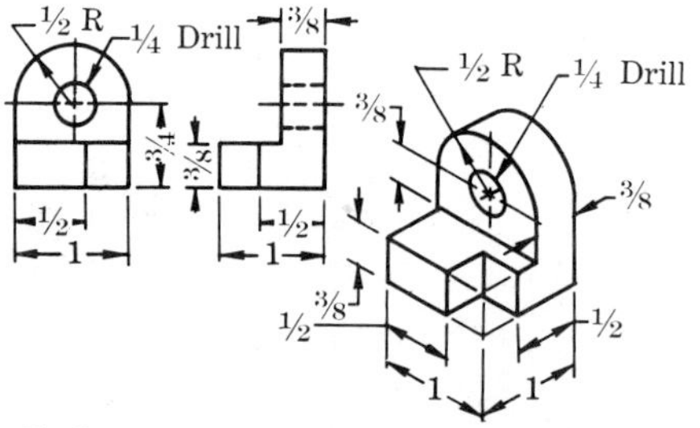

Circles:

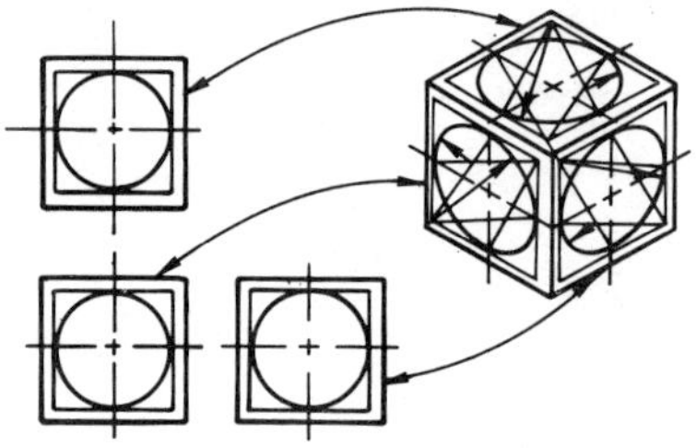

4. Oblique Sketching

Oblique layout offers the advantage that irregular contours may be shown true size when placed parallel to the plane of projection.

Layout:

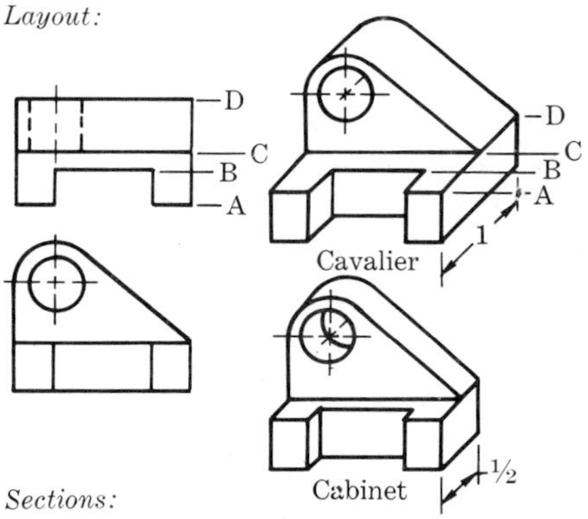

Sections:

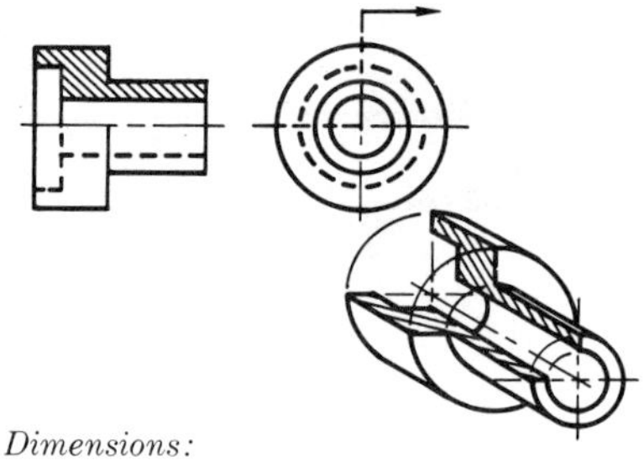

Dimensions:

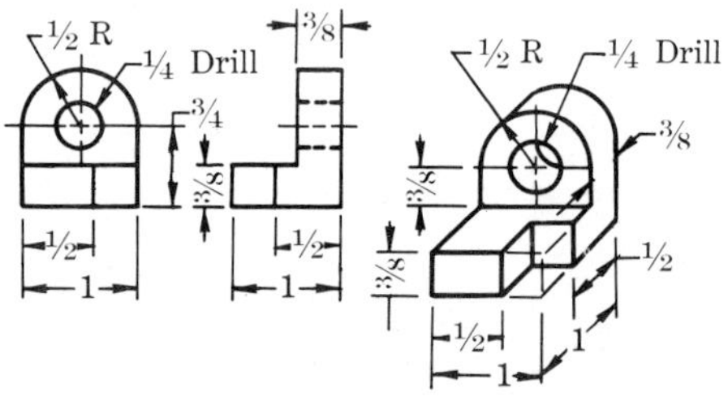

Offset Measurement:

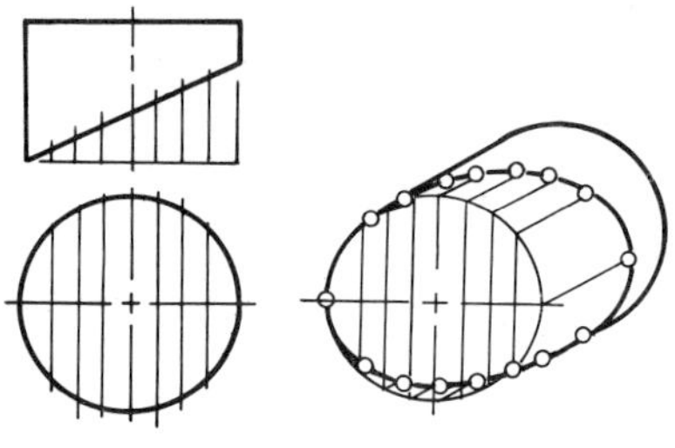

5. Alphabet of a Line: Position, Bearing, Projection

THEORY

Position of a Line: The direction a line passes through space, read alphabetically.

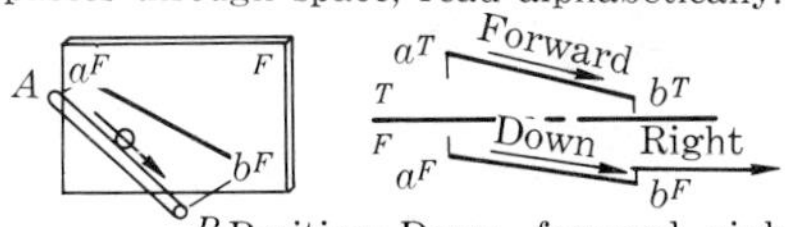

Position: Down, forward, right

Bearing of a Line: The horizontal angle (less than 90°) between the line and the N-S direction.

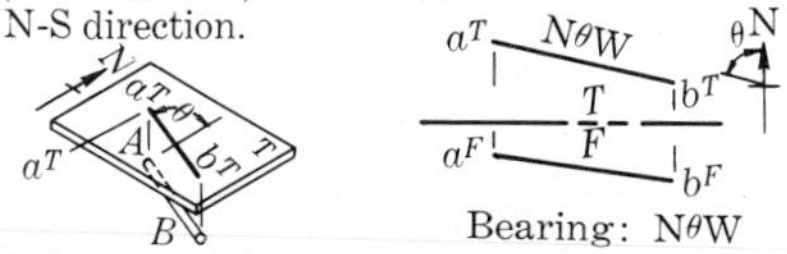

Bearing: NθW

Point on a Line: When a point is on a line, the propections of that point must lie on all respective projections of that line.

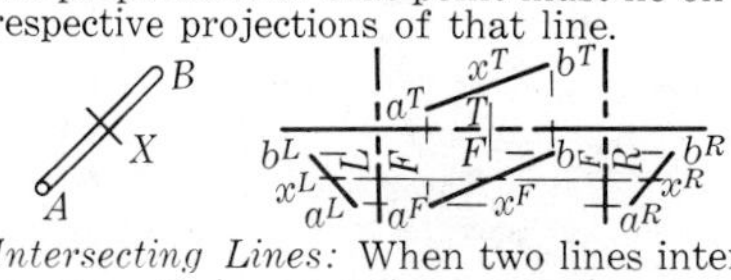

Intersecting Lines: When two lines intersect in space, their respective projections must intersect at a common point.

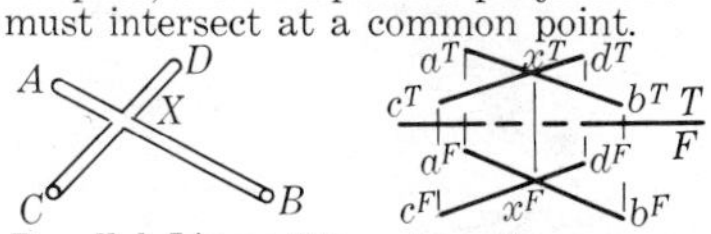

Parallel Lines: When two lines are parallel in space, their respective projections must be parallel or coincide.

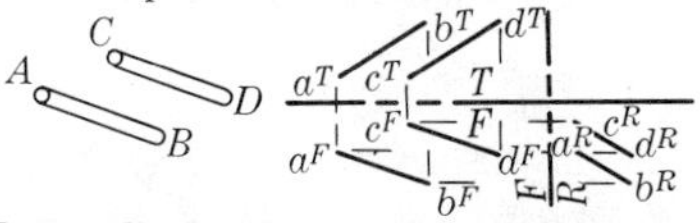

Perpendicular Lines: When two lines are perpendicular in space, a principal projection is perpendicular when one of the lines is parallel to a principal plane.

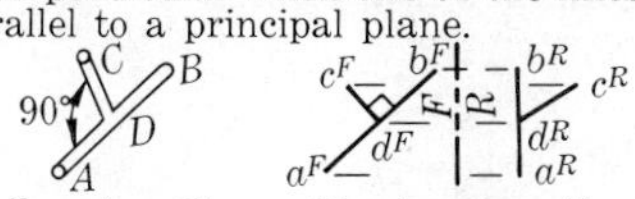

Strike of a Plane: The bearing (horizontal angle less than 90°) of a horizontal line in the plane.

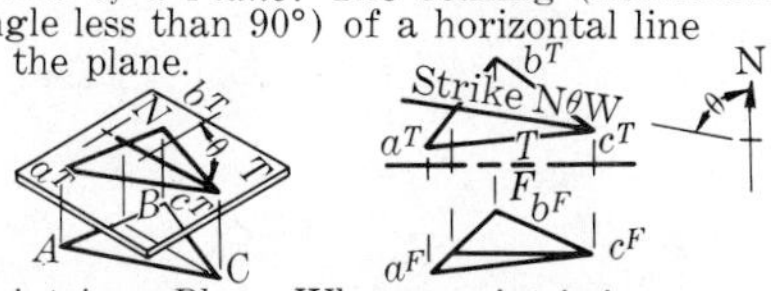

Point in a Plane: When a point is in a plane, the point must be on a line in the plane.

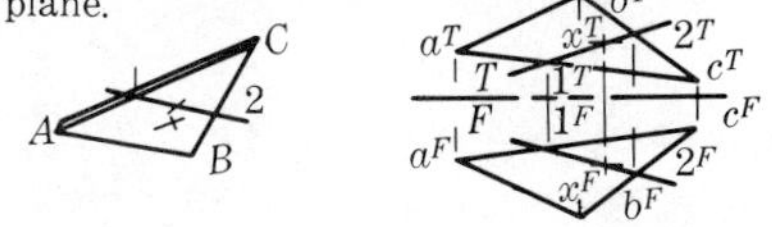

6. True Length by Components and by Projection. Point View of a Line

True Length by X,Y,Z Components:

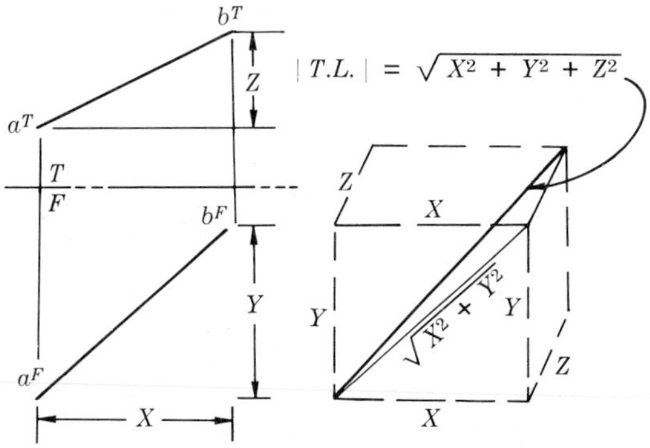

Rule: The true length of a line is the length of a diagonal of a rectangular box whose sides (X, Y, Z) are the components of the line.

$T.\,L. = \sqrt{X^2 + Y^2}$ *Only When* $Z = 0$:

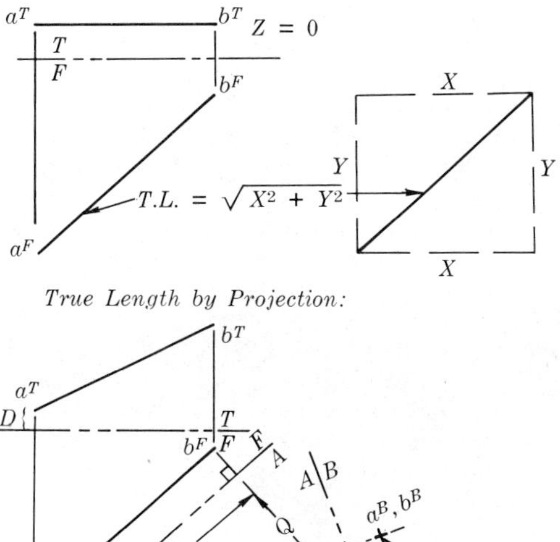

True Length by Projection:

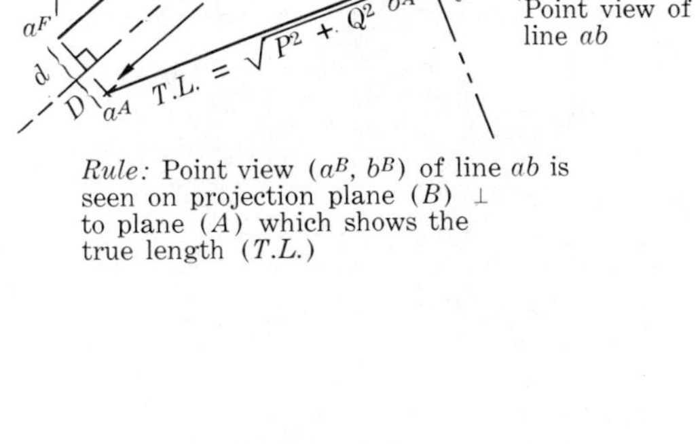

Rule: Point view (a^B, b^B) of line ab is seen on projection plane (B) ⊥ to plane (A) which shows the true length ($T.L.$)

7. True Length and True Size by Rotation

THEORY

True Length of Oblique Lines and Angles with Principal Planes: Line is rotated parallel to a principal plane.

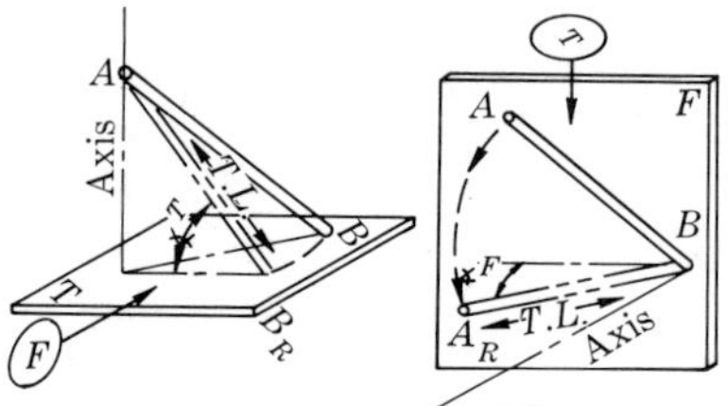

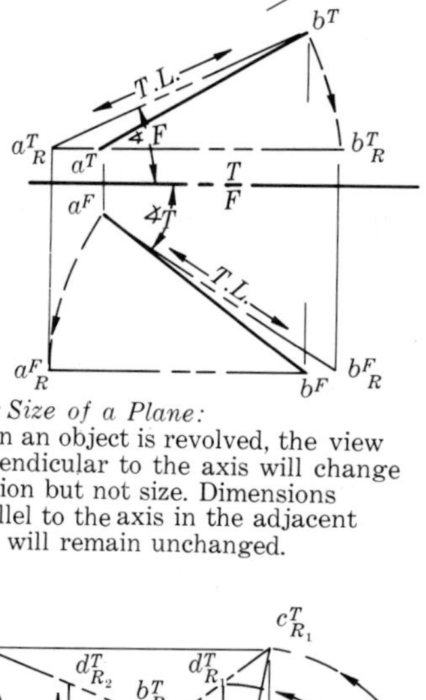

True Size of a Plane:
When an object is revolved, the view perpendicular to the axis will change position but not size. Dimensions parallel to the axis in the adjacent view will remain unchanged.

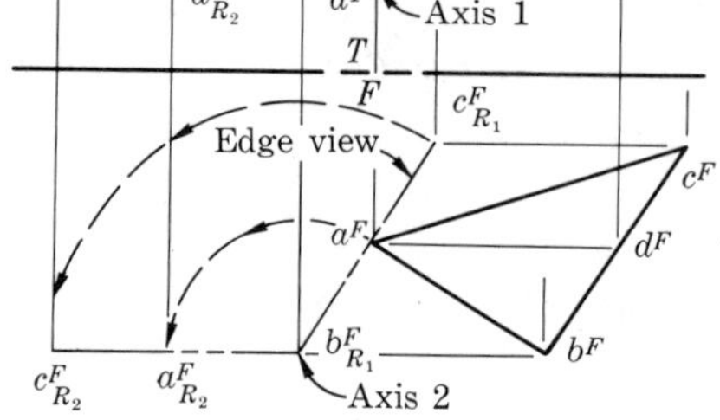

8. Auxiliary Plane; Shortest Distance from a Point

THEORY

Primary Auxiliary Plane:
A plane of projection perpendicular to a principal plane and oblique to the others.

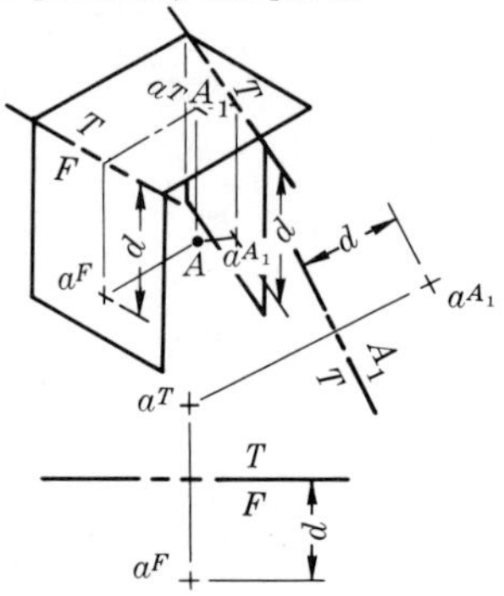

True Length of Oblique Line and Angles with Principal Planes: The primary auxiliary plane is set parallel to the oblique line.

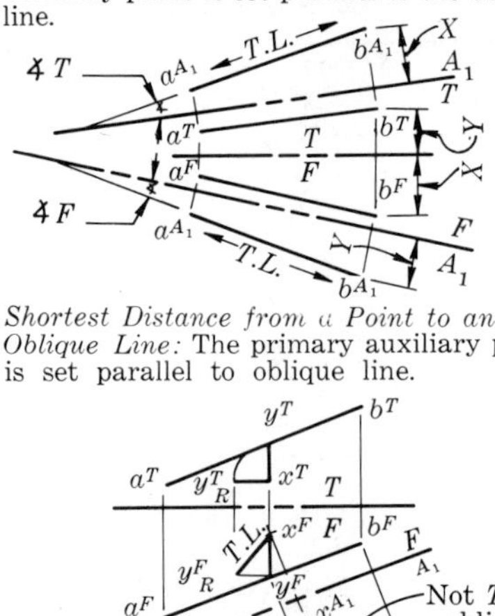

Shortest Distance from a Point to an Oblique Line: The primary auxiliary plane is set parallel to oblique line.

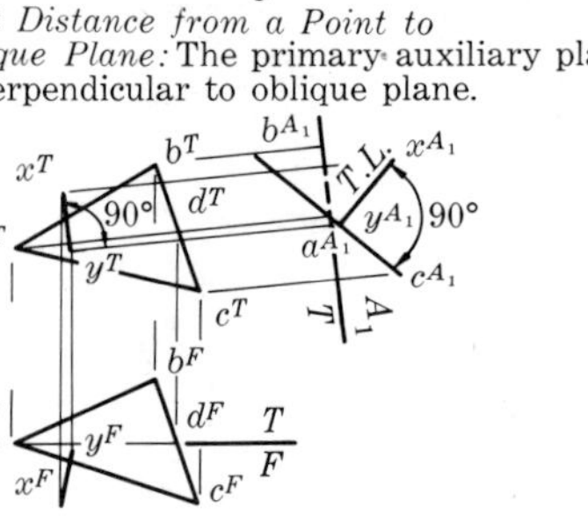

Shortest Distance from a Point to an Oblique Plane: The primary auxiliary plane is set perpendicular to oblique plane.

9. Second Auxiliary Plane

Secondary Auxiliary Plane:
Perpendicular to the primary auxiliary plane.

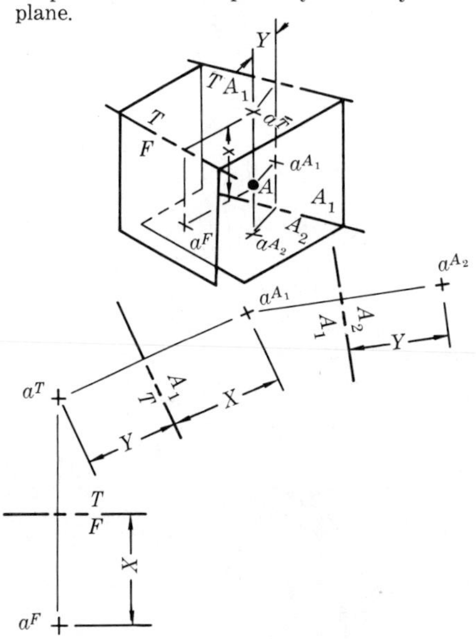

True Size of an Oblique Plane:
Primary auxiliary is set perpendicular to oblique plane to reveal edge view. Secondary auxiliary is set parallel to edge view revealing true size.

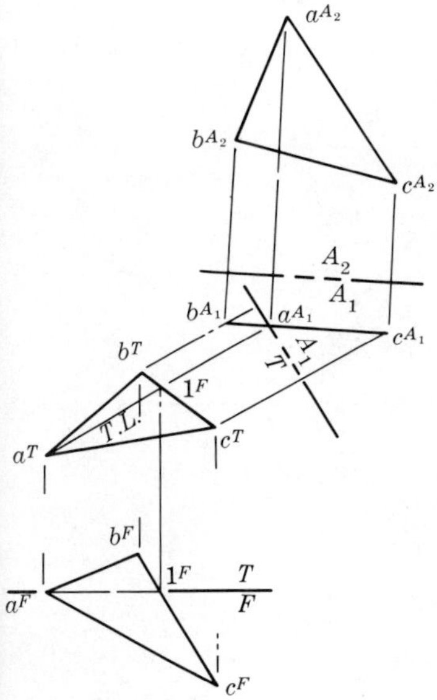

10. Visibility

Rule: Given two lines (A, B) with an apparent intersection in one view (T), that line closest to viewing plane (in adjacent view) is closest to the eye, and hence "visible."

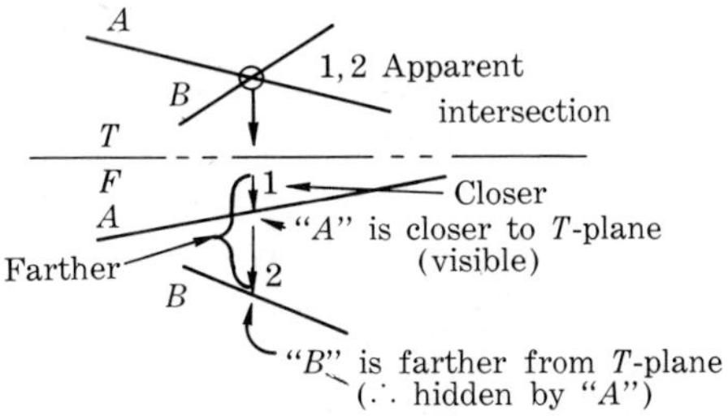

Line Piercing a Plane:

Procedure:
Project 1, 2 in "T" to "F." Line $a^F b^F$ is found closer to T/F than is $A^F B^F$.
Thus a^T lies above $A^T B^T$ and is visible all the way to O^T.

Similarly, $a^F b^F$ lies above $C^F B^F$

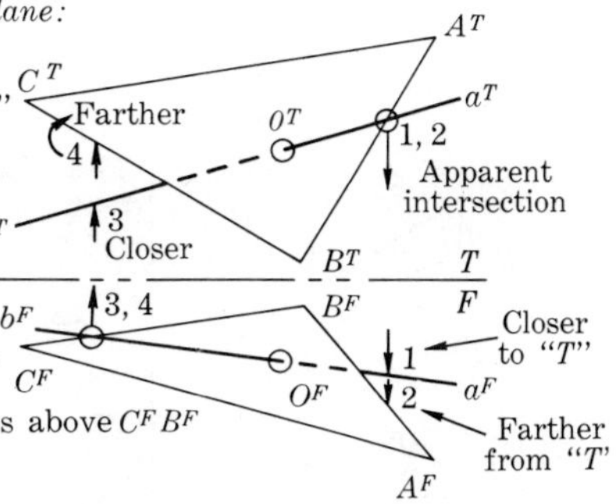

Visible Edges of Solids (Pyramids):

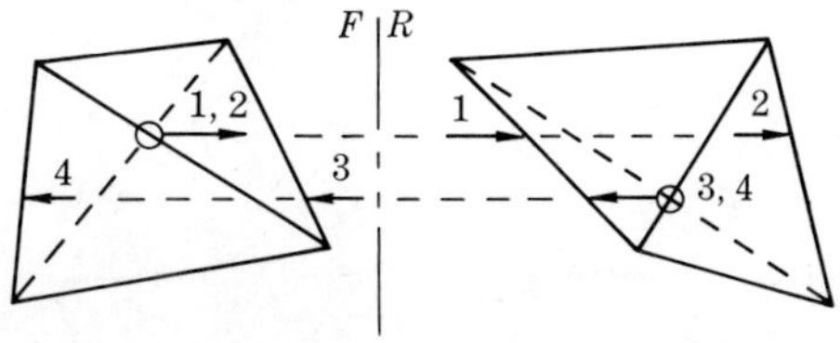

Visibility of Intersecting Planes:

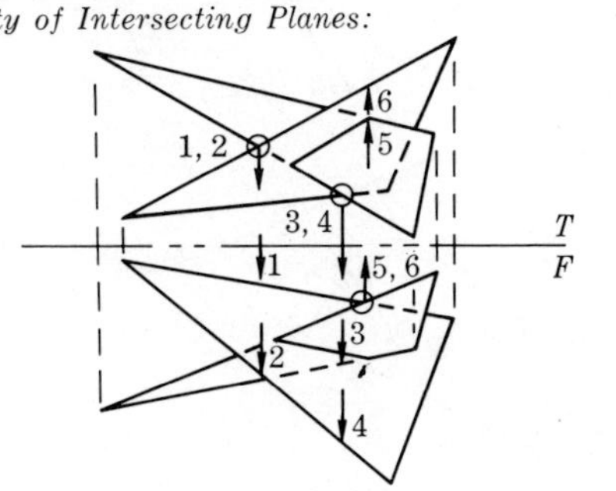

11. Distance between Skew Lines

Skew Lines: Lines that are nonparallel and nonintersecting.

Shortest Distance between Skew Lines: Primary auxiliary plane is set parallel to one line revealing its true length; secondary auxiliary plane is set perpendicular to the true length line.

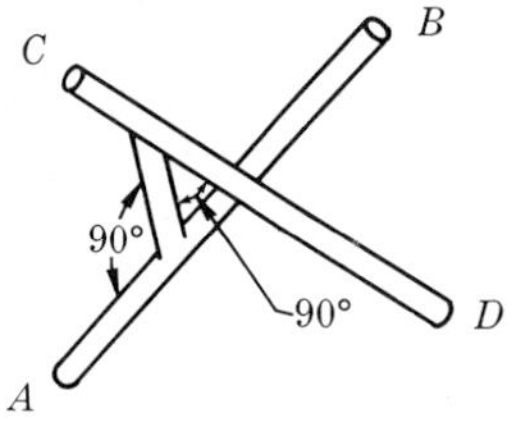

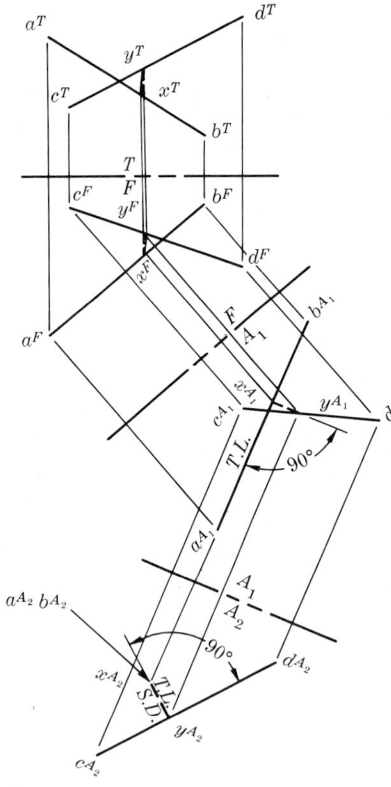

12. Angle between Lines and between Planes

Angle between Intersecting Lines: Since two intersecting lines define a plane, the true angle between the lines may be found through primary and/or secondary auxiliary projection.

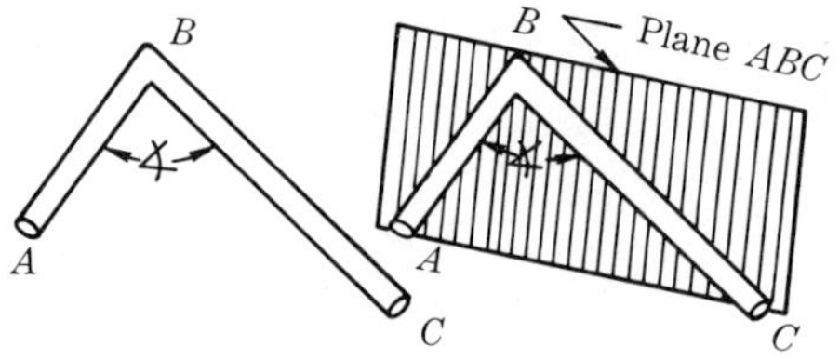

Dihedral Angle: The angle between two planes measured in a plane perpendicular to the line of intersection between the two planes.

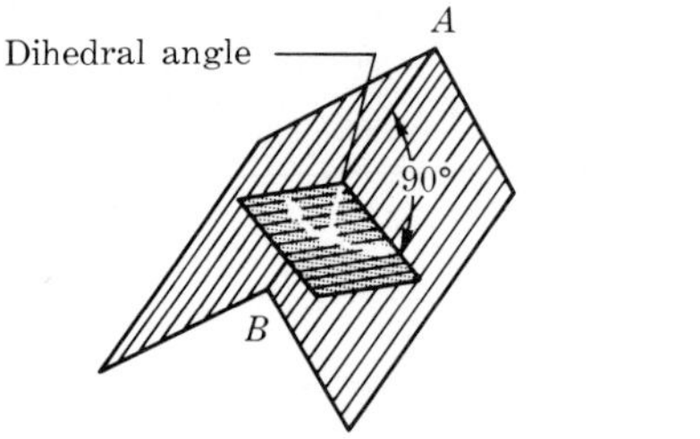

The true size of the dihedral angle may be found by viewing the line of intersection as a point.

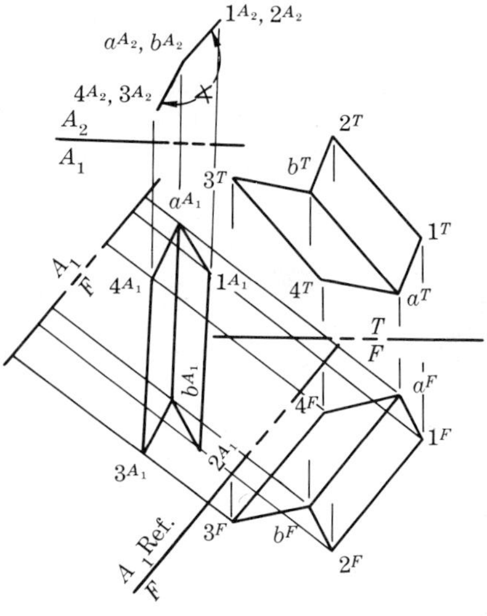

13. Angle between Line and Plane by Projection

The Angle between a Line and a Plane: The angle between the line and its projection on the plane.

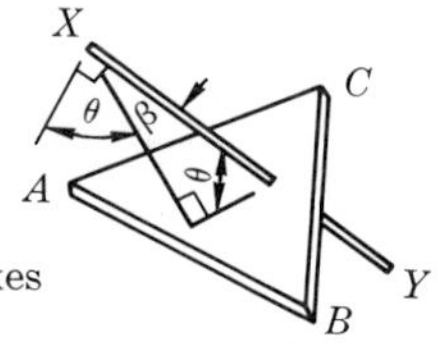

θ: Angle line makes with plane.

β: Complement of angle line makes with plane.

First Method: The true angle may be found in a projection in which the plane projects as an edge and the line in its true length.

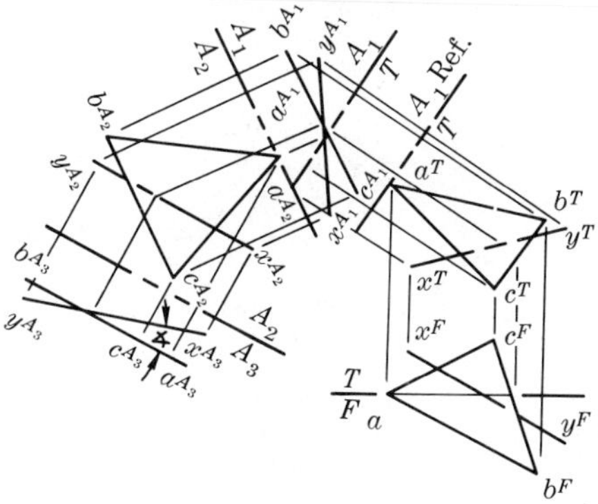

Second Method: The true angle may be found by constructing a perpendicular from the line to the plane.

The angle the perpendicular makes with the line is the complement of the angle the line makes with the plane.

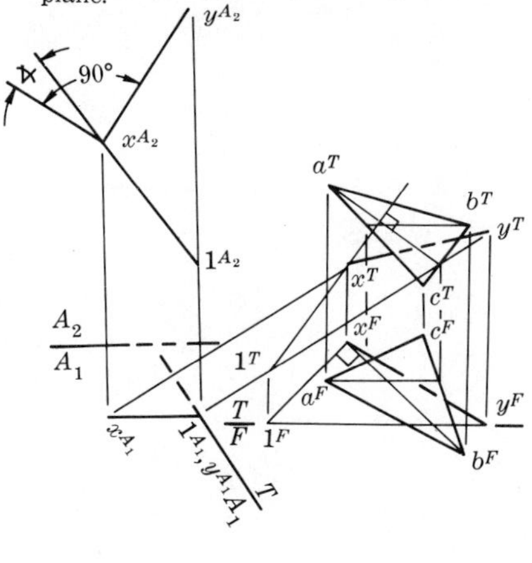

14. Angle between Line and Plane by Cutting Plane

Cutting Plane Method: A plane assumed through line to intersect given plane will reveal a line of intersection that will cross given line at required point.

Edge View Method: A principal view or auxiliary view will reveal edge view of plane and show required point where line intersects plane.

Line and Plane:

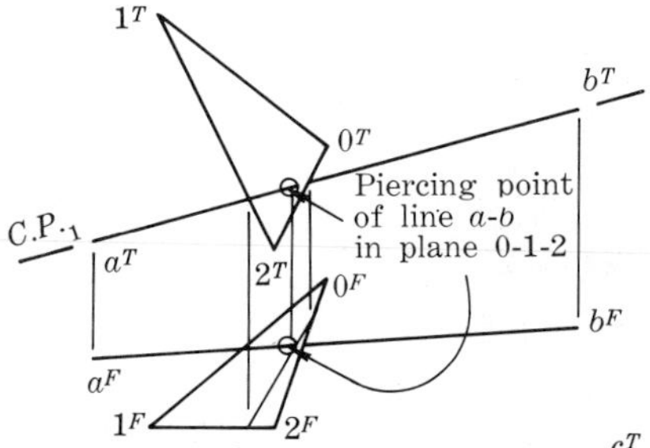

Plane and Plane:

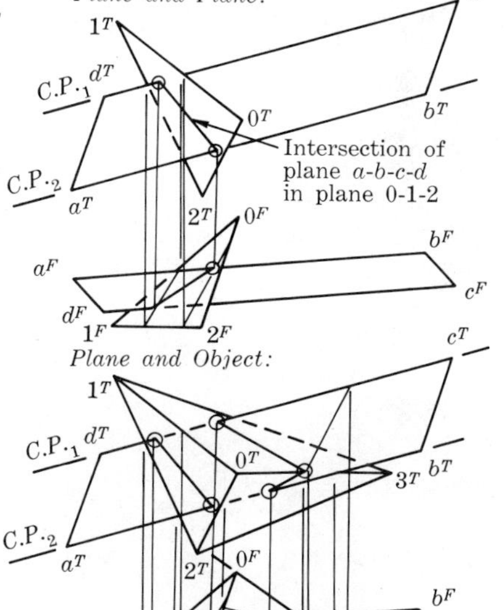

Plane and Object:

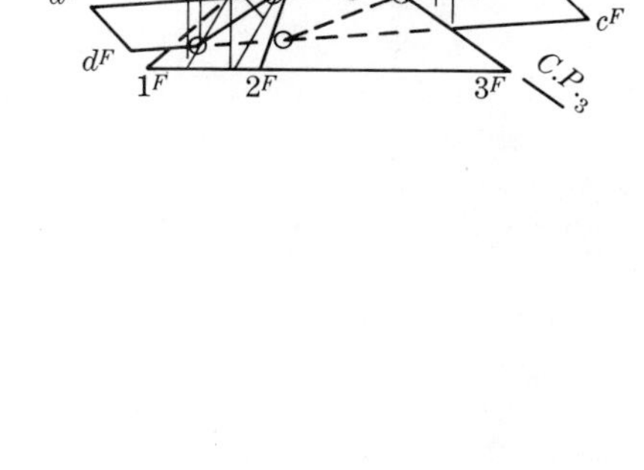

15. Intersections of Solids

Assume a series of planes that will intersect both surfaces along elements or easily determined lines.

The point where respective elements, each in the same plane, cross will be a point on the required line of intersection.

Prism and Cylinder:

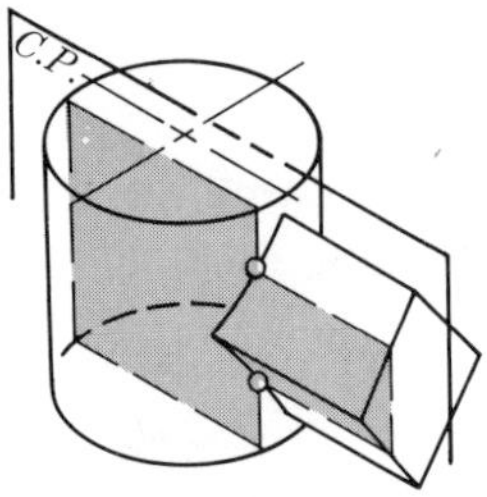

Two Cylinders:

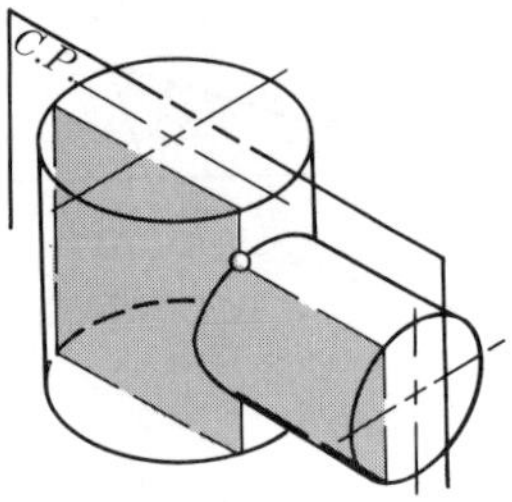

Cylinder and Cone:

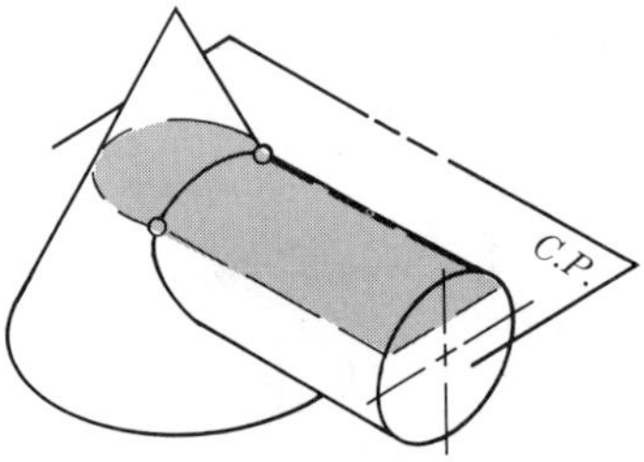

Two Cones:

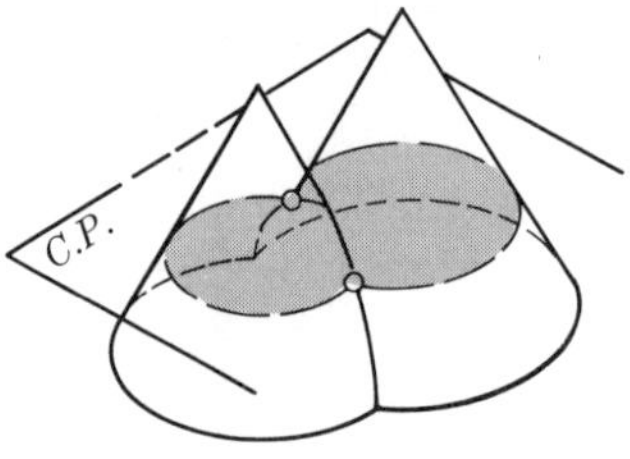

16. Resultant of Force Vectors

DEFINITIONS

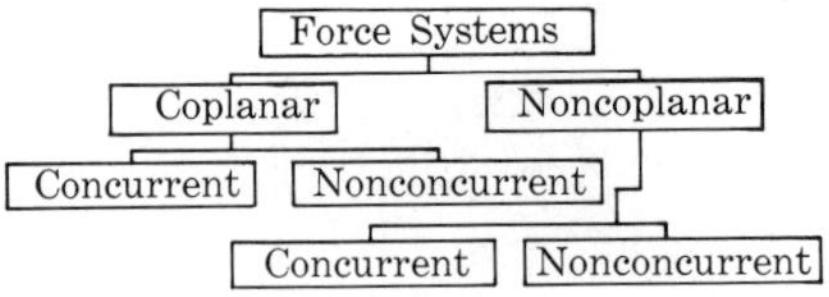

Scalar Quantity: A quantity possessing magnitude only. (Time, temperature, distance, etc.)

Vector Quantity: A quantity possessing magnitude, direction, and sense. (Force, displacement, velocity, etc.)

Force: The quantity that changes or tends to change the state of motion of a body.

Example: Force of 100 lb, acting at 30° with horizontal, and up to right.

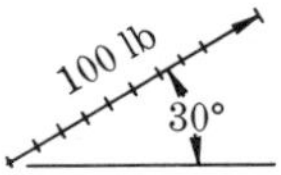

Magnitude = 100 lb
Direction = 30°-H
Sense = Up to right

Resultant: A single force that will have the same effect on a body as the combined effect of a system of forces.

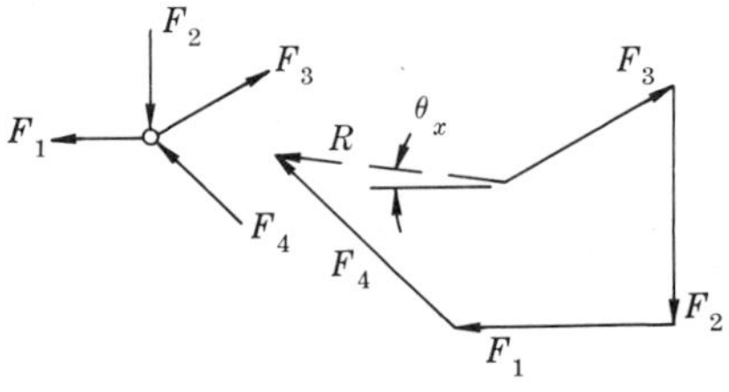

Equilibrium: When the system of forces acting on a body balance one another producing a resultant of zero.

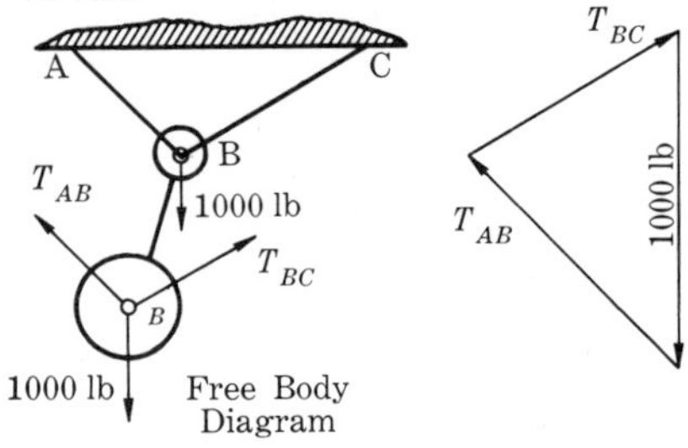

17. Maxwell Force Diagram

THEORY

Maxwell Diagram: A force diagram applied to the solution of stress in members of a truss or bent. Each joint of two unknowns is solved through its application to a composite force diagram.

Bow's Notation: A form of notation for the purpose of designating forces. The space between forces is designated by upper case letters. Any force is designated by the letters of the spaces it separates. Letters are read in a clockwise direction around any free body joint.

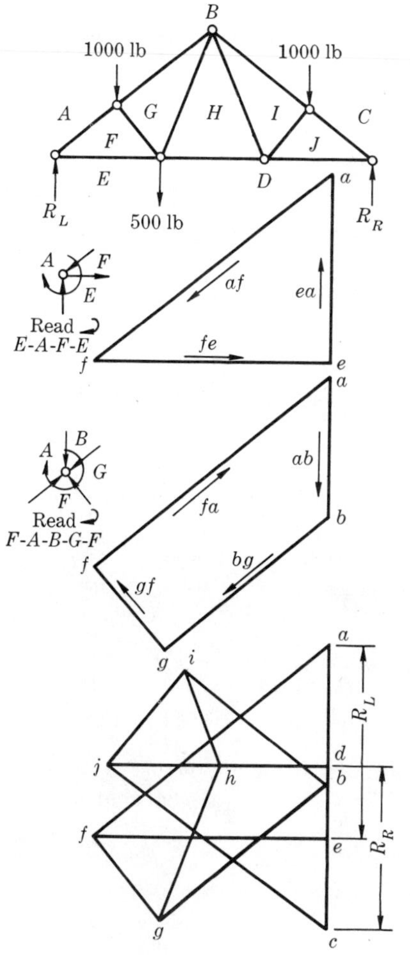

Maxwell Diagram

18. Noncoplanar Force Vectors

THEORY

Resultant: Since a noncoplanar force is positioned in three dimensional space, it must be defined by the three space angles related to the X, Y, and Z axes in addition to its magnitude.

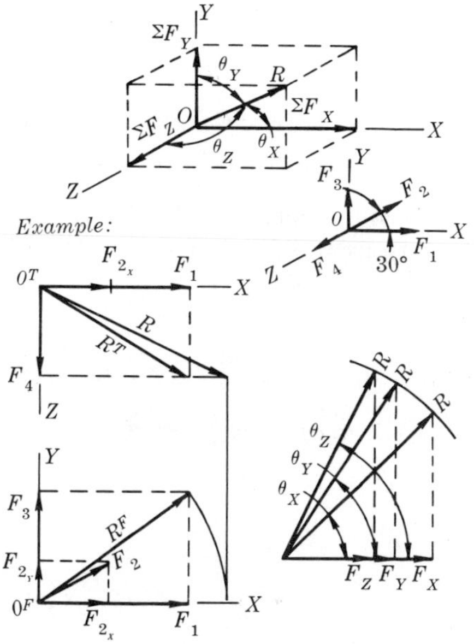

Equilibrium: To determine forces in a concurrent system, project the noncoplanar forces onto one or more planes and solve as related coplanar systems.

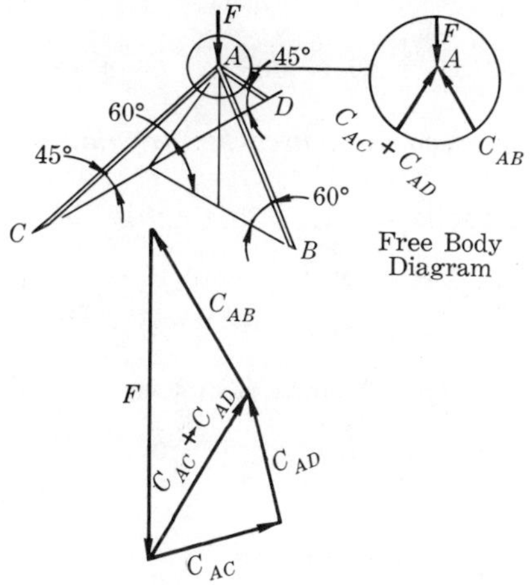

19. Vectors, Components, and Vector Multiplication

GIVEN:

Vectors:

$$\bar{A} = a_1 i + a_2 j + a_3 k$$
$$\bar{B} = b_1 i + b_2 j + b_3 k$$
$$|\bar{A}| = \sqrt{a_1^2 + a_2^2 + a_3^2}$$
$$|\bar{B}| = \sqrt{b_1^2 + b_2^2 + b_3^2}$$

where a_1, a_2, a_3 are coefficients of the three unit vectors i, j, k, which are parallel to the x, y, z axis directions respectively.

θ = Angle between $\bar{A}$ and $\bar{B}$, positive when rotated clockwise, right-hand rule (Figure E.19)

DEFINITIONS:

Vector Addition (used for forces):

$$\bar{A} + \bar{B} = (a_1 + b_1)i + (a_2 + b_2)j + (a_3 + b_3)k \qquad \text{(also a vector)}$$

$$\bar{A} + \bar{B} = \bar{B} + \bar{A}$$

Vector Subtraction:

$$\bar{A} - \bar{B} = (a_1 - b_1)i + (a_2 - b_2)j + (a_3 - b_3)k \qquad \text{(also a vector)}$$

$$\bar{A} - \bar{B} = -(\bar{B} - \bar{A})$$

"Dot Product" (used for computing "work" or angles between vectors):

$$\bar{A} \cdot \bar{B} = a_1 b_1 + a_2 b_2 + a_3 b_3 \qquad \text{(not a vector, only a scalar)}$$

$$\bar{A} \cdot \bar{B} = |\bar{A}||\bar{B}| \cos \theta; \quad \cos \theta = \bar{A} \cdot \bar{B}/|\bar{A}||\bar{B}|$$

"Cross Product" (used for angular velocity, moment of a force):

$$\bar{A} \times \bar{B} = (a_1 i + a_2 j + a_3 k) \times (b_1 i + b_2 j + b_3 k) \qquad \text{(see footnote)}$$

$$\bar{A} \times \bar{B} = (a_2 b_3 - a_3 b_2)i + (a_3 b_1 - a_1 b_3)j + (a_1 b_2 - a_2 b_1)k$$

(also a vector)

Note: The cross product of unit vectors is defined:

$$i \times i = j \times j = k \times k = 0$$
$$i \times j = k = -j \times i$$
$$j \times k = i = -k \times j$$
$$k \times i = j = -i \times k$$

In matrix form

$$\bar{A} \times \bar{B} = \begin{vmatrix} i & j & k \\ a_1 & a_2 & a_3 \\ b_1 & b_2 & b_3 \end{vmatrix} \quad \text{which yields the equation above.}$$

$$|\bar{A} \times \bar{B}| = |\bar{A}||\bar{B}| \sin \theta$$

The direction of this vector is perpendicular to the plane of $\bar{A}$, $\bar{B}$ (positive in the direction of right-hand screw rotating A into B).

GRAPHICALLY:

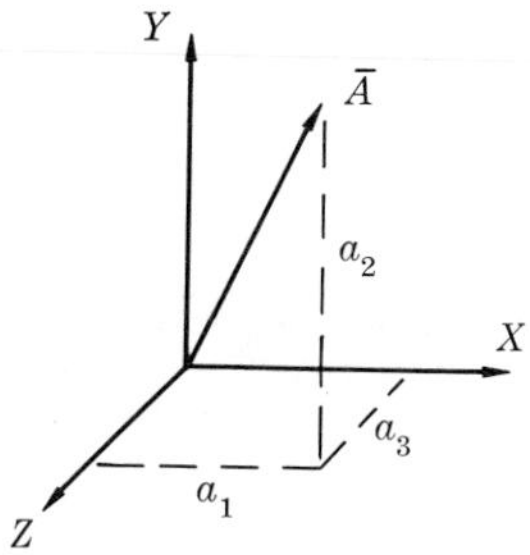

3–Dimensional

$\bar{A} = a_1 i + a_2 j + a_3 k$

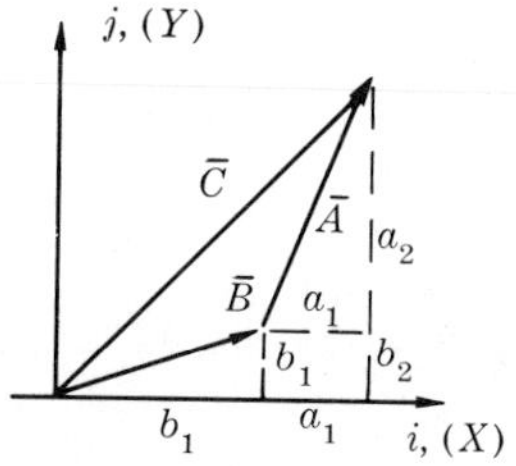

2–Dimensional

$\bar{C} = (a_1 + b_1)i + (a_2 + b_2)j$

Dot Product

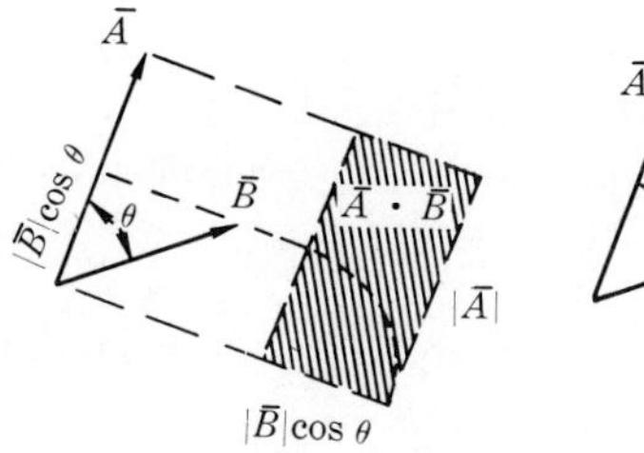

$\bar{A} \cdot \bar{B} = |A||B| \cos \theta$

$\bar{A} \cdot \bar{B} = a_1 b_1 + a_2 b_2 + a_3 b_3$

Not a Vector

Cross Product

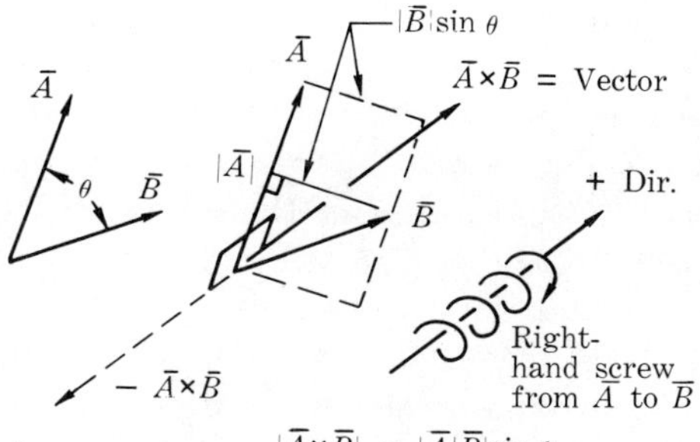

$|\bar{A} \times \bar{B}| = |\bar{A}||\bar{B}| \sin \theta$

$\bar{A} \times \bar{B} =$ Vector $\perp$ plane $\bar{A}, \bar{B}$

Use and Care of the Engineering Notebook*

Appendix F

General Suggestions

Each engineer's laboratory notebook must be the master repository for all his technical planning (including plans not carried out), all his analytical work, all records of experimental work, references to all sources of information used, and all other significant thinking about his project and directly related technical subjects, regardless of the apparent importance of the acquired information. Thus the laboratory notebook becomes the basic record of the engineer's personal understanding.

As design and development progresses, all changes of specifications must be carefully recorded in the lab notebook.

Analyze and plan on paper before doing the work planned.

Record and correlate all results with the plan or specification as they are obtained.

On completion of work required by one step of the plan, evaluate and analyze the results in writing in the lab book. Show how they correspond to the plan and how they differ from it. Describe the next step of the work plan.

* Most of the material for this section is from W. B. Heinz and orginated at the Ampex Corporation, San Jose, California.

When partial results disclose fallacies in the plan, or means to improve it, revise the plan in writing in the book before proceeding.

Start every circuit and mechanism development with the simplest possible diagram in your lab book. Amplify it as work progresses.

Always give quantitative information when possible. Accept no inconsistencies, investigate causes. Explore first the effects of large changes.

Essential Rules for Keeping Lab Notebooks

At the front of your notebook put a table of contents consisting of chronological entries together with page references.

Enter currently (not later) into the lab book all favorable and unfavorable results including things not fully understood at the time. All data must be originals in their primary form, not after recalculation or transformation. They must not be taken on scrap paper for later transcription "to make them neat."

All references to apparatus, places, time, literature, graphs, and people should be sufficiently explained to be understandable years later. It should be possible to take a drawing or a technical report and show just where every figure, description, or statement in it is backed up by original entries in the lab notebook, and exactly why the final and original numbers differ, if they do.

Lab book entries must be coherent, lucid, legible, non-ambiguous, and complete within themselves. Of second-order importance are straightedge-drawn lines, beautiful printing, or figures detailed beyond need for complete understanding. Don't gild lilies!

Enter specific, complete references to books, periodicals, reports, catalogs, patents, drawings, and all other sources of information used from time to time during the work. Make these entries currently, not later.

Rough qualitative graphs can be drawn directly but more careful ones should be prepared on graph paper for entry in the notebook.

The material should not be crowded on the pages. Paper is inexpensive compared with other development expenses.

When patent questions are involved, the entries in the notebook should be recorded in ink or in indelible pencil. They must be witnessed by someone understanding the material other than a co-inventor. Both the basic material and any alterations must be signed, witnessed, and dated.

A statement of purpose of the work and a summary of conclusions are essential to best use of the information. The author should be able to draw conclusions better than anyone else. When he recognizes more work which should be done later, he should describe it in some detail for future reference by himself or others. His thinking should be recorded then and there.

REQUIRED IN LAB BOOKS

1. Analytical work and calculations.
2. Experimental data and analyses thereof.
3. Graphical presentations, selected, arranged, titled, interpreted. *Note*: Coordinate axes must be properly designated and contain scales.
4. Pen recorder charts, selected, arranged, titled, interpreted.
5. References to information sources.
6. References to drawings.
7. Oscilloscope pictures, selected, arranged, titled, interpreted.
8. Technical planning.

FREQUENTLY SUITABLE IN LAB BOOKS

9. Copies of complete technical articles, catalogs, etc., when directly applicable.
10. Copies of tables, graphs, descriptions and other forms of information from other sources, directly related to the work logged, and useful in performing and understanding that work.
11. Original technical correspondence, internal and external, relating solely to the project. If of more general interest the original should be elsewhere (usually a Document Center) and a copy should be in the lab book.

Appendix G

Flow Charting for Digital Computer Solutions

To solve $Ax^2 + Bx + C = 0$ in "Machine Language."

When an algebraic expression, such as the equation above, is numerically evaluated in a computer, the actual detail steps are performed by internal circuits which obey the limited variety of execution codes inherently designed into the machine.* Machines vary, but they all have internal operations corresponding to the following typical commands:

Add	Read (load numbers into memory)
Subtract	Check for negative (is the number negative?)
Multiply	End of data? (check for more input data)
Divide	Print (type out the number)
Store	Punch (punch the number in a new IBM card)
Take square root	Transfer (move number from one memory address to another)
	Stop

* If the operation is not specifically available in "hardware," an appropriate subroutine is called into use by an equivalent operation code.

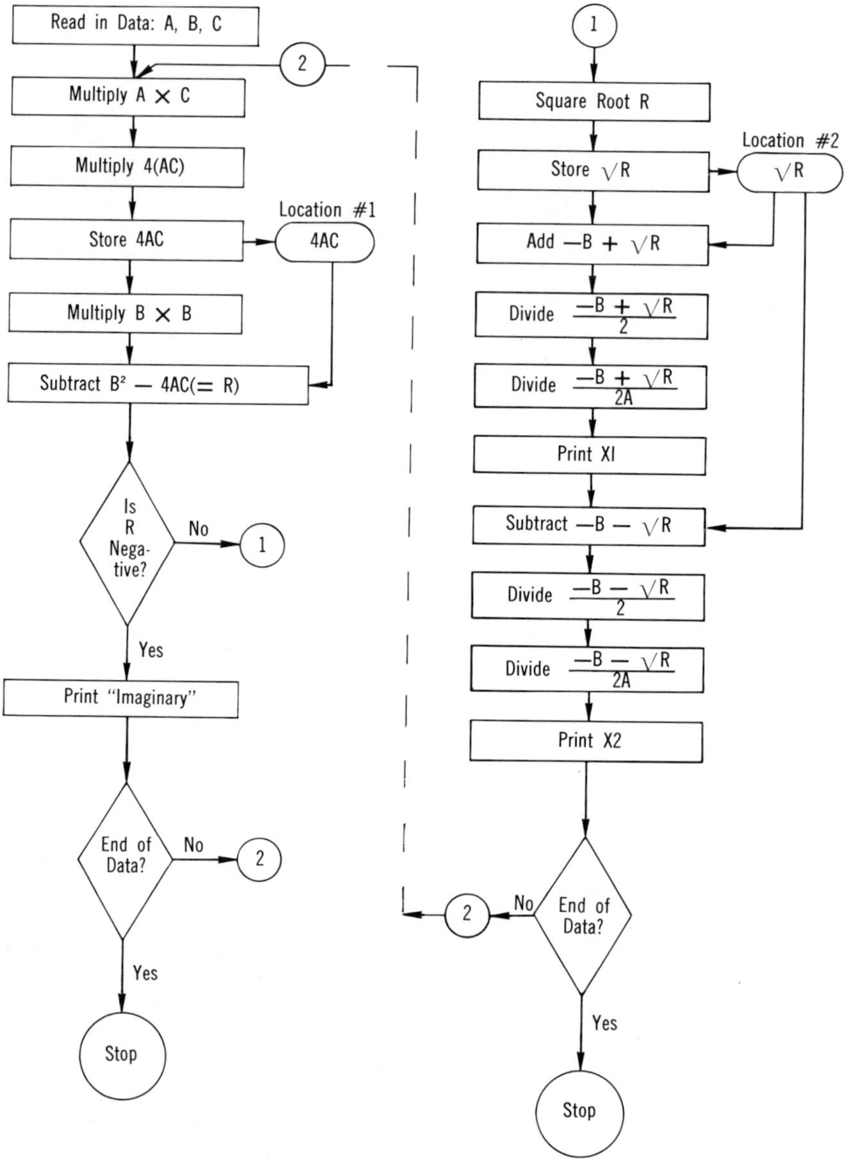

Fig. G.1 Flow chart for computing real roots of quadratic equation $Ax^2 + Bx + C = 0$ for several values of A, B, C.

Thus all computations must be broken down into these above specific typical commands.

A flow chart is one of the best ways of visualizing these steps. Such a chart is shown in Figure G.1, for computing the real roots $X1$, $X2$ of the equation above, when successive different sets of values are supplied for A, B, C.

This chart illustrates several important points:

1. The programmer must anticipate all the possible outcomes (i.e., the imaginary root, if the radical is negative).
2. Looping. The group of operations after the first one and before the last one are repeated for each set of A, B, C before the computer will stop.
3. Specially designated storage locations (#1, #2) must be arranged to hold data temporarily.
4. Program may be graphically broken into pieces, indicated by (1)-(1) and (2)-(2).

The program of Figure G.1 is actually prepared for the computer by 1) substituting appropriate operation-code numbers (e.g., add = 21; multiply = 23, for IBM 1620), 2) selecting the addresses to be used in memory, and 3) punching the program and data in cards. Such a sequence of commands constitutes a *machine language* program.

"Interpretive Language"

Obviously, a programmer must devote much time and effort to writing a machine language program for a problem of any length or sophistication.

To avoid these difficulties and delays, the computer manufacturers and programming specialists have prepared "interpretive" or "translator" languages which allow the engineer to write his program in more or less direct mathematical form called a "source program." The computer is then given instructions by which the "source program" is converted internally into "machine language," yielding an "object" deck of cards. The problem is finally solved by running the latter "object deck" through the computer, for the answers.

The simplicity is illustrated by repeating our above problem in FORTRAN language, first as a FORTRAN flow chart and then in the specific "source" program form. The "object" program takes a full page of machine language commands.

Very briefly, in the FORTRAN program to the right of the Figure G.2 above, the second line computes the radical $R = B^2 - 4AC$, and lines five and six compute the two real roots. The third line sends the computer to

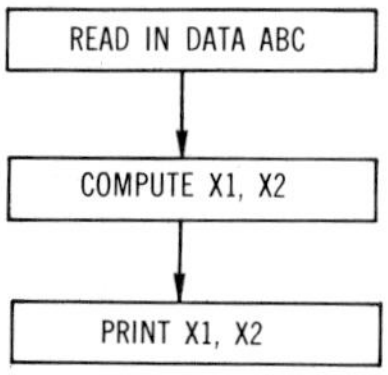


```
2     READ, A, B, C
      R = B**2-4. *A*C
      IF (R) 11, 10, 10
10    SQ = SQRTF (R)
      X1 = (-B-SQ) / (2. *A)
      X2 = (-B+SQ) / (2. *A)
      PRINT 1, A, B, C, X1, X2
1     FORMAT (5F10.6)
      GO TO 2
11    PRINT 3
3     FORMAT (9H IMAGINARY)
      GO TO 2
```

Fig. G.2 Flow chart and FORTRAN program for $Ax^2 + Bx + C = 0$.

lines marked 11 and 3 to print "imaginary" if R is negative, and to line marked 10 if R is zero or positive.

A few of the FORTRAN symbols are repeated below to assist in understanding some of the programs illustrated in the book.

SYMBOLS

+ means add

− means subtract

/ means divide

* means multiply

** means raise to the power of

( ) means quantity of

COMMANDS

READ means put data in memory

PRINT means type out the number

PUNCH means punch the number in IBM card

FORMAT defines columns and spaces and words in answer

DO defines a loop with initial, increment and limit

IF transfers control if negative, zero, or positive

GO TO transfers control unconditionally

Appendix H

Large System Modeling: Supersonic Aircraft Transport (SST) for 1970-80

In 1964, the President's Advisory Committee on the Supersonic Transport recommended (and Congress agreed) that the Federal Aviation Agency authorize two airframe companies to examine the effect of several variables upon the proposed aircraft's purchase price, its direct operating cost, and the controversial sonic boom. The basic question was, "What should be the speed, size, and flying range of the new commercial aircraft which the United States will build?" At this writing, the first such manufacture is expected in 1972.

Lockheed, as one of the two companies under contract, engineered 24 realistic designs, and, interspersing subsonic and supersonic wind-tunnel tests, exercised these proposed designs on its sophisticated Airline Simulation System. By means of this computerized system model, the proposed new SST was pitted against present and proposed competitive aircraft under the expected conditions of the 1970–80 world transport market. The competitive aircraft were the new British Concorde (1500 mph), the present models 707-DC8 (600 mph), and a proposed 700 passenger subsonic jet (600 mph).

The inputs to and outputs from the computer are shown in condensed form in Figure H.1. The inputs comprise the markets, the planes, the routes, the passenger preferences, the expected competition, and desired return on investment. Following a step-by-step analysis, the computer produced the outputs of route decisions, aircraft assignment, and loading and the resulting system profit, loss, and investment summary.

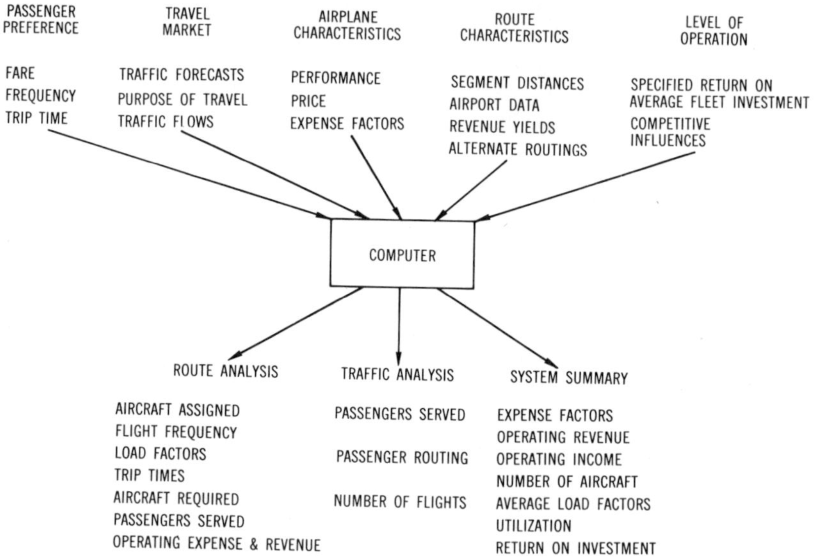

Fig. H.1 Airline simulation system. Computer inputs and outputs. (Courtesy Lockheed-California Company, Burbank, California)

Trade-offs between criteria, such as sonic boom, passenger route preference and flying range were made more or less continuously during the many stages of development and optimization; although most trade-offs or compromises were made by decisions between stages rather than by inclusion in the equation of the Criterion Function. For instance, it was found that 0.2 lb/sq ft added sonic boom (over the recommended 1.5 value) permitted 800 more miles of flying range. Decisions to accept or modify such exchanges of values were necessary in the midst of the project.

The major parameters, listed in Table H.1, were at different times treated as constraints, constants, or variables. The adjacent numbers in parentheses are the tentatively optimized values.

Table H.1 Major Parameters in the Optimization of the Supersonic Transport* for 1970–80 (Tentatively selected values in parentheses).

x_1	Gross take-off weight	(500,000 lb max.)
	Weight of fuel	(240,000 lb max.)
x_2	Number of seats (domestic)	(227)
x_3	Wing area	(8,400 sq ft)
	Tail area	(980 sq ft)
x_4	Wing sweep angle	(60°)
x_5	Wing aspect ratio	(1.9)
	Wing span	(116 ft)
	Fuselage length	(221 ft)
	Fuselage diameter	(11 ft)
x_6	Engine air flow (size-rating)	(700–800 lb/sec)
x_7	Cruise speed	(Mach 3 $\cong$ 2100 mph)
x_8	Climb speed	(Mach 2 $\cong$ 1400 mph)
	Climb sonic boom over-pressure	(2.0 psf)
	Cruise sonic boom over-pressure	(1.5 psf)
	Landing approach speed	(145 knots)
	Take-off runway length	(2 mi, max.)
x_9	Cruising range	(4000 mi)
	Total payload	(50,000 lb)
	Direct operating cost per passenger unit	(1¢/seat mi)

* Courtesy Lockheed-California Company, Burbank, California.

During the phase of design called "airplane characteristics" in Figure H.1, the criterion function was expressed in dollars, but these costs came from functions of the identified parameters in Table H.1. That is,

$$CF = \text{direct operating cost, to be minimized}$$

or,

$$CF = f(x_1, x_2, \ldots, x_9)$$

In summary, the currently recommended designs culminate more than six years of large-scale effort, building upon the previous art, upon extensive research, testing, and numerical modeling.

Two views of the aircraft as now conceived are shown in Figure H.2 on the following page.

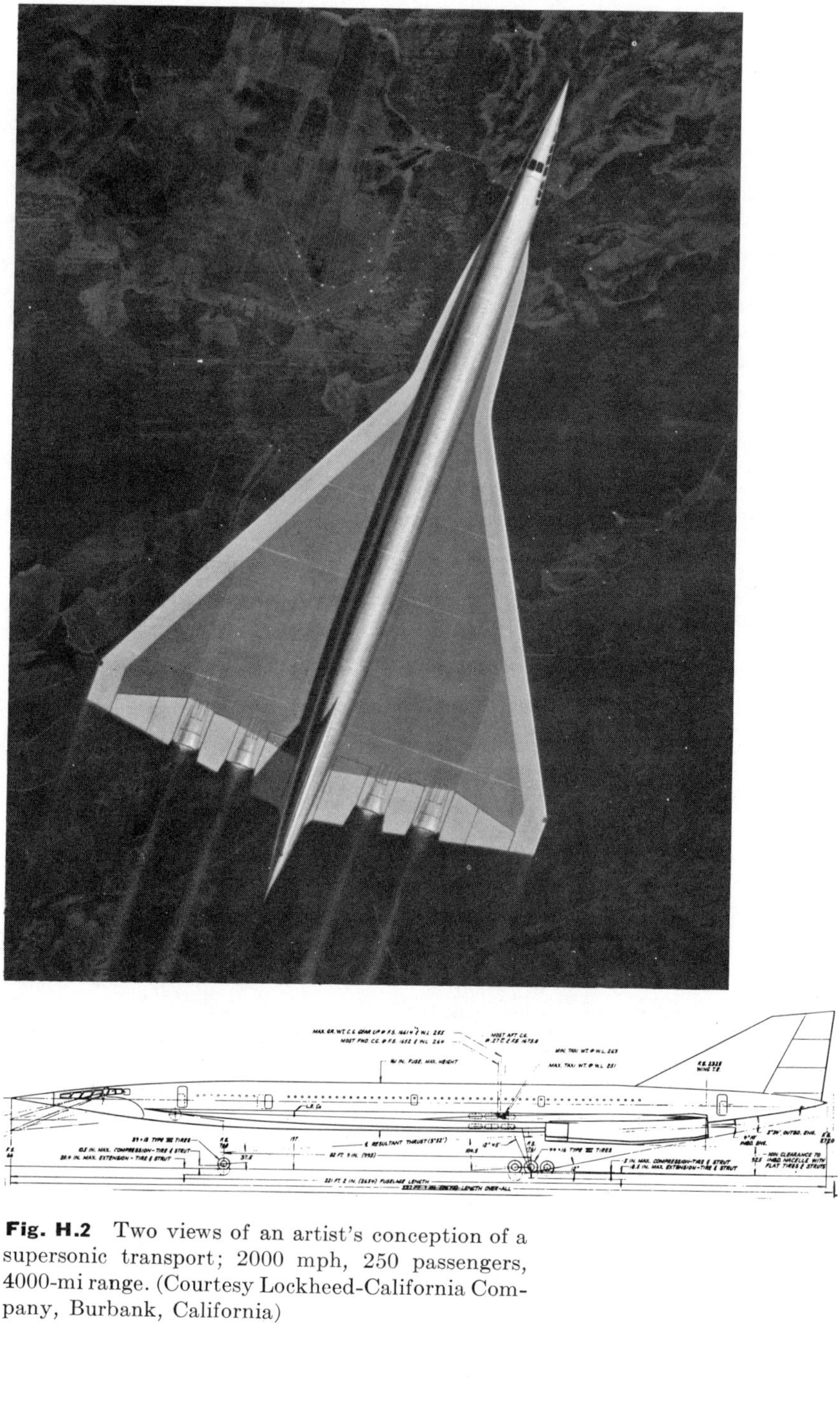

Fig. H.2 Two views of an artist's conception of a supersonic transport; 2000 mph, 250 passengers, 4000-mi range. (Courtesy Lockheed-California Company, Burbank, California)

Index

Abstraction, mathematic, 156
semantic, 156
Aesthetics, in city planning, 366
from start of a project, 221
Airplane, analog model for, 147
flight simulating model for, 151
lift-to-drag ratio, 106
supersonic transport, 423–426
Algorithm, definition, examples, 315
for the square-root, 316
Analog(s), analogy between technical fields, 153
computer (electrical, mechanical), 153, 155
model, definition of, 143, 146
from nature, 81
scanning tables for, 75–80, 81, 82
Analysis, economic, 116
of need, primitive statement, 48
order-of-magnitude, definition and use, 99
examples, 103–106
summary, 107
of structure (pole guy-wire), 277
vs. synthesis, 61
Auto, body design schedule, 28
drag on braking, 253
engine starter, 39
ignition invention, 90
license plate example, 62

Biotechnology and engineering, 369
Black-box, as engineering project, 112
as system in input/output analysis, 60
Book project examples, Chap. 1, introduction, 14
Chap. 2, applied morphology, 32
Chap. 3, needs analysis, 53
Chap. 4, problem statement, 65
Chap. 5, (1) information types, 68
(2) information sources, 74
Chap. 6, sand loader invention, 94
Chap. 9, sand conveyor, 130
Chap. 10, three types of models, 158
Chap. 12, (1) water tank model, 198
(2) water tank model revised, 209
Chap. 15, bucket conveyor optimization, 270
Chap. 16, plant location algorithm, 317
Brazil, development project, 372
tropical house for, 388, 389
Business unit, capital decisions in, 118
loans for, 121
management of, 122
money flows in, 118
national economy, 120
operating portion, 118
similarity to engineering project, 117

Cash flow, 115
Cause-and-effect chart, 82
Checking, by experiment and experience, 254
by logic and mathematics, 242, 243
by order-of-magnitude analysis, 99
role and necessity, 239
summary, 255
by symmetry and limiting value, 249, 252
by units and dimensions, 245
Cities, environment, 365
land use, 365
metabolism, 366
transportation, 368
Common sense, 98
Communication, definition, 322
importance to engineer, 321
and noise, 325
redundancy, 325
and report writing, 327
system and diagram, 323
Compatibility analysis, general, in preliminary design, 125
pole guy-wire example, 129
Computation, algorithmic, 315
iterative, 303
massive arrays, 300
Computer, analog, 153, 155
combination of old parts, 31
digital, antecedents, 287
blocks and components, 289, 295
data and command numbers, 294
flow charting, 420
IBM 1620, 149
for information handling, 369
looping, 298
memory addresses, 293

Computer, programs, interpretive language, 291
machine language, 291
solving equations, 419
simulation of aircraft system, 424
table of characteristics of, 319
use in design, 299
algorithms, 315
iteration, 303
massive computation, 300
Concrete, mix and aggregate, 94
plant project, 14
Constraint(s), expressions for conveyor, 275
functional (in optimization), 268
home intercom example, 220
in input/output analysis, 60, 62
regional, as linear program, 279
in optimization, 268
resources, 3
Construction job costs, 377
Consumer, 48–50
Conveyor, bucket, optimization of, 270
sand, hydraulic analog, 159
Conveyor belt, optimization of, 130
Corporation, business unit, 117
management, 122
taxes, income, 116
other, 113
Cost, criterion for information, 69
of construction items, 377
of energy and fuels, 375
of factory product, 380
incremental, generalization on, 216
of materials, 376
of water, 363
Cost criteria, differing, 233
Creativity and innovation, characteristics of creative person, 87
emotional source of, 84
examples, 87
flash of insight, 87
genius vs. layman, 87
hindrances to, 89
how to develop, 89
occurrence, 88
summary, 93
synthesis and, 85
Criterion (criteria), formulation of, arguments recorded in, 224
check list, 212
illustrative evaluation table, 201
quantities vs. qualities in, 192
Criterion (criteria), formulation of, relative to value system, 204
water tank example of, 211
weighting coefficients in, 206
function, arguments to establish, 202
bucket conveyor surfaces, 273
computation by computer, 300
evaluation summary, 237
expansion, for intercom, 222
for pencil, 206
for rake, 207
for toy, 230
for water tank, 212
general shape of surfaces, 269
generalities, 203, 215
for multiple criteria, 217, 219, 261
normalized variables, 206
permutation of alternatives for, 227
purpose of, 214
scale factors between variables of, 219
for single criterion, 216
symbolic equation, 203
synonym (objective function), 203
weighting coefficients, 206, 208
in preliminary design, for conveyor belt, 132
generalities, 125
for pole guy-wire, 127
Curves, array of fitting functions, 180
highway standards for, 390
plotting standards for, 338–340

Decision(s), and the engineer, 205
making generalities, 3
in optimization, 261
Depreciation allowance, 116, 120
Design, anatomy of, 61
art in, 3
basic equation of, 202, 215
and checking, 239
and conversion of resources, 3
and creativity, 85, 87
and decisions, 3
detail, 25, 27
definition of, 3
and estimation, 98
and input/output analysis, 59
iteration, 3, 30
morphology of, 24
detail design, 25, 27
feasibility study, 24, 26
preliminary design, 24, 26

Design, morphology of, revision, 25, 27
optimization, 258
phases of, 24–27
plans for system, 4
prices paid in trade-off, 4
process (Asimow), 29
project organization, 21
reporting, 327–337
vs. research, 60
tools, 61
Design project, 24
changes in criteria of, 24
economic analysis of, 112
end objective, goals of, 31, 51, 52, 197
goals definition, 23, 197
information for, 41, 42
input/output analysis of, 60
phases, 24
vs. "problem," 21, 23, 25
student, information on, spacecraft accessory, 384, 385
anthropometric data, 387, 388
highway data, 390, 391
nuclear container, 383
tropical construction costs of, 389–391
waste water reclamation, 392–396
suggestions for, 17, 18, 382, 383
successful operation of, 114–116
Differentiation, graphical, 178
Dimensions of quantities, in checking, 246
table, 403
Discipline of self, by checking, 241
by self-knowledge, 13

Economic(s), analogous operation of, 122
basis, of information, 69
of project, 111, 112
conclusions, 116
and criterion function, 215
dividends, 120
division of cash flows, 116
in engineering, 122
gaining experience in, 111
loans to finance, 121
retained earnings, 120
summary, 122
working capital, 120
Energy and power, costs, 375
definition (units), 374
needs, 360
potential (kinetic), 132
Energy and power, solar constant of, 363
trends (U.S.A. and world), 361
Engineer(s), as apprentice, historically, 2
change in viewpoint with project, 24
and checking, 67
and communication, 67
confidence of, 97
and continuing education, 67
as converter of information, 37
as decision maker, 205
as designer, 4, 13
emotional drive of, 84
as employee in industry, 8
and estimation, 57, 98
evaluation of self, 13
and failure, 57
as generalist, 4
goal oriented, 14
job satisfaction, 6
no longer in engineering, 7
and optimization, 261
orientation on job, 8
participation in goals definition, 23
as professional, 8
responsibility, 205
as "service," 13
variation in analytic approach, 125
Engineering, and biotechnology, 369
branches, 6
challenges, 357
and the city, 365
departments, 10, 11
and developing nations, 371
fields, 6
functions, 7
and health, medicine, 370
and information handling, 369
jobs and expectations, 5, 8, 13
notebook, 416
objectives (of excellence), 87
presentations, 347
product vs. process, 6
reference data, 374
reports, 329
research problems, 358
vs. science, 14, 52
societies, 399
work evaluation, 14
Engineering reports, 329
conclusions within, 343
essentials of, 330
and graphics, 337

Engineering reports, idea development in, 336
  oral types of, 347
    advantages of, 348
    facilities for, 353
    preparation for, 350, 353
    presentation of, 355
    rehearsal for, 353
    speaker for, 354
    summary of, 356
  purpose and content, 333
  written, preparation of, 334
      summary of, 334
      writing draft or outline for, 335
  written vs. oral, 348
Environment, in input/output analysis, 60
  urban, 365
Equation, basic design, 202
  criterion function, 203
  differential, solution of, 314
  general optimizing form of, 208
  quadratic, 419
  recursive form of, 303
  simultaneous, iterative solution of, 311
Estimation, general, 98
  mechanical and heat examples, 103–106
  parking lot example, 99
  preliminary, 140
  preliminary steps for water tank, 210
  significant figures in, 107
  summary, 107
Evaluation in the design process, 30
Evidence, by checking, 241
  in legal analogy, 45
Eye-and-Mind, graphic capabilities, 165, 168

Failure, checking to guard against, 240
  and the engineer's reputation, 57
Feasibility study, definition, 56
  elements and glossary, 26, 57, 58
  solutions to start preliminary design, 125
  as stage in design, 24, 26
Flow diagram of information, 39
Force diagram, pole guy-wire, 126
  vectors, 412, 413, 415
FORTRAN, computer language, 291
  sample programs, 292, 302
  solution of quadratic equation, 421
  symbols and operations for, 422

Goals, engineering excellence, 87
  of project, 51, 52
Goals, of project, converting to "problem," 62
    via "desired outputs," 60
      intercom example, 221
      toy example, 230
    rechecking starting point, 84
Graphics, alphabet of lines, 405, 407
  as communication, 163
    in reports, 337, 344
  for computation, 173–181
    arithmetic proportions, 174
    curve fitting, 178
    differentiation, 178
    integration, 175
    interpolation, extrapolation, 173, 174
    nomograms, 175
    optimization, 180
    solution to equations, 179
    trigonometry, 174
    vector calculations, 174, 415
  descriptive geometry cases, 407–412
  and engineering, 162, 166
  forces, resolution of, 412
  isometrics, 406
  layout, 172
  optimization, 264
  and oral presentation, 350, 356
  orthographic projection, 407–409
  sketching, 181, 182, 405, 406
  skills and practice, 186
  for specification, 168, 172
  true length, shape, angle, 407–411
  variants-invariants, 181
  and visual aids, 183

Handbooks, useful, list of, 378
Health, medicine, and engineering, 370
Humanities, importance of, 12, 13, 204, 221

Iconic model (*see* Models, iconic)
Illustration(s) in reports, 337
  standards, 338–340
  summary, 344
Industry, business aspects, 117
  development indices, 359
  engineering organization in, 9–11
  income in U.S.A. for, 8
Information, availability, 69
  catalogs, general, 103
  conversion, 37–39
  criteria of rationality of, 45
  definitions of, 41

Information, economic basis of, 69
feedback, 29
flow, 39
general, 30, 37
input/output analyses of, 59
Library of Congress classifications, 397
measured by probability, 323
methods of obtaining, 71, 72
numerical, 47
processing by computer, 369
and production-consumption cycle, 42
and project, 42
real problems of, 41
source addresses, 398
theory, 323
uncertainty of, 44
varieties of, 43–45
hard, soft, necessary, and dispensable, 44
variety of sources of, 68–71, 378, 397–401
Input/output transformation, analysis, general, 59
for intercom, 220
of materials in cities, 368
physical and nonphysical, 40
primary and secondary functions, intercom, 222
toy, 230
system (product) vs. project, 60
from value system, 205
varieties of inputs to, 59
Intangibles, decisions on, 204
evaluation of, 221
Integration, graphical, 175, 177
Intercommunication system case, 219–227
input/output analysis for, 220
Inventor and invention, for optimizing, 276
patents, 90, 91
records, 92
rights, protection of, 90
Isometrics, examples, 169, 406
for free-hand sketching, 182
(*See also* Graphics)
Iteration, convergence of, 30
criteria for, 30, 31
in design, 3
end objective, 31
in solution of equations (recursion), 300, 303

Laws, in checking, 254
and effects, analogies from nature, 81
Laws, and effects, input/output tables of, 82
scanning tables of, 75–80
empirical, 194
and evidence, 46
implicit answers by, 195
perfect, 194
physical vs. logical, 194
Layout, graphic, 172–181
Legality, as a criterion, 261
of evidence, 46
Library of Congress classification, 397
Linear programming, concrete-mix example, 279
general uses, 283
Linkage, four-bar, optimization, 266
Logic, in checking, 242
constructed, 194
definition, 194
in design decisions, 205, 208, 237

Materials, for cities, 366
consumption in U.S.A., 363
costs and properties, 376
in production-consumption cycle, 43
Mathematics, in checking, 243
implicit answers through, 195
laws of, 194
models and modeling, 125, 156
attributes, 157
as criterion function, 203
general optimizing form, 268
in preliminary design, 125
conveyor belt example, 132
pole guy-wire example, 127
symbolic or, 156
system description by, 192
water-tank example, 197
Models and modeling, analog, 146–153
of airplane, 147
of arithmetic, 148
of heat system, 150
of power system, 148
attributes, summary of, 157
engineering problems, 138
formal, of system, 205
iconic, 142–146
definition, 143
graphics, 162–190
Mariner structure, 185
limitations, 139

Models and modeling, mathematical (*see* Mathematics, models and modeling)
purposes of, 142, 214
symbolic, 156–158, 191–238
characteristics, 198
definition, 143, 156
as design equation, 203
general optimizing form, 268
starting, 197
value of, 138
variety of, 143
Money, amount used in engineering, 120
and economic analysis of project, 112
and engineering economics, 122
and engineers, 109
flows, 115, 116
investment of, 113
questions vs. technical questions, 110
Morphology of design, 24
Asimow, 22, 24, 25
Buhl, 22, 25
as check list, 31

Nation(s), development indices of, 359
economy of, capital reservoir, 118, 121
components of, 118
money flows in, 117
purchasing power reservoir, 118
education and development of, 371
engineers employed in U.S.A., 8
needs, 357–373
Needs, analysis of, 48–51
and information, 42
starting point of, 48
changes with study of, 52
conflicting statements of, 43
designing for, 1, 2, 14
of engineers vs. customers, 51
future, 50
to goals, 51
human, 4, 50
past, present, and future, 50
primitive statement of, 24, 48
psychological, 50
summary of, 53
variety, 50
Nomogram, examples, 175
Notebook for engineering records, 416
Numbers, checking for errors in, 240, 243
in floating point notation, 293

Optimization, 258–286
of criterion function, intercom, 226
toy, 237
of electric cable insulation, 309
in general, 32, 258
graphic, 180
by invention, 276
by linear programming, 279
mathematical, purpose of, 130
methods, analytic, 267
general principles, 263
graphic, 264
subjective, 263
in new vs. old product, 32
optimum vs. optimal, 260
outputs as features to optimize, 60
in preliminary design, of conveyor belt, 132
general, 125
of pole guy-wire, 128
structural, rule for, 278
summary of, 283
of supersonic transport aircraft, 425
systems and criteria for, 259
when to optimize, 262
Order-of-magnitude analysis (*see* Analysis, order-of-magnitude)
Organization of engineering group, 9
by product, 10
by specialty, 11
Outcome, of design process, 30
of engineering successes, 87
Owner, entrepreneur, and stockholder, 112
equity funds of, 113
and the successful project, 116

Parameters, expressions for, bucket conveyor, 271–274
educational toy, 231
home intercom, 224
ranges of value, 208
water tank, 200
Parking lot, estimation example, 99–102
Patents, categories patentable, 91
disclosure, 92, 93
recording idea for, 92
rights, conditions, and law, 90
Perspective, examples of, 164, 169, 170, 182
PERT diagram, as analog model, 152
in system schedule reference, 192
Philosophy and value systems, 13
Physical realizability, 58

Pole guy-wire examples, 125–130, 277
Pollutants, atmospheric, 367
Prediction of performance in design, 125
Preliminary design, 24, 26
  stages in, 125
Primitive statement of need, 24, 48
  (*See also* Needs)
Probability and statistics, in checking, 242
  in needs analysis, 47
  normal distribution, 48
  standard deviation in, 48
Problem(s), estimation to reduce size of, 98
  perfect solutions for, 127
  possible outcomes of, 141
  scientific vs. engineering kinds of, 52
  solving, 21–23
    in anatomy of design, 61
    sub-problems leading from, 63
  statement from goals, 62
  technical statement of, 58, 60
Production-consumption cycle, 42
  importance of consumer in, 49
  subgroups within, 49
Professional engineer, 8*n*.
  attitudes of, 9, 13
    change during project, 24, 25
    in checking, 241
  vs. commercial man, 13
  engineering excellence goals of, 87
  and problems of world, 358
  societies, 399
  summary, 373
Project, design (*see* Design project)
Prosthetics, 370

Quantities vs. qualities, reconcilement in criterion equation, 209
  in systems, 192

Railway track standard gauge, 47
Records, 416–418
  drawings, 183
  patent disclosure, 183
  project notebook, 416–418
Recursion expression, 303
Reports, engineering (*see* Engineering reports)
Research vs. design, 60
Research expenditures in U.S.A., 358
Resources, 3
  in input/output analysis, 60
Revision in design morphology, 25, 27

Scalar (vs. vector), 412, 414
Scaling, constants in criterion function, 225
  in iconic model, 184
Science and technology, expenditures in U.S.A., 358
  fields of, 73, 74
  Library of Congress classification of, 397, 398
Semantic(s), abstraction, 156
  noise, 326, 327
Sensitivity analysis, 125
  check(s), in general, 209, 227
    for toy, 236
  of pole guy-wire variables, 129
Significant figures, 99, 244
Simulation of aircraft system, 424
Sketching, 181, 182, 405, 406
Social science, importance of, 12, 13
Societies, professional engineering, names of, 399
Solar power constant at earth's surface, 363
Solar power satellite, 106
Solutions, by algorithm, 315
  alternative, 58, 84
    combinatorial, 86
    how to generate, 89
    by input/output analysis, 60
  of differential equations, 314
  by iteration, 303
  perfect, 127
  possible, 141
  of quadratic equation (computer), 419
  recursive expression for, 303
  of simultaneous equations, 311
Sources of information, procedures to use, 71, 72
  public vs. private, 70
  scanning tables for ideas, 75–80
  variety of, 68, 378, 398
Spacecraft, photographs of, 334–336
  satellite variety of (solar power), 106
Specification by words, graphics, etc., 167
Statistics (*see* Probability and statistics)
Structures, general construction principle, 278
  pole guy-wire example, 125–130
  preference for tension members, 126
Student, confidence level, 97
  general knowledge, 98
  project alternatives, 17, 18, 382, 383
Supersonic transport, 423

Symbols, definition of, 196
FORTRAN, 422
graphic, 164
mathematic, 196
(*See also* Models and modeling, symbolic)
Synthesis, of alternative solutions, 204
vs. analysis, 60, 61
combinatorial nature of, 86
and creativity, 85
in feasibility study, 58
sequence of events in, 88
of systems, 85, 86, 193
and visualization, 168
Systems, biological, 369
description, analytical, 191–193
end result of design, 61
inputs as limitations to, 60
and optimization, 259
potential variety, 58, 162
Systems Command Manual, Air Force, 205

Telephone handset dimensions, 47
Toy, educational, criterion function for, 236
criterion variables for, 227
proposed design for, 229
Trade-offs, 3, 4, 60
in optimization, 283
in preliminary design, 125
Transport and transportation, commuter, 368
proposed supersonic aircraft, 423
regional needs for, 369
tentative SST design values, 425

United States, annual income by industry, 8
comparison by indices, 359
materials consumption in, 363
population trends in, 361
problems facing, 358
science spending in, 358
Units and dimensions, 402–403
in checking, 245
fundamental values of, 246
of quantities, 402
for work and power, 374

Value system, and criterion function, 204
example for intercom, 224
in optimization, 262
Variables, parameters, and constants,
in criterion function, 215, 223
permutation of, 237
in preliminary design, 134
ranges of value predictions, 208
conveyor, 274, 275
intercom, 223
toy, 230
water tank, 198, 199
Vectors and scalars, definition and operations, 414
force resolution by, 412, 413
graphic computation of, 174, 175
Visual aids, and graphics, 183
for oral presentations, 350, 356
and reports, 337
summary of, 344
types of, 351
Visualization and synthesis, 168, 171

Washing machine agitator power computation, 177
Wastes, from city, 366
in material extraction, 363
reclamation of water, 392–396
Water, costs, 363
desalination, 364
reclamation project, 392–396
Water-heater recovery time, 103
Weight minimization, 277
of steel tension-member, 127
Weighting coefficients, evaluation, 206, 208
for intercom, 224
for toy, 228
World, energy trends, 361
indices of industrial development, 359
population trends, 361
problems, 357
Wright brothers' flight tests, 44